Trigonometric Identities

I0070133

Reciprocal Identities

1. $\sec\theta = \dfrac{1}{\cos\theta}$
2. $\csc\theta = \dfrac{1}{\sin\theta}$
3. $\cot\theta = \dfrac{1}{\tan\theta}$

Ratio Identities

1. $\tan\theta = \dfrac{\sin\theta}{\cos\theta}$
2. $\cot\theta = \dfrac{\cos\theta}{\sin\theta}$

Pythagorean Identities

1. $\sin^2\theta + \cos^2\theta = 1$
2. $1 + \tan^2\theta = \sec^2\theta$
3. $1 + \cot^2\theta = \csc^2\theta$

Sum and Difference Identities (Addition and Subtraction Formulas)

1. $\cos(A - B) = \cos A\cos B + \sin A\sin B$; **4.** $\sin(A - B) = \sin A\cos B - \sin B\cos A$

2. $\cos(A + B) = \cos A\cos B - \sin A\sin B$; **5.** $\tan(A + B) = \dfrac{\tan A + \tan B}{1 - \tan A\tan B}$

3. $\sin(A + B) = \sin A\cos B + \sin B\cos A$, **6.** $\tan(A - B) = \dfrac{\tan A - \tan B}{1 + \tan A\tan B}$

Double Angle Identities

1. $\sin 2\theta = 2\sin\theta\cos\theta$; **2.** $\cos 2\theta = \cos^2\theta - \sin^2\theta$; **3.** $\cos 2\theta = 2\cos^2\theta - 1$;

4. $\cos 2\theta = 1 - 2\sin^2\theta$; **5.** $\tan 2\theta = \dfrac{2\tan\theta}{1 - \tan^2\theta}$

Half Angle Identities

1. $\cos\frac{1}{2}\theta = \pm\sqrt{\dfrac{1 + \cos\theta}{2}}$; **2.** $\sin\frac{1}{2}\theta = \pm\sqrt{\dfrac{1 - \cos\theta}{2}}$;

3. $\tan\frac{1}{2}\theta = \dfrac{1 - \cos\theta}{\sin\theta} = \dfrac{\sin\theta}{1 + \cos\theta}$

The ± sign indicates the sign to use depending on the quadrant location of $\dfrac{\theta}{2}$

Product or Product as a Sum Formulas

1. $\cos(A - B) + \cos(A + B) = 2\cos A\cos B$
2. $\cos(A - B) - \cos(A + B) = 2\sin A\sin B$

{Derived from sum and difference identities}

3. $\sin(A + B) + \sin(A - B) = 2\sin A\cos B$
4. $\sin(A + B) - \sin(A - B) = 2\cos A\sin B$

{Derived from sum and difference identities}

Sum Identities

1. $\cos A + \cos B = 2\cos\dfrac{A + B}{2}\cos\dfrac{A - B}{2}$
2. $\cos A - \cos B = -2\sin\dfrac{A + B}{2}\sin\dfrac{A - B}{2}$
3. $\sin A + \sin B = 2\sin\dfrac{A + B}{2}\cos\dfrac{A - B}{2}$
4. $\sin A - \sin B = 2\sin\dfrac{A - B}{2}\cos\dfrac{A + B}{2}$

Cofunction Relationships

$\sin A = \cos(90^\circ - A)$
$\cos A = \sin(90^\circ - A)$ } cofunctions.

$\tan A = \cot(90^\circ - A)$
$\cot A = \tan(90^\circ - A)$ } cofuntions.

$\sec A = \csc(90^\circ - A)$
$\csc A = \sec(90^\circ - A)$ } cofuntions.

Other identities (Useful when finding functional values of negative angles)

For any θ,

1. $\cos(-\theta) = \cos\theta$
2. $\sin(-\theta) = -\sin\theta$
3. $\tan(-\theta) = -\tan\theta$
4. $\sec(-\theta) = \sec\theta$
5. $\csc(-\theta) = -\csc\theta$
6. $\cot(-\theta) = -\cot\theta$

If you can remember a needed information, you make decisions faster, you learn faster, you work faster, and you are more productive.

Yes, you can memorize them.

Squares of Natural Numbers

$1\times1=1$	$26\times26=676$
$2\times2=4$	$27\times27=729$
$3\times3=9$	$28\times28=784$
$4\times4=16$	$29\times29=841$
$5\times5=25$	$30\times30=900$
$6\times6=36$	$31\times31=961$
$7\times7=49$	$32\times32=1024$
$8\times8=64$	$33\times33=1089$
$9\times9=81$	$34\times34=1156$
$10\times10=100$	$35\times35=1225$
$11\times11=121$	$36\times36=1296$
$12\times12=144$	$37\times37=1369$
$13\times13=169$	$38\times38=1444$
$14\times14=196$	$39\times39=1521$
$15\times15=225$	$40\times40=1600$
$16\times16=256$	$41\times41=1681$
$17\times17=289$	$42\times42=1764$
$18\times18=324$	$43\times43=1849$
$19\times19=361$	$44\times44=1936$
$20\times20=400$	$45\times45=2025$
$21\times21=441$	$46\times46=2116$
$22\times22=484$	$47\times47=2209$
$23\times23=529$	$48\times48=2304$
$24\times24=576$	$49\times49=2401$
$25\times25=625$	$50\times50=2500$

FREMPONG'S STEP-BY-STEP SERIES IN MATHEMATICS

College Trigonometry

Includes SAMPLE PROBLEMS WITH

STEP-BY-STEP SOLUTIONS

plus

PRACTICE PROBLEMS WITH ANSWERS

A.A. FREMPONG

College Trigonometry

Published by Yellowtextbooks.com. N.Y.

ISBN 978-1-946485-18-2

Printed in the United States of America

In Memory of My Parents

Mom:
She was a devoted mother, sharing, kind, kinder to strangers and generous to a fault. She never cursed, she never hated; she never cheated, and she never envied. She never lied, and she never got angry. Once, she nursed an almost dying stranger renting a room in her house back to good health to the extent that the relatives of this renter later travelled one hundred miles just to thank mom. She was always peaceloving and forever forgiving.
An angel once lived on this earth to serve others.

Dad:
A great dad, kind, generous and forgiving. He emphasized and was an example of both formal education and self-education. A veterinarian, a bacteriologist, an Associate of the Institute of Medical Laboratory Technology (UK), a Fellow of the Royal Society of Health (UK); an incorruptible civil servant; his book on ticks has always inspired me to write whenever the need arises.

NOTE TO THE STUDENT

This book was written with you in mind at all times. You may use this book as the course textbook or as a review book since the book gets to the point quickly on all relevant topics and yet covers these topics in detail.

Begin to master the definitions and the solutions of the sample problems thoroughly. (You have mastered a sample problem if you can solve the sample problem and similar problems without any reference to this book or any other source. For some problems, two or more methods are presented. Read the various methods and decide which methods you would like to remember; but always be aware of the existence of the other methods, in case the need arises. After having mastered the sample problems, try the exercise problems. The answers to these problems are presented immediately after the problems. You may cover the answers with paper before you attempt these problems, if the answers are too obvious. You may refer back and forth to the solved problems when you do not remember how to proceed.

You may also attempt some of the sample problems first, if you have been exposed to the topics previously, before reading the solution methods, and in this approach, the sample problems become more practice problems for you.

As a reminder, in any book, do not dwell on the few inadvertent errors you may find, but rather concentrate on what is useful to you.

For this book to be useful both as the course textbook, as well as review for exams, it is **important** to **Understand, Remember, Apply**, and **Remember** the material covered.

Wishing you Good Luck on all the exams
A.A.Frempong

Books in the series by the author: Integrated Arithmetic; Elementary Algebra; Intermediate Algebra, Elementary Mathematics; Intermediate Mathematics; Elementary & Intermediate Mathematics (combined); College Algebra; **College Trigonometry**; College Algebra & Trigonometry and Calculus 1 & 2.

PREFACE

This book is the trigonometry part of the parent book entitled "College Algebra & Trigonometry. The first four chapters review the basic concepts of functions with reference to polynomials and rational functions, since the concepts in these chapters can be applied to trigonometric functions. If you are familiar with these chapters , you may skip them and begin with chapter five.
The idea of producing this book first came to me some time when I was working with students on one-to-one basis. My experience at that time was that I could not find that one book which was easily read and understood on all the required topics for College Trigonometry. On occasion, I could recommend a book only for one or two topics, and then I had to search in the library for other recommendable books (to explain the other topics) but without much success.
Another observation of mine was that there was a group of students who inspite of all the hard work could not understand their textbooks well. These students had done well at the elementary and intermediate levels. Often, after having helped a student understand a material, the student would remark" Why doesn't the book state it so?".

Even though the author has published the companion larger volume" College Algebra & Trigonometry, some programs may offer trigonometry separately from college algebra,

This book is an attempt to help such a student. This book could be used for self-study, especially by the working student or the continuing student with so much demand on his or her time. It could also be used as a reference book since each topic has been covered very well. This is not a book for the lazy student. It is a book for the serious student who in spite of all the hard work finds his or her textbooks rather difficult to understand or follow.
This book could be used as a textbook for a trigonometry course. To the instructor, this book should be a relief so that more time could be spent on the applications of the definitions and principles.
This book was written with the student in mind at all times, and at times, some of the explanations may seem redundant, however, this is intentional. Analogies from everyday life are presented whenever they help to explain a principle. At times, the book may be found to be rather informal, but this also is intentional, because the main objective is to communicate.
The following concepts have been treated in more detail than most current textbooks do at this level:
4
Decreasing functions, **increasing** functions, **continuous** functions, discontinuous functions, positive functions, negative functions, **translation** of axes, contraction and expansion of curves, symmetry, **asymptotes,** critical points, and **maximum** and minimum points. The author recommends that as soon as possible, the above mentioned concepts be mastered. The early mastery of these concepts will help the student read and understand subsequent material much more quickly than otherwise. The student would also be able to see the beauty in studying mathematics as a system and would be able to appreciate the unifying concepts among the various functions and relations.
A new concept of relating simple continuous functions and their reciprocals has been presented, perhaps for the first time (by the author).
In **trigonometry,** the concept of the period of a function is explained as the simultaneous occurrence of two events.
A step-by-step approach is used throughout the book.
This book can also be used for short term programs such as mini-sessions, immersion programs, workshops, as well as in distance learning programs.

My sincere appreciation goes to those who wrote before me, and without whom this book would have been inconceivable.

A. A. Frempong
New York, October, 2012

CONTENTS

Page

CHAPTER 1 1

Lesson 1: **Definitions, Terminology, and Types of Numbers** 1

CHAPTER 2 4

Basic Review of Functions I

Lesson 2: **Sets, Relations, Functions, Comparison of Relations and Functions** 4

Lesson 3: **Functional Notation; Defined Functions; Excluded Values, Domain and Range** 8

Lesson 4: **One-to-One Functions, Composite Functions** 20

Lesson 5: **Inverse Functions and Inverse Relations** 24

CHAPTER 3 33

Basic Review of Functions II

Lesson 6: **Continuous and Discontinuous Functions** 33

Lesson 7: **Asymptotes** 36

Lesson 8: **Graphs of Rational Functions** **46**

CHAPTER 4 60

Basic Review of Functions III

Lesson 9: **Positive, Negative, Increasing, and Decreasing Functions** **60**

Lesson 10: **Concavities of Curves; Critical Points** 64

Lesson 11: **Reflection of Points, Lines and Curves** 69

Lesson 12: **Transformations: Translation** of Points, Relations, Functions, Axes 75

Lesson 13: **Contraction and Expansion of Curves** 82

Lesson 14: **Symmetry; Even and Odd Functions** 96

CHAPTER 5 91

Introduction to Trigonometry, Right Angle Trigonometry

Lesson 15 **Basic Review for Geometry** 92

Lesson 16: **Right Triangle Trigonometry** 97

Lesson 17: **Applications of Trigonometric ratios: Angles of Elevation and Depression; Bearing;Linear Interpolation** 113

CHAPTER 6 119
Trigonometric Functional Value of any Angle

Lesson 18: **Definitions; Functional Value Given the Measure of the Angle** 119

Lesson 19: **Given a Point on the Terminal Side of the Angle** 125

Lesson 20: **Trigonometric Functional Values of Quadrantal Angles** 129

CHAPTER 7 135
Finding Other Trigonometric Functional Values

Lesson 21: **Given a Functional Value and the Quadrant of the Angle** 135

Lesson 22: **Given a Functional Value and no Specification of the Quadrant of the Angle** 137

CHAPTER 8 139
Trigonometry of Oblique Triangles

Lesson 23: **The Law of Cosine; The Law of Sines** 139

Lesson 24 **Applications of the Laws of Cosine and Sines** 145

CHAPTER 9 150
Application of Trigonometry to Vectors

Lesson 25: **Basic Definitions; Representation of Vectors** 150

Lesson 26 **Addition (Sum, Resultant, or Composition) of Vectors** **153**

CHAPTER 10 163
Trigonometry of Real Numbers

Lesson 27: **Definitions; Radian Measure; Arc Length; Reference Number** 163

Lesson 28: **Trig Functional Values of Angles and of Real Numbers** 172

CHAPTER 11 174
Graphs of Trigonometric Functions

Lesson 29: **Introduction; Labeling the Coordinate Axes;Illustration of the Periodicity of Trigonometric Functions** 174

Lesson 30: **Sketching the Graph of** 179

Lesson 31: **Sketching the Graph of** 184

Lesson 32: **Graphs of Discontinuous Trigonometric Functions:** $y = \tan x$; $y = \csc x$; $y = \sec x$; $y = \cot x$ 187

Lesson 33: **Direct Procedure for Sketching the Graphs of** $y - k = a\sin(bx - h)$ and $y - k = a\cos(bx - h)$ 195

CHAPTER 12 216

From Trigonometric Functions to

Inverse Trigonometric Functions

Lesson 34: **Introduction; Definitions; Operations Involving Inverse Trigonometric Functions** 216
Lesson 35: **How to Sketch the Graph of** $y = \text{Arcsin}\, x$ 223
Lesson 36: **How to Sketch the Graph of** $y = \text{Arccos}\, x$ 227
Lesson 37: **How to Sketch the Graph of** $y = \text{Arctan}\, x$ 230

CHAPTER 13 232

Lesson 38: **Trigonometric Identities** 232
Lesson 39: **Proving Trigonometric Identities; Applications of Sum and Difference Identities** 236

CHAPTER 14 241

Lesson 40: **Solutions of Trigonometric Equations** 241

Appendix A 247

How to change a terminating decimal to a rational number 248
How to change a repeating decimal to a rational number 248

Appendix B 251

Definitions of Inverse Trigonometric Functions 252

Appendix C 253

Area and Perimeter of a Circle; Arc Length and Sector of a Circle; Area and Perimeter of Composite Figures 254

Appendix D 259

About Measurements

Standard Unit, Error, Rounding-off Numbers, Significant Digits, Scientific Notation

EXTRA
Appendix E
Complex Numbers

Lesson 41: **Basic Definitions; Basic Operations with Complex Numbers** 270

Lesson 42: **Equality of Complex Numbers; Roots of Complex Numbers; Equations Involving Complex Numbers** 278

Lesson 43: **Graphical Representation and Addition of Complex Numbers** 282

Lesson 44: **Polar (Trigonometric) Form of Complex Numbers** 286

Lesson 45 **Powers of Complex Numbers; De Moivre's Theorem; Roots of Complex Numbers** 294

Appendix F

Lesson 46: **Graphing Polar Coordinates** 302

Lesson 47: **Graphing Polar Equations** 308

INDEX

317

CHAPTER 1

Lesson 1

Definitions, Terminology, and Types of Numbers

The basic elements we deal with in our study of mathematics are numbers. It is important that we obtain a good understanding of the types of numbers we deal with in mathematics. It is also important that we are able to distinguish between the different kinds of numbers and their associated terminology. A very good grasp of the terminology will help us read and understand subsequent material much more quickly than otherwise.

As shown in the number flow chart below, all the numbers we deal with in mathematics can be divided into two main sets, namely the set of **real numbers** and the set of **non-real numbers**. (We must note that the terms "real", "non-real", and similar terms are only names we use in mathematics to distinguish between numbers and we must note that the real numbers are **not literally** more real than the non-real numbers. These terms are only names for convenient distinction between some numbers.) The real numbers are also divided into two main sets, namely the **rational numbers** and the **irrational numbers.** The rational numbers are further subdivided into **integers** (strictly integers) and **fractions** (strictly fractions). The integers consist of the **negative integers, zero,** and the **positive integers** (natural numbers); and the positive integers consists of the **prime numbers** and the **composite numbers**.

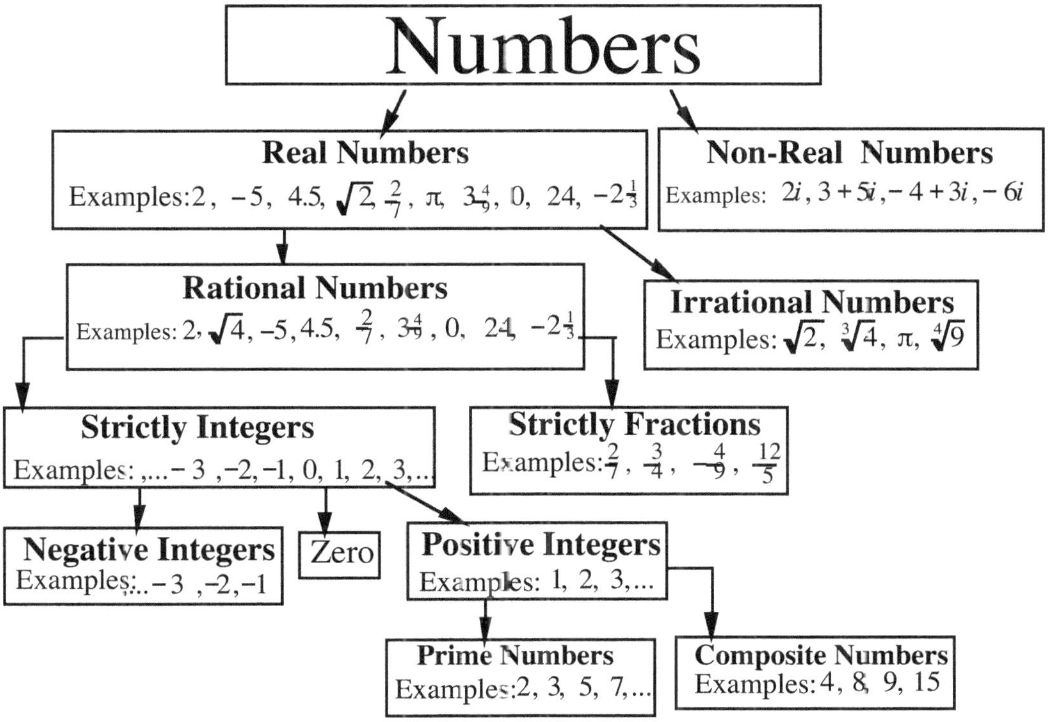

Number Flow Chart

We shall now define some terms mentioned in the number flow chart, above.

We define a **set of numbers** as a well-defined collection of numbers.

The set of the **natural numbers** consists of the numbers 1, 2, 3, 4, 5, 6, 7, 8, 9, 10, 11, 12, 13,...., If we know a natural number, to obtain the next natural number we add 1. The smallest natural number is 1, but we do not know the largest natural number, since given any large natural number, we can always obtain the next natural number by adding 1. The natural numbers are also known as the **counting numbers** or the **positive integers**.

The set of **whole numbers** consists of the numbers 0, 1, 2, 3, 4, 5, 6, 7, 8, 9, 10, 11, 12, 13,..., If we know a whole number, to obtain the next whole number we add 1. The smallest whole number is 0, but we do not know the largest whole number, since given any large whole number, we can always obtain the next whole number by adding 1.

If we take the opposites of the set of natural numbers also called positive integers, we obtain a set of numbers called **negative integers** (such as -1, -2, and -12). If we combine the whole numbers with the negative integers we obtain a set of numbers called the **integers**. The set of integers therefore consists of the set of numbers ...,-7, -6, -5. -4, -3, -2, -1, 0, 1, 2, 3, 4, 5, 6, 7,...
(The three dots preceding the -7 on the left indicates that the numbers continue to decrease to the left and the three dots after the 7 on the right indicates that the numbers continue to increase to the right)

Ratio: The ratio of a is to b is the fraction $\frac{a}{b}$. **Example:** The ratio of 3 is to 4 is the fraction $\frac{3}{4}$.

Rational number: A rational number (a fraction) is a number which **can** be written as the ratio of two integers. The word **rational** pertains to the word **ratio.**

Examples are (a) $\frac{2}{3}$; (b) $\frac{1}{5}$; (c) 4 (since $4 = \frac{4}{1}$)

(d) 0 (since $0 = \frac{0}{7} = \frac{0}{3}$ ... or $0 = \frac{0}{b}$, where b is an integer and $b \neq 0$)

(e) $\sqrt{4}$ (because $\sqrt{4} = 2 = \frac{2}{1}$)

A rational number can also be written either as a terminating decimal or as a repeating decimal.

Examples of terminating decimals: $\frac{1}{4} = .25$; $\frac{13}{2} = 6.5$; $\frac{37}{8} = 4.625$.

Examples of repeating decimals: $\frac{1}{3} = .333...$ or $.\bar{3}$ and $\frac{1}{6} = .1666...$ or $.1\bar{6}$; $\frac{2}{3} = .66...$ or $.\bar{6}$.

Note the bar (vinculum) placed over the repeating digit or block of digits.
We may also regard a terminating decimal as non-terminating if we attach zeros to the right of the decimal. Examples are .25 = .250000...., .5 = .5000...

We define an **irrational number** as a number which **cannot** be written as the ratio of two integers. However, we can approximate irrational numbers as closely as we wish by rational numbers or decimals.

Examples of irrational numbers are $\sqrt{2}, \sqrt[3]{4}$, and π (pi). We can for example, approximate $\sqrt{2}$ by 1.414, and π by $\frac{22}{7}$ or 3.142.

When written in decimal form, an irrational number is non-repeating and non-terminating.

If we combine the natural numbers, the fractions, decimals, irrational numbers, the negatives of these numbers, and zero, we obtain the set of **real numbers**. Simply, the real numbers consists of the rational numbers and the irrational numbers.

We can represent real numbers by points on a horizontal line called the **real number line** (Fig.1)

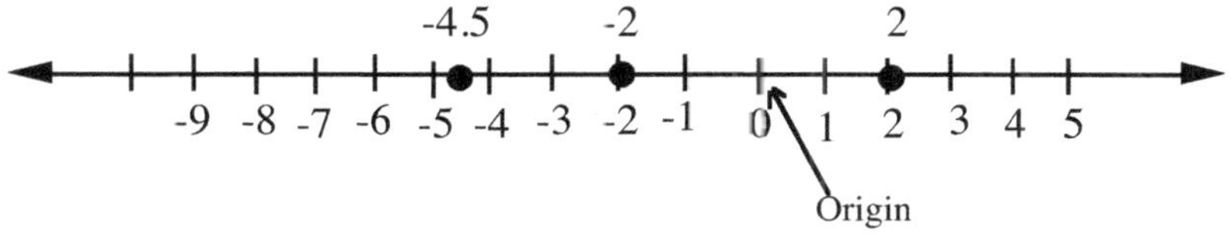

Figure 1

The **real number line** is a horizontal straight line with equally spaced intervals as in Figure 1 above. We label a point called the origin, 0 (zero). Points to the right of the origin are labeled positive and points to the left of the origin are labeled negative. The numbers increase as one moves from the left to the right on the real number line. Roughly speaking, a real number is a number that can be r epresented by a point on the real number line. The real numbers consists of the integers, fractions, mixed numbers, decimals, and radicals. In Figure 1, if the real numbers, -4.5, -2, and 2 are of interest, we can represent them by the dots shown. Every point on this line is associated with a real number; and every real number is associated with a point on this line. We can also say that the set of real numbers consists of the signed numbers and zero.

Summary for Some Number Terminology

Positive integers, or natural numbers or counting numbers $= \{1, 2, 3, 4, ...\}$

Negative integers $= \{..., -4, -3, -2, -1\}$

Non-negative integers or **whole numbers** = $\{0, 1, 2, 3, 4, ...\}$

Non-positive integers = $\{..., -4, -3, -2, -1, 0\}$

Non-negative real numbers consist of 0 and the positive real numbers.
Examples are $0,\ 4, 7.5,\ 3\frac{1}{2},\ \frac{1}{4},\ 6,$ and $\sqrt{2}$

Non-positive real numbers consist of 0 and the negative real numbers.
Examples are $-\sqrt{11}, -4, -3\frac{1}{2}, -2, -1, -\frac{1}{4}, -.126,$ and 0.

We must understand thoroughly the above terms because they will be used over and over in the future. Anytime we meet any of these terms, we should try to form a quick mental picture of representative examples.

Signed Numbers

A signed number is a number with either a plus sign " +" or a minus sign "-" preceding it.
If there is no sign preceding a number, we will assume that the number has a plus sign.

Absolute Value

The absolute value of a signed number may be defined as the number without its sign. The absolute value of zero is zero.

Note: The absolute value of a signed number is also its distance from zero on the number line.

Absolute value defined more formally:

The absolute value of a real number x is x if x is a positive number or zero, but it is $-x$ if x is negative number . (i.e. the negative of a negative number).

CHAPTER 2

Basic Review of Functions I

Lesson 2: **Sets, Relations, Functions, Comparison of Relations and Functions**

Lesson 3: **Functional Notation; Defined Functions; Excluded Values, Domain and Range**

Lesson 4: **One-to-One Functions, Composite Functions**

Lesson 5: **Inverse Functions and Inverse Relations**

Lesson 2

Sets, Relations, Functions, Comparison of Relations and Functions

Ordered Pair

An **ordered pair** of numbers is an arrangement of two numbers in a specified order. In an x-y rectangular coordinate system of axes, the first element (or component) is the x-value and the second element is the y-value.

Example 1 (*a*) (1, 2) <--- ($x = 1, y = 2$)
(*b*) (2, 3) <---- ($x = 2, y = 3$)
(*c*) (5, -1) <---- ($x = 5, y = -1$)

Note that each ordered pair represents a point in an x-y coordinate system of axes.

Set of numbers

A **set of numbers** is a well-defined collection of numbers. The numbers are called the elements or members of the set.

Example 2: If we denote the set of the numbers 2, 5 and 6 by A, then we may write $A = \{2, 5, 6\}$

Example 3: The set B of the ordered pairs (1, 2), (2, 3), and (5,-1) is given by
$B = \{(1, 2), (2, 3), (5, -1)\}$.

Relation

A **relation** is a set of ordered pairs. (A collection of ordered pairs of numbers)

Example 4: The set $E = \{(2, 3), (2, 5), (4, 6)\}$ is a relation, <---There are three ordered pairs.

Example 5: The set $C = \{(6, 2), (7, 4), (11, 5)\}$ is a relation.

Definition of a **Function**

A function may be defined in a number of ways, namely,

(a) in terms of ordered pairs; (b) in terms of a rule involving two variables;
(c) in terms of a rule for inputs and outputs; (d) in terms of correspondence of two sets; (e) as a graph

Definition 1: In terms of ordered pairs

A **function** is a relation in which no two distinct ordered pairs have the same first component; or a function is a set of ordered pairs in which for any two different ordered pairs, the first elements are different. The set in Example 5, above, is a function but the set in Example 4 is not a function, because the first two ordered pairs have the same first element, namely 2.

The set of all the first elements of the ordered pairs is called the **domain** of the function; and the set of all the second elements is called the **range** of the function.

Example 6: In the function $C = \{(6, 2), (7, 4), (11, 5)\}$. The domain, D = {6, 7, 11} (first elements)
The range, R = {2, 4, 5}. (second elements)

Other definitions of a function

Definition 2: If x and y are two variables. then we say that y is a function of x if there is a rule which gives just one corresponding value of y for **each** value of x. The variable x is called the independent variable, and a variable y is called the dependent variable. The rule may be specified in the form of a set, in the form of a graph, in the form of a table, or in the form of an equation or formula.

The **domain** of a function is the set of numbers that can be assigned to x (the independent variable).

The **range** of a function is the set of all the corresponding numbers y (the dependent variable) associated by the function (rule) with the numbers, x, in the domain .

We symbolize that f is a function of x by $f(x)$, where x is called the independent variable, y is called the dependent variable. **Note**: $f(x)$s is read as f of x.

The following are examples of how the rules for functions may be specified:

(a) In the form of an equation or a formula: $y = 2x$.
(b) In the form of a set: $\{(2, 3), (1, 4), (7, 5)\}$.
(c) In the form of a table for x and y: See Table 1.
(d) In the form of a graph. See Figure

Table 1:: y = $2x$

$x =$	0	1	2	3	4
$y =$	0	2	4	6	8

Figure: Graph of $y = 2x$

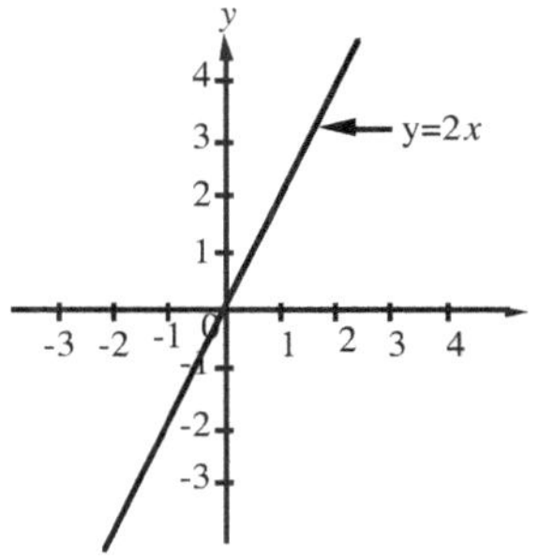

Definition 3 In terms of the correspondence of two sets (Fig. 1)
A function is a correspondence between a first set, say set A and a second set, say Set B such that **each** element of set A corresponds to exactly one element of set B. The set of all the elements of set A is a called the **domain** of the function, and the set of all the corresponding elements of set B is called the **range** of the function.

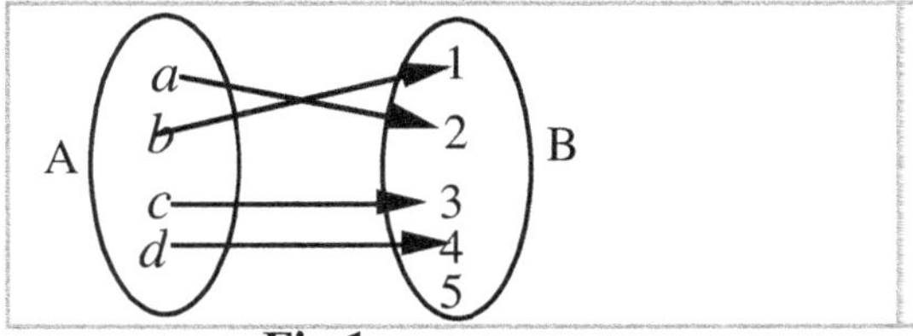

Fig 1

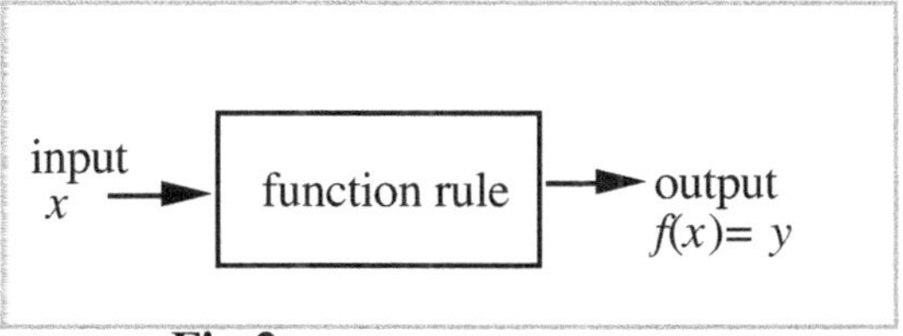

Fig 2

Definition 4 In terms of a rule for inputs and outputs (Fig. 2)
A function is a rule which assigns to each input number exactly one output number. The set of all input numbers that the rule is applicable to is called the **domain** of the function; and the set of all the corresponding output numbers is called the **range** of the function.
A variable representing an input number is called the independent variable, and a variable representing an output number is called the dependent variable.

Given a graph (Vertical line test)
A given graph is that of a function if every possible vertical line drawn to intersect the graph intersects (cuts) the graph exactly once (i.e., at one point only).

Comparison of a Function and a Relation

Similarities: Each is a set of ordered pairs.

Differences: In a relation, two or more ordered pairs may have the same first component; but in a function, no two distinct ordered pairs may have the same first component.

Example: The set $D = \{(1,6),(3,4),(3,5),(4,6)\}$ is only a relation and **not** a function. because the second and third ordered pairs have the same first component, which is 3.

Example: The set $E = \{(1,2),(2,3),(4,5),(7,5)\}$ is a function (even though the second components of the third and fourth ordered pairs are the same).

A function is a relation, but a relation is not necessarily a function.

Determining if a given graph is a relation or a function

We will use the so-called **vertical line test.**

Procedure

Step 1: Draw as many vertical lines as possible (This can be done visually.) to intersect the graph.

Step 2: If any of the possible lines intersects (cuts) the graph at more than one point, then the given graph is not a function but a relation. However, if each of the possible vertical lines intersects the graph only once (at one point only), then the graph represents a function. Figure.. is a graph which is a relation but not a function.

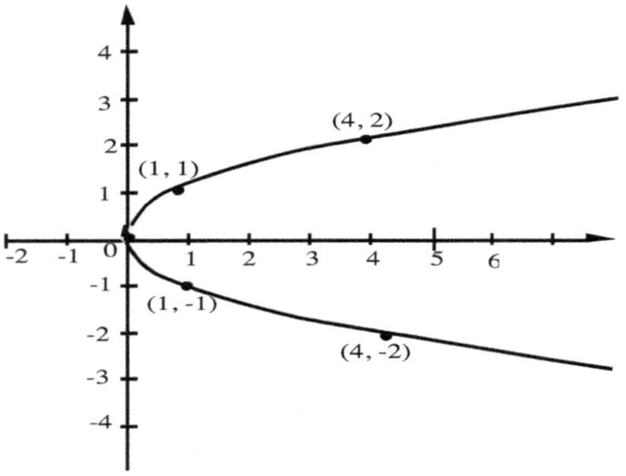

Figure: Graph of $y = \pm\sqrt{x}$ or $x = y^2$

This graph is a relation but not a function

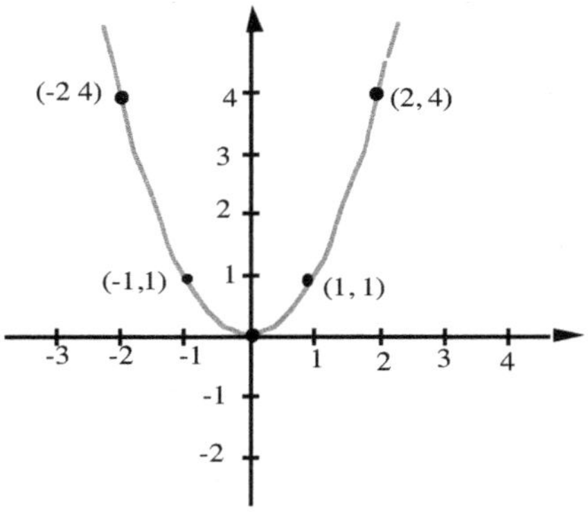

Figure: Graph of $y = x^2$.

This graph is that of a function.

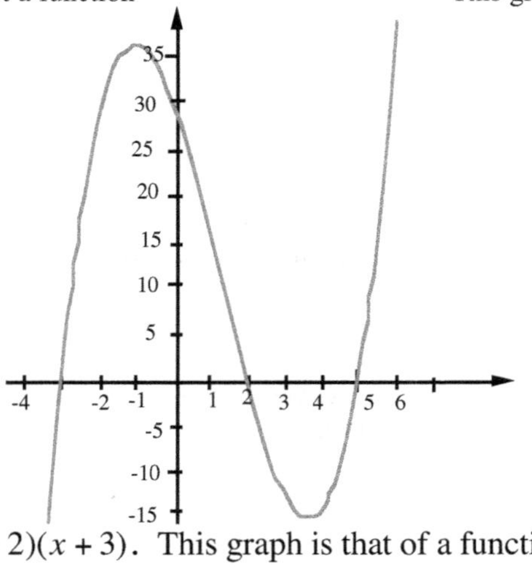

Figure: Graph of $y = (x-5)(x-2)(x+3)$. This graph is that of a function.

Lesson 2 Exercises

Determine which of the following are graphs of functions.

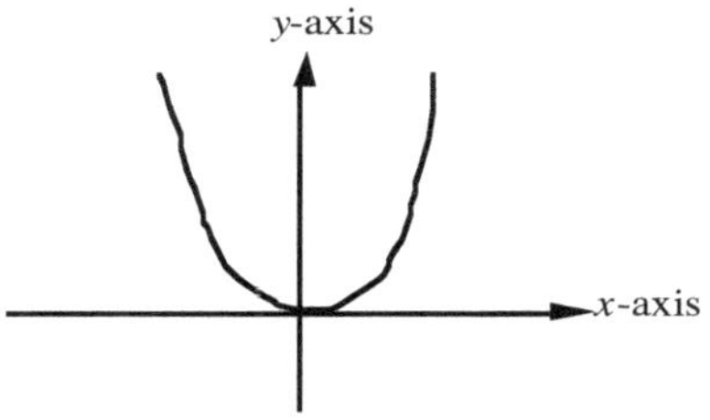

Figure (a)

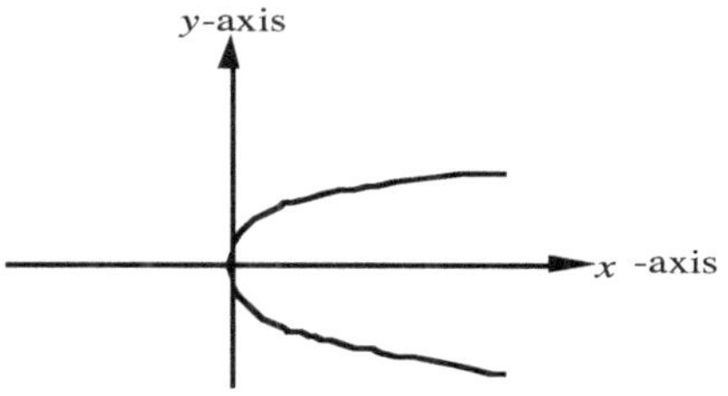

Figure (b)

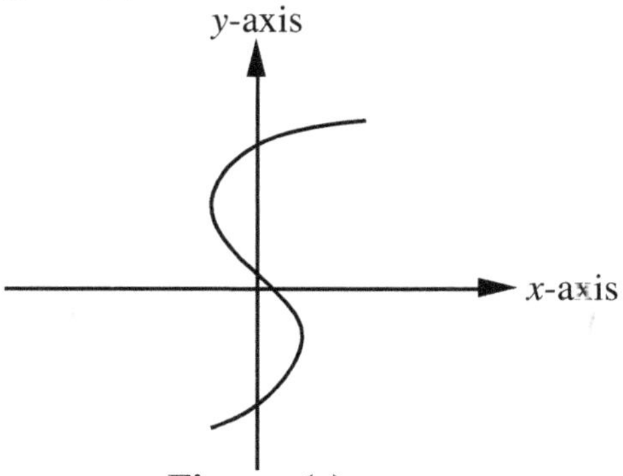

Figure (c)

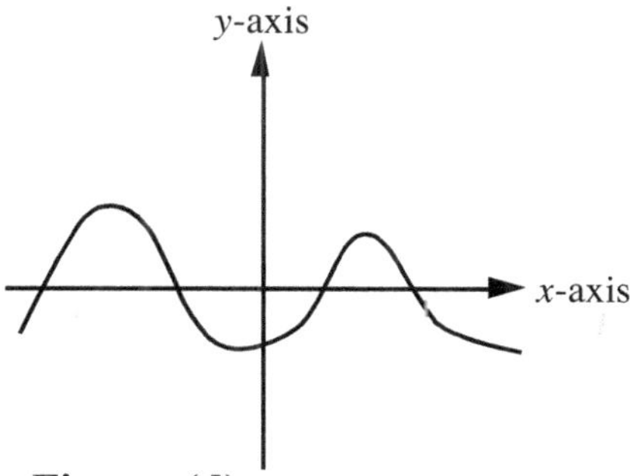

Figure (d)

Answers: (a) A function; (b) Not a function; (c) Not a function; (d) A function

Lesson 3

Functional Notation; Defined Functions; Excluded Values; Domain and Range

Functional Notation

Let a function $f(x)$ be specified by the rule $f(x) = x^2 + 3$. (1)

In equation (1), $f(x)$ is read "*f* of *x*" or *f* is a function of *x*.

To evaluate a function for a particular value of *x*, we substitute that value of *x* in the rule that defines $f(x)$. Note that $f(x)$ does not mean *f* times *x* but that $f(x)$ is written as a symbol.

Example 1 Given that $f(x)= x^2 + 3$, find (a) $f(-1)$.; (b) $f(x_0 + h) - f(x_0)$

Solution: (a) $f(x) = x^2 + 3$

$f(-1) = (-1)^2 + 3$ (replacing *x* in the given equation by -1)

$= 1 + 3$

$f(-1) = 4$

(b) (Replace *x* in the given equation by $(x_0 + h)$ and x_0, accordingly in $f(x)= x^2 + 3$

$$f(x_0 + h) - f(x_0) = [(x_0 + h)^2 + 3] - (x_0{}^2 + 3)$$

$$= [x_0{}^2 + 2hx_0 + h^2 + 3] - x_0{}^2 - 3$$

$$= x_0{}^2 + 2hx_0 + h^2 + 3 - x_0{}^2 - 3$$

$$= 2hx_0 + h^2$$

Example 2 Find $f(2)$, given that $f(x) = x + 7$

Solution

$f(x) = x + 7$

$f(2) = 2 + 7$

$= 9$

Example 3 If $f(x) = 2 - \frac{1}{x - 4}$, find (*a*) $f(-3)$; (b) $f(x_0 + h)$; (*b*) $f(-x)$.

Solution (*a*) $f(x) = 2 - \frac{1}{x - 4}$

$f(-3) = 2 - \frac{1}{(-3) - 4}$ (replacing *x* in the given equation by -3)

$= 2 - \frac{1}{-7}$

$= 2 + \frac{1}{7}$

$= 2\frac{1}{7}$

(c) (Replace x in the given equation by $x_0 + h$

$$f(x) = 2 - \frac{1}{x-4} \quad \text{<---- given equation}$$

$$f(x_0 + h) = 2 - \frac{1}{(x_0 + h) - 4} \quad \text{<---replacing } x \text{ by } x_0 + h$$

$$= 2 - \frac{1}{x_0 + h - 4}$$

$$= \frac{2(x_0 + h - 4) - 1}{x_0 + h - 4}$$

$$= \frac{2x_0 + 2h - 8 - 1}{x_0 + h - 4}$$

$$= \frac{2x_0 + 2h - 9}{x_0 + h - 4}$$

(*c*) (Replace x in the given equation by $-x$)

$$f(x) = 2 - \frac{1}{x-4} \quad \text{<---- given equation}$$

$$f(-x) = 2 - \frac{1}{(-x) - 4} \quad \text{<-----}(replacing\ x\ by - x)$$

$$= 2 - \frac{1}{-x - 4}$$

$$= \frac{2(-x - 4) - 1}{-x - 4}$$

$$= \frac{-2x - 8 - 1}{-x - 4}$$

$$= \frac{-2x - 9}{-x - 4}$$

$$= \frac{-(2x + 9)}{-(x + 4)} \quad (factoring\ out - 1)$$

$$= \frac{2x + 9}{x + 4}$$

Note above that in (*b*) the final result contains x. This is so, because we replaced x by $-x$. In the case of (*a*), we replaced x by the integer -3 , and the final result was purely a numerical value.

Furthermore, in Example 3, $f(-a) = \dfrac{2a + 9}{a + 4}$

Defined Real-Valued Function

Meaning of a defined function of x

A real-valued function $f(x)$ is said to be defined for a variable x if the following conditions are satisfied:

1. The x and $f(x)$ must be real (i.e., x and $f(x)$ should not be the square root or an even root of a negative number). Thus, a value such as $\sqrt{-4}$ or $2i$ is not allowed.

   For example, in $f(x) = \sqrt{x-4}$, we have to make sure that $x \geq 4$, since otherwise, we obtain imaginary numbers.

2. The value of x when substituted in the functional equation should yield specific real numbers. (i.e., the value of x when substituted in the functional equation should **not** make the denominator become zero.)

Condition (2) implies that the function should not become undefined when the value of x is substituted in the functional equation. When the function involves a denominator, we have to make sure that the denominator is not allowed to be zero. A particular example of this function occurs when the given function is the ratio of two polynomial functions. (We call such functions rational functions.)
In this book, it is agreed that a function is real-valued unless otherwise specified.

Examples of rational functions are: $(a)\ f(x) = \dfrac{x^2+4}{x-1}$; $(b)\ f(x) = \dfrac{1}{(x-3)(x+4)}$

Excluded Values, Domain and Range

The **excluded values** are (usually) the values which when substituted in the functional equation make the function either undefined or imaginary.

The **domain** (say , D) of a function $f(x)$ consists of those real values of the independent variable say, x, for which $f(x)$ is real and defined.

The **range** (say, R) of the function consists of the corresponding values which $f(x)$ assumes for x in the domain of the function .

Specification of Domain and Range

The domain and the range of a given function may be specified in several ways, namely in the form of a set, in the form of a table, in the form of an equation, in the form of an inequality, or in the form of a graph.(In some of the examples that follow, we will use a graphing calculator or a computer grapher to generate graphs which will be useful in determining the range of some of the functions. Determining the range of some of the functions analytically may require advanced methods (which we do not cover in this book) , and therefore, we use graphing as an aid.)

Example 1: In the form of a set

Find the domain and range of the function specified by $\{(1,2),(3,2),(4,4),(5,5)\}$.

Solution: The domain consists of the first components of the ordered pairs.
The domain, $D = \{1,3,4,5\}$.
The range consists of the second components of the ordered pairs.
The range, $R = \{2,4,5\}$.

Example 2: In the form of a table of values

Find the domain and range of the function specified by the table of values below.

x	y
1	2
3	2
4	4
5	5

Solution: The domain consists of (the x-values) 1, 3, 4, and 5.

The range consists of the (y-values) 2, 4 and 5.

Note that this form is the set form written in a different format.

Example 3: In the form of an equation:

The domain consists of those real values of x for which the function is defined. The corresponding values of $f(x)$ form the range of the function.

Example 4: In the form of a graph

Find the domain and range of the function specified by the graph below.

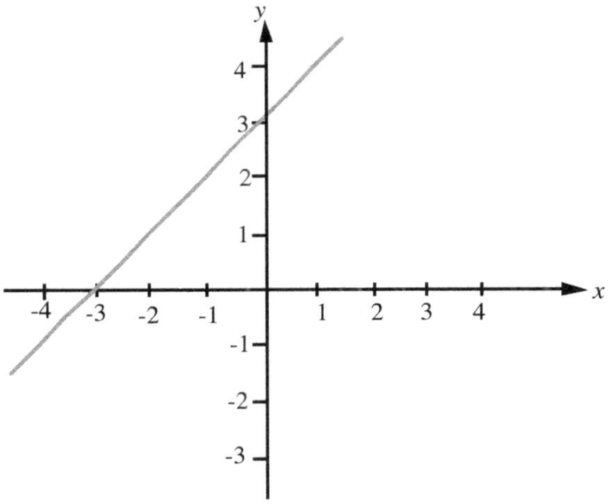

Figure: The graph of $y = x + 3$

Solution: From the graph, the function is defined for all real values of x.

The domain consists of all real x-values. The range consists of all real y-values.

Implicit and Explicit Specification of the Domain of a Function

The domain of a function may be specified either implicitly or explicitly.

Consider the function $f(x) = x^2$

As it stands, the domain is implicitly specified. The above function is that of a polynomial and as such the domain consists of all real numbers. Here, we assume the largest possible domain.

Now, consider $f(x) = x^2 \qquad 0 \le x \le 5$

In this case, the inequality written to the right of the function specifies explicitly and restricts the domain. The domain of this function is such that x is between 0 and 5, including 0 and 5.

If the inequality to the right had not been indicated, the domain would have consisted of all real x-values.

The restrictions on the domains are very important and useful in sketching the graphs of functions.

Other functions with explicitly specified domains are

$(a)\ \ y = \sin x \qquad 0 \le x \le 2\pi$

$(b)\ \ y = \cos x \qquad -2\pi \le x \le 2\pi$

Determining the Excluded Values, Domain and Range of a Function

Case 1: Polynomial functions

Example 1 (a) For what values of x is the following function not defined? (b) what is the domain?

(c) what is the range? $f(x) = x^2 - 3x + 1$

Solution: All polynomial functions are defined for all real values of the independent variable.

(a) Since the given function is a polynomial function, it is defined for all real values of x. We may note that the right-hand side of the equation does not involve denominators or square roots (or even roots) of the independent variable. There are **no** excluded values.

(b) The domain consists of all real values of x. Set-builder notation: Domain = $\{x \mid x \text{ is a real number}\}$

Using interval notation: Domain $=(-\infty, +\infty)$

(c) Generally, determining the range of a polynomial function may require advanced methods. However, since the given function is a quadratic function , we may apply the range inequality formula, $y \ge \dfrac{4ac - b^2}{4a}$

From the function, $a = 1, b = -3$, and $c = 1$.

Substituting these values,

$$y \ge \frac{4(1)(1) - (-3)^2}{4(1)} = -\frac{5}{4} \quad (\text{i.e., } y \ge -\tfrac{5}{4})$$

Set-builder: $\{y \mid y \ge -\frac{5}{4}\}$

The range consists of all real numbers greater than or equal to $-\frac{5}{4}$.

(Note that any horizontal line drawn below the point $(\frac{3}{2}, -\frac{5}{4})$ will **not** intersect the curve, but any horizontal line through or above this point will intersect the curve)

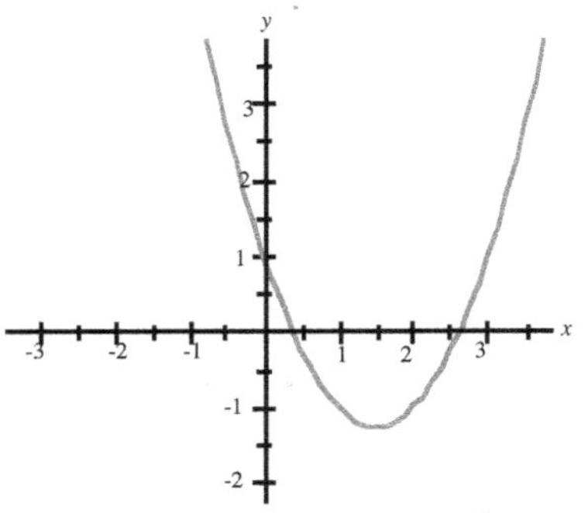

Figure: Graph of $f(x) = x^2 - 3x + 1$

Case 2: Rational functions

Note: A rational function is a function which is the ratio of two polynomial functions.

Example 2 (a) For what values of x is the following function not defined? (b) what is the domain? (c) what is the range?

$$f(x) = \frac{3x - 2}{x - 1}$$

Solution Step 1: Setting the denominator to zero,

$$x - 1 = 0$$

Step 2: Solving for x, $x = 1$.

(a) The function is not defined when $x = 1$. (The excluded value of x is 1.)

(b) The domain is all real values of x, except 1. same as $\{x \mid x \text{ is a real number and } x \neq 1\}$

(c) The range (from graph) if found by being guided by the horizontal asymptote, $y = 3$. (see p.36)

The range is given by the set $\{y \mid y < 3 \text{ or } y > 3\}$ or simply $\{y \mid y \neq 3\}$

(Note that a horizontal line drawn through (0, 3) will **not** intersect the curve)

Checking for $x = 1$, $f(1) = \frac{3(1) - 2}{1 - 1} = \frac{3 - 2}{0} = \frac{1}{0}$ which is undefined .(The right-hand side is division by zero).

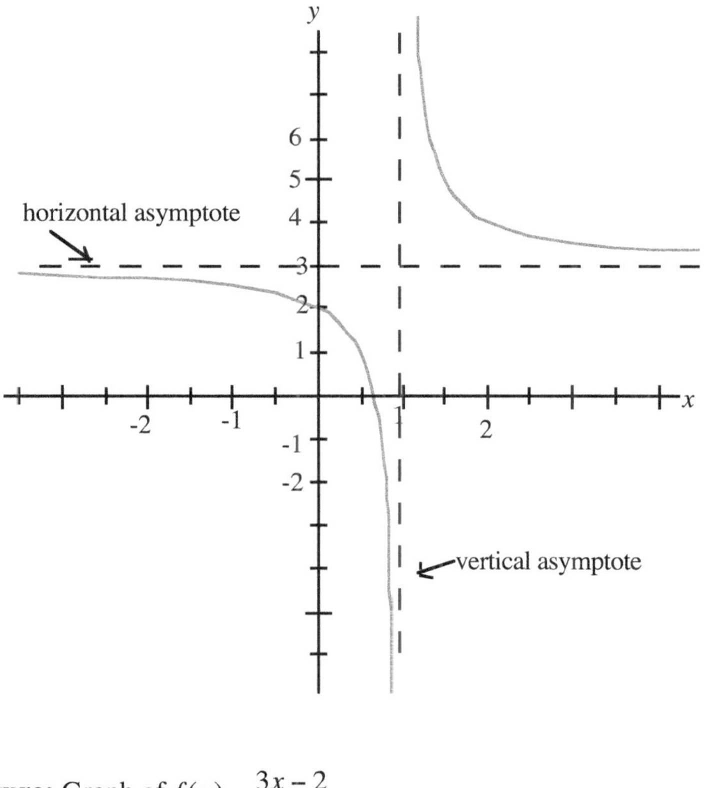

Figure: Graph of $f(x) = \dfrac{3x-2}{x-1}$

Example 3 (a) For what values of x is the given function not defined? (b) what is the domain?
(c) what is the range?

$$f(x) = \frac{2(x - 1)}{(x - 2)(x + 4)}$$

Solution Step 1: Setting the denominator to zero,

$$(x - 2)(x + 4) = 0$$

Step 2: Solving for x, $x = 2$, or $x = -4$.

(a) The function is not defined when $x = 2$ and -4. (The excluded values of x are 2 and -4.)

(b) The domain is all real values of x, except 2 and -4.

Set-builder notation: $\{x \mid x \text{ is a real number and } x \neq -4,\ x \neq 2\}$

(c) The range (from graph) is all real y.

Set-builder notation: $\{y \mid y \text{ is a real number}\}$

(Note that any horizontal line drawn through the y-axis will intersect the curve)

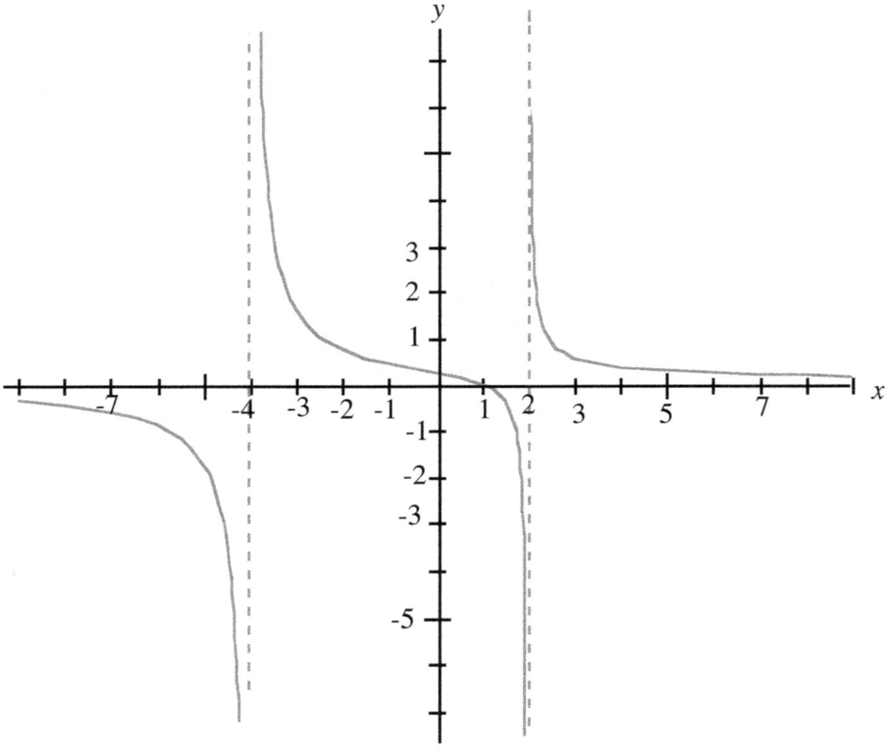

Figure: Graph of $f(x) = \dfrac{2(x-1)}{(x-2)(x+4)}$

Example 4 (a) For what values of x is the given function not defined? (b) what is the domain? (c) what is the range?

$$f(x) = \frac{1}{x}$$

Solution Setting the denominator to zero,
$x = 0$

(a) The function is not defined when $x = 0$. (The excluded value of x is 0.)

(b) The domain is all real values of x, except 0.
Set-builder notation: $\{x \mid x \text{ is a real number and } x \neq 0\}$

(c) The range (from graph) is given by the set $\{y \mid y < 0 \text{ or } y > 0\}$ or simply $\{y \mid y \neq 0\}$
(the horizontal asymptote is $y = 0$)
(Note that any horizontal line drawn through the y-axis, except through the point $(0, 0)$, will intersect the curve.

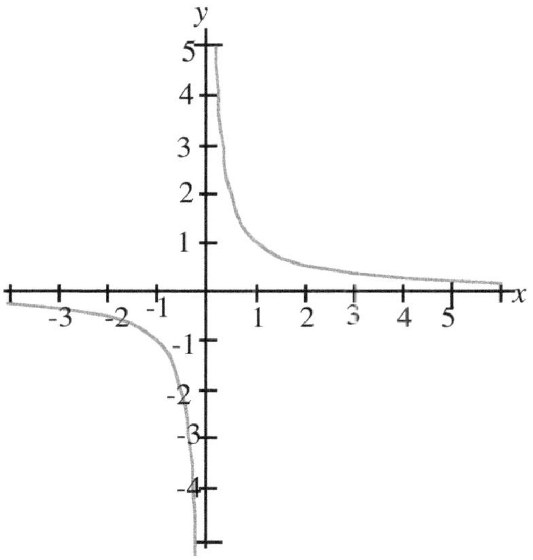

Figure: Graph of $f(x) = \frac{1}{x}$

Example 5 (a) For what values of x is the given function not defined? (b) what is the domain? (c) what is the range?

$$f(x) = \frac{x^3 + x^2 + 2}{x^2 - 16}$$

Solution: Setting the denominator to zero and solving,

$$x^2 - 16 = 0$$

$$(x + 4)(x - 4) = 0$$

$$x = -4 \text{ or } x = 4$$

The function is not defined when $x = -4$ or 4. (The excluded values of x are -4 and 4.)

(a) The domain is all real x except -4 and 4.

$\{x \mid x \text{ is a real number and } x \neq -4,\ x \neq 4\}$

(b) The range (from graph) is all real values of y. same as $\{y \mid y \text{ is a real number}\}$

(Note that any horizontal line drawn through the y-axis will intersect the curve)

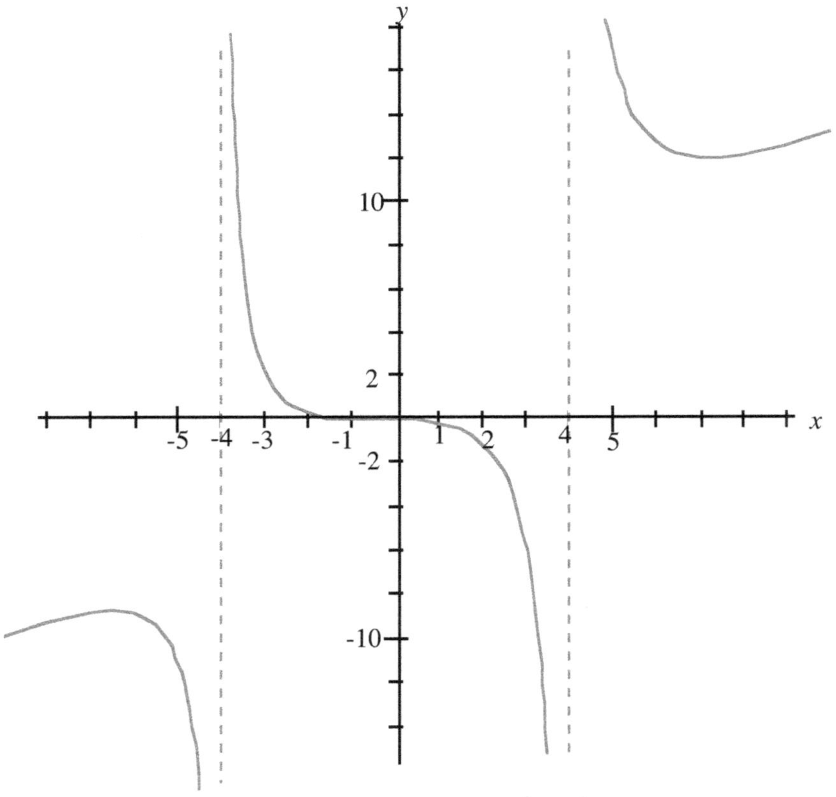

Figure: Graph of $\frac{x^3 + x^2 + 2}{x^2 - 16}$

Example 6 (a) For what values of x is the given function not defined? (b) what is the domain? (c) what is the range?

$$f(x) = \frac{8}{x^2 - 4}$$

Solution Step 1: Setting the denominator to zero,

$$x^2 - 4 = 0$$

$$(x + 2)(x - 2) = 0.$$

Step 2: Solving for x, $x = 2$, or $x = -2$.

(a) The function is not defined when $x = 2$ and -2. (The excluded values of x are 2 and -2.)

(b) The domain is all real values of x except 2 and -2.

Same as $\{x \mid x \text{ is a real number and } x \neq -2\ ,\ x \neq 2\ \}$

(c) The range (from graph) is given by the set $\{y \mid y \leq -2 \text{ or } y > 0\}$

(Note that any horizontal line drawn through or below $y = -2$ or above $y = 0$ will intersect the curve)

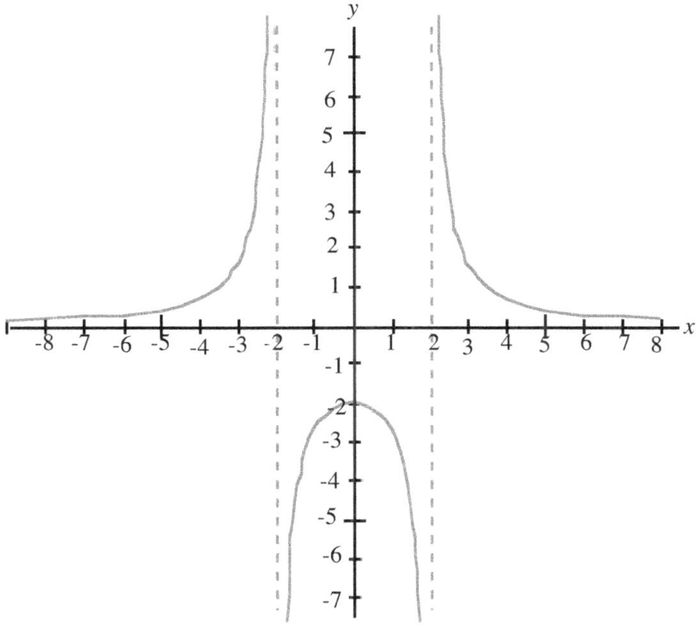

Figure: Graph of $f(x) = \frac{8}{x^2 - 4}$

Example 7 (a) For what values of x is the given function not defined? (b) what is the domain?
(c) what is the range?

$$f(x) = \frac{8}{x^2 + 4}$$

Solution Step 1: Setting the denominator to zero, and solving, we obtain non-real values. Since we are dealing with real-valued functions, we conclude that

Step 2: (a) There are **no** excluded values.
$x^2 + 4$ is positive for all real values of x and never zero, since the square of any nonzero real number is always positive.

(b) The function is defined for all real values of x.
Domain: = $\{x \mid x \text{ is a real number}\}$

(c) The range (from graph) is given by the set $\{y \mid 0 < y \leq 2\}$

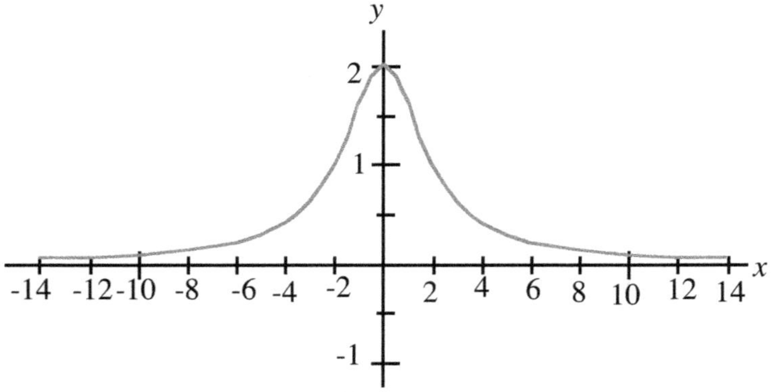

Figure: Graph of $f(x) = \dfrac{8}{x^2 + 4}$

Lesson 3 Exercises

1. If $(x) = x^2 - 5x + 2$, find (a) $f(-2)$; (b) $f(-1)$, (c) $f(-x)$, (d) $f(a + h)$, (e) $\dfrac{f(a+h) - f(a)}{h}$;

A: 2. If $(x) = 4 - \dfrac{1}{x - 2}$, find (a) $f(-3)$; (b) $f(-x)$; (c) $f(a)$.

3. If $(x) = x^3 - x^2 - x - 1$, find $f(-1)$.

Answers: **1.** (a) 16; (b) 8; (c) $x^2 + 5x + 2$; (d) $a^2 - 5a + 2ah - 5h + h^2 + 2$; (e) $2a + h - 5$

2. (a) $4\frac{1}{5}$; (b) $\dfrac{4x + 9}{x + 2}$; (c) $\dfrac{4a - 9}{a - 2}$; **(3)** -2

B **1**.What is meant by the domain of a function?

2.What conditions do you look out for in determining the domain of a function?

3.What is meant by the range of a function?

Find the domain and range in each of the following:

4. $A = \{(2,4),(3,9),(4,16),(5,25)\}$; **5.** $B = (2,1),\ (5,1),\ (6,1),\ (7,3)$

6. $y = x + 2$; **7.** $y = x^2 + 3x + 7$; **8.** $y = -x - 4$; **9.** $y = x^3 + x^2 - 5$

Answers: **4**. Domain : $\{2,3,4,5\}$;, range: $\{4,9,16,25\}$

5. Domain : $\{2,5,6,7\}$;, range: $\{1,\ 3\}$

6. Domain : All real values of x: range: all real values of y.

7. Domain : All real x; range: all real y such that $y \geq 4.75$.

8. Domain : All real values of x: range: all real values of y.

9. Domain : All real values of x.

Lesson 4

One-to-One Functions, Composite Functions

One-to-One Functions

(In Lesson 5, we will learn that if a function is one-to-one, then it has an inverse function.)
We consider three main cases according to how the function is specified.

Case 1: **Given a set of ordered pairs** (or a table of x- and y-values)

A one-to-one (1-1) function is a set of ordered pairs in which for any two ordered pairs, the first elements are different from each other and the second elements are also different from each other.

Example The set, $A = \{(3,2),(4,7),(1,5),(2,3)\}$ is a one-to-one function. However.
The set, $B = \{(3,2),(4,7),(1,5),(5,2)\}$ is **not** a one-to-one function because the first and the last ordered pairs have the same second elements, namely, 2.

Case 2: Given the graph of the function

By the so called **horizontal line test**, the graph of a function is that of a one-to-one function if every possible horizontal line drawn to intersect the graph cuts (intersects) the graph only once (at one point only). **Figures** 2 and 3, p. 26, are graphs of one-to-one functions but **Figure 1** is not one-to-one.

Case 3: Given the equation of the function

A function $f(x)$ is one-to-one if whenever $f(x_1) = f(x_2), x_1 = x_2$. If we let $y_1 = f(x_1)$, and $y_2 = f(x_2)$, then $y_1 = y_2$ **must** imply that $x_1 = x_2$, otherwise, $f(x)$ is not one-to-one.

Example 2: Determine if $f(x) = 2x + 3$ is one-to-one.

Solution

Step 1: $f(x_1) = 2x_1 + 3$

$f(x_2) = 2x_2 + 3$

Equate RHS of $f(x_1)$ to RHS of $f(x_2)$
(That is, let $f(x_1) = f(x_2)$):

$$2x_1 + 3 = 2x_2 + 3 \qquad (1)$$

Step 2: Solve for x_1. (You may also solve for x_2)

$$2x_1 + 3 = 2x_2 + 3$$
$$2x_1 = 2x_2$$
$$x_1 = x_2 \qquad (2)$$

Since from above, whenever $f(x_1) = f(x_2)$ (from equation (1)) $x_1 = x_2$ (from equation (2))
$f(x) = 2x + 3$ is one-to-one. (You may check by sketching its graph and using the horizontal line test)

Example 3 Determine if $f(x) = \sqrt{25 - x^2}$ is one-to-one.

Solution

Step 1: $f(x_1) = \sqrt{25 - x_1^2}$

$f(x_2) = \sqrt{25 - x_2^2}$

Equate RHS of $f(x_1)$ to RHS of $f(x_2)$
(That is, let $f(x_1) = f(x_2)$.)

$$\sqrt{25 - x_1^2} = \sqrt{25 - x_2^2} \qquad (1)$$

Step 2: Solve for x_1.

$$\sqrt{25 - x_1^2} = \sqrt{25 - x_2^2}$$
$$25 - x_1^2 = 25 - x_2^2$$
$$-x_1^2 = -x_2^2$$
$$x_1^2 = x_2^2$$
$$x_1 = \pm\sqrt{x_2^2}$$
$$x_1 = +x_2 \text{ or } -x_2 \qquad (2)$$

Since from above, whenever $f(x_1) = f(x_2), x_1 = -x_2$, (That is for the same y-value, x is not unique: $x_1 = +x_2$, and **also** $x_1 = -x_2$ (from equation (2))
$f(x) = \sqrt{25 - x^2}$ is **not** one-to-one. (You may check by sketching its graph and using the horizontal line test)

Example 4: Determine if $f(x) = |x - 2|$ is one-to-one.

Solution

Step 1: $f(x_1) = |x_1 - 2|$

$f(x_2) = |x_2 - 2|$

Equate RHS's of $f(x_1)$ to $f(x_2)$

(That is, let $f(x_1) = f(x_2)$):

$$|x_1 - 2| = |x_2 - 2| \qquad (1)$$

Step 2: Solve for x_1:

If both $x_1 - 2$ and $x_2 - 2$ are positive,

$x_1 - 2 = x_2 - 2$ and from which $x_1 = x_2$

(same result as if both $x_1 - 2$ and $x_2 - 2$ are negative)

However, if $x_1 - 2$ is positive and $x_2 - 2$ is negative,

$$x_1 - 2 = -(x_2 - 2)$$

$$x_1 - 2 = -x_2 + 2$$

$x_1 = -x_2 + 4$. That is, $x_1 \neq x_2$

Since from above, whenever $f(x_1) = f(x_2)$, $x_1 \neq x_2$

$f(x) = |x - 2|$ is **not** one-to-one. (You may check by sketching its graph and using the horizontal line test)

Another method for determining if a function is one-to-one

The author proposes the following definition for a one-to-one function.

Definition : A function is one-to-one if its inverse relation is a function.

This definition provides another method for determining if a given function is one-to-one.

Example 5

Determine if $f(x) = 2x + 6$ is one-to-one

Solution

Given: $f(x) = 2x + 6$

Required: To determine if $f(x) = 2x + 6$ is one-to-one.

Plan: If it can be shown that $f(x) = 2x + 6$ has an inverse relation which is function, then $f(x) = 2x + 6$ is one-to-one.

Determination:

Step 1: Let $f(x) = y$ to obtain $y = 2x + 6$.

Step 2: Interchange x and y to obtain the inverse relation $x = 2y + 6$

Step 3: Solve for y.

$$2y = x - 6$$

$$y = \tfrac{1}{2}x - 3$$

Since clearly, for a given value of x there is exactly one corresponding value of y.

the inverse relation, $y = \frac{1}{2}x - 3$, is a function and therefore, $f(x) = 2x + 6$ is one-to-one.

(You may also check that $y = \frac{1}{2}x - 3$ is a function by sketching its graph and applying the vertical line test)

Example 6

Determine if $f(x) = x^2$ is one-to-one.

Solution

Given: $f(x) = x^2$

Required: To determine if $f(x) = x^2$ is one-to-one.

Plan: If it can be shown that $f(x) = x^2$ has an inverse relation which is function, then

$f(x) = x^2$ is one-to-one.

Determination

Step 1: Let $f(x) = y$ to obtain $y = x^2$

Step 2: Interchange x and y to obtain the inverse relation $x = y^2$

Step 3: Solve for y.

If $y^2 = x$

$y = \pm\sqrt{x}$ (that is, $y = +\sqrt{x}$ or $y = -\sqrt{x}$

Clearly, for a given value of x there are two different corresponding y-values. Therefore, y is **not** a function of x, and the inverse relation $y^2 = x$ is **not** a function and therefore , the given function $f(x) = x^2$ is **not** one-to-one.

(You may also check that $y^2 = x$ or $y = \pm\sqrt{x}$ is **not** a function by sketching its graph and applying the vertical line test)

Composite Functions

Some authors refer to composite functions as "product functions". This alternative terminology may be misleading because a reader might be inclined to multiply the given functions. Perhaps, a better alternative terminology for a composite function is " a function within another function".

Use of Composition of Functions:
The principle of composition of functions can be used to determine algebraically if two given functions are inverses of each other.

Definition: If two functions f and g are such that the range of g is in the domain of f, then the composite function of f with g, symbolized $f \circ g$, is specified by $(f \circ g)(x) = f[g(x)]$ which is read "f of g of x". That is, the output of g becomes the input for f.

Similarly, if the range of f is in the domain of g, then the composite function of g with f, symbolized $g \circ f$ is specified by $g \circ f = g[f(x)]$ which is read "g of f of x." That is, the output of f becomes the input for g.

We must **note** that generally, $f \circ g \neq g \circ f$ (i.e., generally, $f \circ g$ is not equal to $g \circ f$)

Example 1

If $f(x) = x^2$, and $g(x) = x + 1$, find (*a*) $f[g(x)]$; (*b*) $g[f(x)]$.

Solution

Recall that if,for example,

$f(x) = x^2$

then $f(3) = (3)^2$.

Similarly, (a) $f[g(x)] = f[x + 1]$

$= [x + 1]^2$

$f[g(x) = x^2 + 2x + 1$

Thus, (in the above problem) wherever there is x in the equation for $f(x) = x^2$, write (substitute) $x + 1$.

(*b*) $g[f(x)]$

$= g[x^2]$

$= (x^2) + 1$ (substitute x^2 for x in the equation for $g(x) = x + 1$)

$= x^2 + 1$

Thus, wherever there is x in the equation for $g(x) = x + 1$, we substitute x^2.

Example 2: If $f(x) = 2(x + 10)$, and $g(x) = x - 2$, find (*a*) $f[g(x)]$; (*b*) $g[f(x)]$.

Solution

(*a*) $f[g(x)] = 2[(x - 2) + 10$

$= 2[x - 2 + 10]$

$= 2[x + 8]$

$f[g(x)] = 2x + 16$

(Substituting x - 2 in the equation for $f(x)$.

(*b*) $g[f(x)] = 2(x + 10) - 2$

$= 2x + 20 - 2$

$g[f(x)] = 2x + 18$

(Substituting $2(x + 10)$ in the equation for $g(x)$.

Domains of Composite Functions

The approach here is similar to that for domains of algebra of functions. However, here, in the first step, we check only for the domain of the "inside" function. In the second step, we check for the domain of the resulting composite function. We add any excluded values to those from Step 1 and if there is an overlap in the domains we use the intersection of the domains from Step 1 and Step 2. Pay attention to radical functions such as $f(x) = \sqrt{x-2}$.

Given $f(x) = \frac{3}{x}$ and $g(x) = \frac{3x-4}{x+2}$, find the domains of the following:
find (a) $f[g(x)]$; (b) $g[f(x)]$.

(a) Step 1: We check the domain of the "inside function", $g(x) = \frac{3x-4}{x+2}$

Domain of $g: \{x \mid x \text{ is a real number and } x \neq -2\}$.

Step 2: Form the composite function and check its domain.

$$f[g(x)] = \frac{3}{\frac{3x-4}{x+2}} \qquad (f(x) = \tfrac{3}{x})$$

$$= \frac{3}{1} \bullet \frac{x+2}{3x-4}$$

$$f[g(x)] = \frac{3(x+2)}{3x-4}$$

Set $3x - 4 = 0$ to obtain $x = \frac{4}{3}$. Therefore $x \neq \frac{4}{3}$.

Combining the domains from Step 1 and Step 2, the excluded values are -2, and $\frac{4}{3}$.; and the domain of $f[g(x)] = \frac{3(x+2)}{3x-4}$ is

$\{x \mid x \text{ is a real number and } x \neq -2 \text{ and } x \neq \frac{4}{3}\}$

(a) Step 1: We check the domain of the "inside function", $f(x) = \frac{3}{x}$

Domain of $f: \{x \mid x \neq 0\}$.

Step 2: Form the composite function and check its domain.

$$g[f(x)] = \frac{3(\frac{3}{x})-4}{\frac{3}{x}+2} \qquad (g(x) = \tfrac{3x-4}{x+2})$$

$$g[f(x)] = \frac{-4x+9}{2x+3}$$

Set $2x + 3 = 0$ to obtain $x = -\frac{3}{2}$.

Therefore $x \neq -\frac{3}{2}$.

Combining the domains from Step 1 and Step 2, the excluded values are 0, and $-\frac{3}{2}$.; and the domain of $g[f(x)] = \frac{-4x+9}{2x+3}$ is

$\{x \mid x \text{ is a real number and } x \neq -0 \text{ and } x \neq -\frac{3}{2}\}$

Application of Composition of Functions

If two functions f_1 and f_2 are inverses of each other, then, the following two conditions must be satisfied simultaneously. **1.** $f_1[f_2(x)] = x$ and **2.** $f_2[f_1(x)] = x$. For examples, see page 31. .

Lesson 4 Exercises

A Show and determine which of the following functions are one-to-one

1. The set $\{(4,5), (3,4), (1,2)\}$; 2. The set $\{(2,2), (4,3), (5,7)\}$; 3. $f(x) = 3x - 2$;

4. $f(x) = \sqrt{x+3}$; **5.** $f(x) = \sqrt{16-x^2}$; **6.** $f(x) = |x|$; **7.** $f(x) = |x-2|$; **8.** $f(x) = x^3 + 2$.

Answers: **1.** Yes; **2.** Yes; **3.** Yes; **4. Yes**; **5.** No; 6. No; **7.** No.; 8. Yes

B **1.** Given that $f(x) = x + 2$, $g(x) = x - 1$, find (a) $f[g(x)]$; (b) $g[f(x)]$

2. Given that $f(x) = 3(x+2)$, $g(x) = \frac{1}{x}$ find (a) $f[g(x)]$; (b) $g[f(x)]$

3. Given that $f(x) = (x+1)^2 + 3$, $g(x) = -x$, find (a) $f[g(x)]$; (b) $g[f(x)]$

Answers: **1.** (a) $x + 1$; (b) $x + 1$; **2.** (a) $\frac{3}{x} + 6$; (b) $\frac{1}{3x+6}$; **3.** (a) $x^2 - 2x + 4$; (b) $-x^2 - 2x - 4$

Lesson 5

Inverse Functions and Inverse Relations

(Exchange is no Robbery)

Recalling the definitions of a relation and a function (page 4), a relation is a set of ordered pairs and function is a relation in which no two distinct ordered pairs have the same first elements (components).

Inverse Relation

Given a relation which is specified by the set of ordered pairs (x, y), the inverse relation of this given relation is the set of ordered pairs (y, x). This inverse relation is obtained by interchanging the first and second elements of each ordered pair.

Example 1: Find the inverse relation of the function specified by the set A:

$$A = \{(1, 3), (3, 2), (5, 7)\}$$

Solution The inverse relation is obtained by interchanging the first and second elements of each ordered pair

The inverse relation of A is the set

$$\{(3, 1), (2, 3), (7, 5)\}$$

From the definition of the inverse, we can conclude that when we form the inverse of a relation, the domain and the range of the original relation are interchanged. Thus, the domain of the original relation becomes the range of the inverse, and the range of the original relation becomes the domain of the inverse. The inverse may be a relation or a function. (See page 4)

Inverse function

Let a function be specified by the set of ordered pairs $\{(x, y)\}$. Then the inverse relation of this function is the set of ordered pairs $\{(y, x)\}$. If this inverse relation is also a function, then for the inverse relation ,we symbolize $f^{-1}(x)$, which is read "*f* inverse of *x*" and we say that $f(x)$ and $f^{-1}(x)$ are inverse functions of each other.

Example 2 Find the inverse function of the function specified by the set B,:

$$B = \{(a, b), (c, d), (e, f)\}$$

Solution: The inverse function of B is the set, denoted by B^{-1}, is given by

$$B^{-1} = \{(b, a), (d, c), (f, e)\}$$

Sometimes, authors ask the question :" Does this function have an inverse?" Such a question may sometimes be misunderstood, since the inverse is found by interchanging the roles of x and y, which is always possible. What is implied in such a question is whether or not the inverse relation obtained is **also** a function. Perhaps, an unambiguous form of the question should be "Does this function have an **inverse function**?" If the inverse relation of a given function is also a function, then we also say that the given function is invertible

Note that it is possible that a given relation which is not a function may have an inverse relation which is a function.

Let us elucidate how the terms, " relation, function and inverse", are connected by using the terms "inverse relation" and "inverse function" in the following statements:

Every relation has an inverse relation. This inverse relation may or may not be a function.
Every function has an inverse relation. Some functions have inverse relations which are (inverse) functions.
A function is also a relation , but a relation is not necessarily a function.

Determining if a function has an inverse function

Necessary and sufficient condition for a function to have an inverse function: A function has an inverse function if and only if it is a one-to-one function.

We will consider the different forms in which the rules for specifying a function may be given and determine if the function has an inverse function. We will consider the set form, the tabular form, the graphical form, and the equation form.

Case 1: Given the set form of the function

A function has an inverse function if for any two different ordered pairs, the second elements are different. If a function has an inverse function it is said to be invertible.

The necessary and sufficient condition is a consequence of the definitions of a function (page 4) and of an inverse function. We may note here that the condition for invertibility refers to the differences in the second components while the condition for being a function refers to the differences in the first components.

Example 3 Given the function specified by the set

$A = \{(1,2),\ (2,3),(4,5),(6,7)\}$, determine if the inverse relation of this set is a function.

Solution Since the second components are all different from one another, the inverse relation is a function.

In fact, the inverse of the set A is given by $A^{-1} = \{(2,1),(3,2),(5,4),(7,6)\}$, which is clearly a function, since the first components are all different from one another.

An example of a function whose inverse relation is **not** a function is the set $\{(3,4),(5,4),(6,5)\}$, because the first and the second ordered pairs have the same second component , 4.

Case 2: Given the tabular form of the function

If any two y-values are the same, then the inverse relation is not a function (but only a relation) otherwise, it is a function. This case corresponds to the set form of a function.

Example 4 **Table 3** represents a function whose inverse relation is a function, while **Table 4** represents a function whose inverse relation is **not** a function.

Tables 3

x	y
1	2
2	3
3	4
4	5
5	6
6	7

Table 4

x	y	
1	2	
2	**3**	} same two y-values
3	**3**	
4	5	
5	7	
6	8	

Case 3: Given the graphical form of the function

The horizontal line test

If any **horizontal line** drawn to intersect the graph intersects the graph at only one point, then the inverse relation of the function (graph) is a function. However, if a horizontal line meets the graph in more than one point, then the inverse of the graph is not a function but only a relation.

See Figs 1, 2 and 3 below/ .

Example 3 Figure **1** is **not** the graph of a one-to-one function, Figures **2** and **3** are the graphs of one-to-one functions.

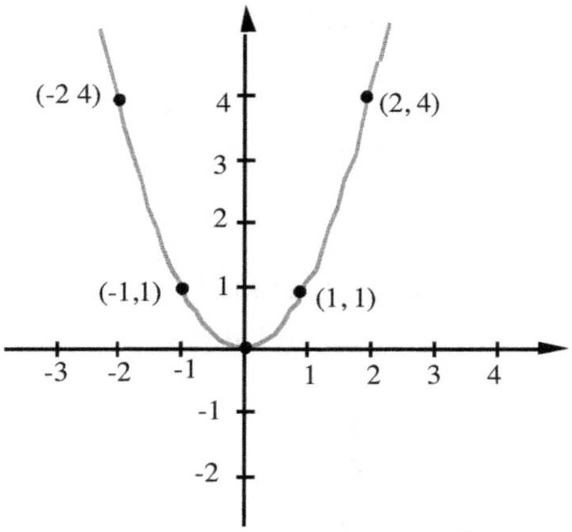

Figure 1: The graph of $y = x^2$,
The inverse relation of this graph is **not** a function

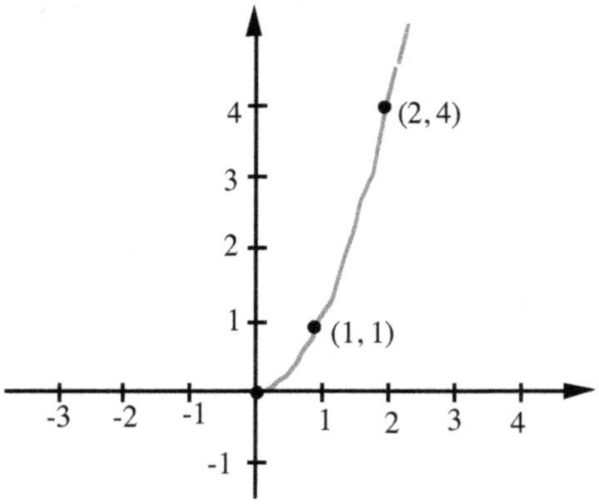

Figure 2: The graph of $y = x^2,\ x \geq 0$.
The inverse relation of this graph is a function
(This is the graph of a one-to-one function)

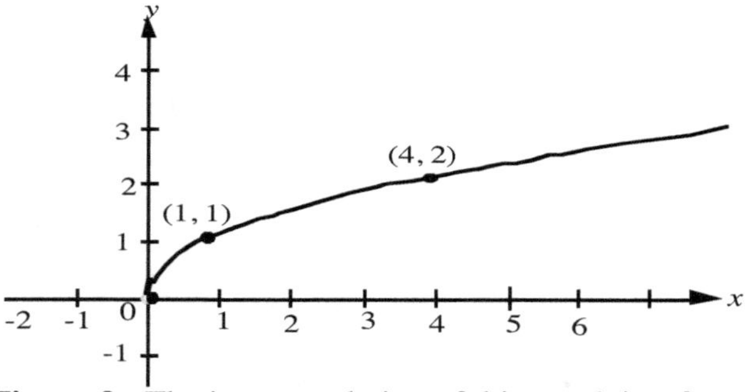

Figure 3: The inverse relation of this graph is a function. (This is the graph of a one-to-one function)

Case 4: Given the equation of the function

By definition, y is a function of x if there is a rule which gives only one corresponding value of y for a given value of x.

Example Determine if the inverse relation of $y = x^2$ is a function.

Method 1: By interchanging the roles of x and y in the given equation, we obtain the inverse relation

$x = y^2$ or $y^2 = x$. Solving $y^2 = x$ for y, we obtain, $y = +\sqrt{x}$ or $y = -\sqrt{x}$

Since for the same x-value (say, 4) we have two different y-values (+2 and -2),. the inverse relation of $y = x^2$ is not a function. However, we could make the inverse relation become a function by restricting the domain of this function (see Figures 1 and 2, above)

For some functions, we will graph the function and use the graphical method.

Method 2: Let $y = f(x)$

Step 1: $f(x_1) = x_1^2$

$f(x_2) = x_2^2$

Equate RHS of $f(x_1)$ to RHS of $f(x_2)$ (That is, let $f(x_1) = f(x_2)$):

$x_1^2 = x_2^2$ (1)

Step 2: Solve for x_1. (You may also solve for x_2)

$x_1^2 = x_2^2$

$x_1 = \pm\sqrt{x_2^2}$

$x_1 = +x_2$ or $-x_2$ (2)

Since from above, $f(x_1) = f(x_2)$, does **not** imply $x_1 = x_2$ (from equation (2)), $f(x)$ is **not** one-to-one and therefore does **not** have an inverse function.

Finding the inverse of a function specified by an equation

Example 3. Find the inverse function of $f(x) = 3x + 2$. (1)

Solution

Step 1: Let $f(x) = y$. Then $y = 3x + 2$.

Interchange x and y in the given equation.

Then, we obtain $x = 3y + 2$. (2)

By tradition, we want to keep the x-axis horizontal and express y as a function of x.

Step 2: Solve equation (2) for y.

Then from $x = 3y + 2$.

$$\frac{x-2}{3} = y$$

$$\text{or } y = \frac{x-2}{3}$$

The inverse $f^{-1}(x) = \dfrac{x-2}{3}$

Alternatively,

Step 1: Solve $y = 3x + 2$. for x.

Then $x = \dfrac{y-2}{3}$

Step 2: Interchange x and y (by definition of the inverse)

$$y = \frac{x-2}{3}$$

The inverse $f^{-1}(x) = \dfrac{x-2}{3}$

We can observe from above that Steps 1 and 2 are interchangeable.

Since the inverse $y = \dfrac{x-2}{3}$ is also a function, we can say that $y = 3x + 2$. and $y = \dfrac{x-2}{3}$ are inverse functions (of each other). We may also add that $f(x) = 3x + 2$. is invertible.

Example 4 Find the inverse function of $f(x) = x^2 + 6$.

Solution

Step 1: Let $f(x) = y$

Step 2: Solve for x.

$y = x^2 + 6.$

$x = \pm\sqrt{y-6}$ (2)

Step 3: Interchange x and y in equation (2).

Then $y = \pm\sqrt{x-6}$ (same as $y = +\sqrt{x-6}$ or $y = -\sqrt{x-6}$) (3)

Clearly, equation (3) does not represent a function, since for a given value of x, there are two y-values. We can test this by graphing and using the so called **vertical line test** (see page 6).

We can say that the given function has an inverse relation specified by $y = \pm\sqrt{x-6}$. However, the given function does not have an inverse function.

In this example, if we were asked to find $f^{-1}(x)$, we would say that there is no $f^{-1}(x)$ (no inverse function) for the given function.

Example 5 Find the inverse function of $f(x) = x^2 + 6$ $x \geq 0$

Solution

Step 1: Let $f(x) = y$

Then $y = x^2 + 6$ $x \geq 0$

Step 2: Solve for x.

Then $x = \pm\sqrt{y-6}$ (That is, $x = +\sqrt{y-6}$ or $x = -\sqrt{y-6}$) (1)

Since we are only interested in the domain $x \geq 0$, we reject the negative part of equation (1).

Then, we obtain $x = +\sqrt{y-6}$ (2)

Step 3: Interchange (to obtain the inverse) x and y in equation (2)

Then $y = +\sqrt{x-6}$ (3)

Equation (3) is the inverse of $y = x^2 + 6$ $x \geq 0$

Therefore, $f^{-1}(x) = \sqrt{x-6}$ $x \geq 6$

The graph of this inverse (Figure **2**) indicates that the inverse is a function.

Similarly, if the given function were $f(x) = x^2 + 6$ $x \leq 0$,

the inverse relation would be $f(x) = -\sqrt{x-6}$ and this also is a function.

ALTERNATIVELY

Step 1: Let $f(x) = y$. Then $y = x^2 + 6$ $x \geq 0$

Step 2: Interchange x and y and solve for y.

$x = y^2 + 6$ $y \geq 0$

$y = \pm\sqrt{x-6}$ $y \geq 0$

$y = \sqrt{x-6}$ since $y \geq 0$

Therefore, $f^{-1}(x) = \sqrt{x-6}$ $x \geq 6$

We may note from Examples 2 and 3 that, sometimes, by redefining (or restricting) the domain of a given function, a function which does not have an inverse function can be made to have a n inverse function. See Figures 1, 2 and 3.

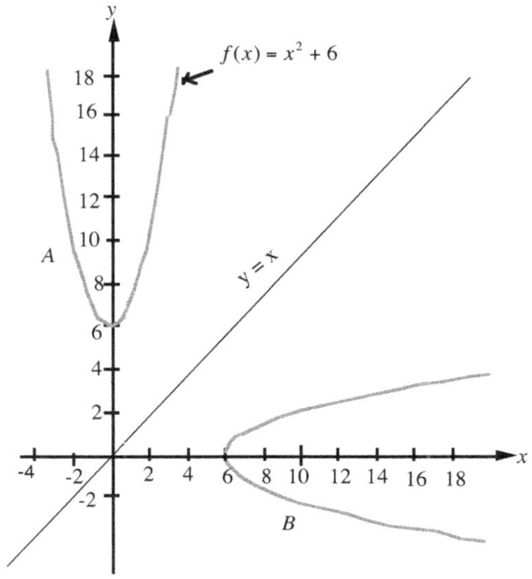

Figure 1:B, the inverse relation of A, is **not** a function .

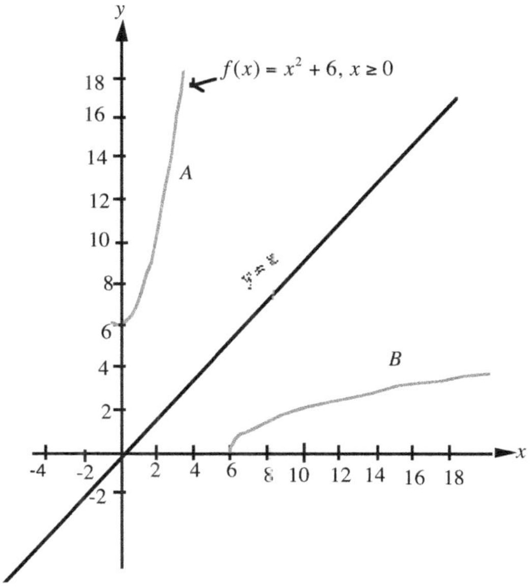

Figure 2: B, the inverse relation of A, is a function (by the vertical line test.)

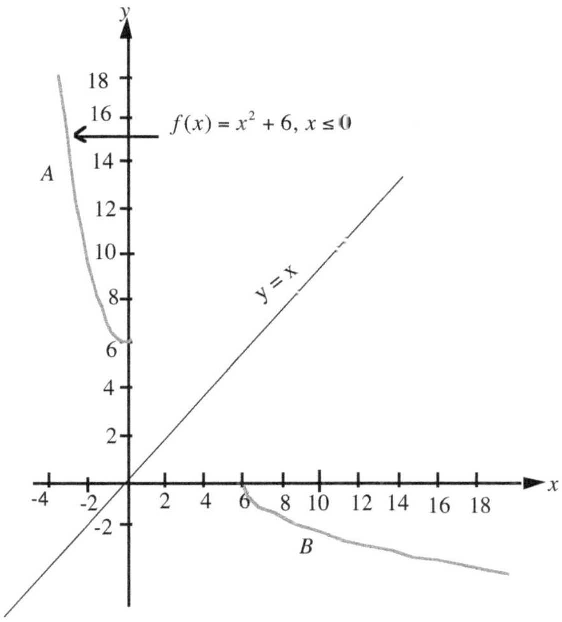

Figure 3: B, the inverse relation of A, is a function (by the vertical line test)

Given the graph of a function, how to sketch the inverse of the graph

Example 5 Given the graph of $y = 3x + 2$, sketch the graph of the inverse of this function.

Geometrically, a function (or relation) and its inverse are symmetric with respect to the line $y = x$. To obtain the inverse graph, we will reflect (see page 73) the given graph about the line $y = x$. The given function is a straight line and so, we will reflect two points about the line $y = x$, and then connect these points by a straight line.

Step 1: Interchange the x- and y-coordinates of any two points on the given line, say, (1, 5) becomes (5, 1); and (-2, -4) becomes(-4, -2).

Step 2: Plot the points (5, 1) and (-4,- 2) on the same coordinate system of axes (as the given graph)

Step 3: Connect the points from Step 2 by a straight line to obtain the inverse graph (Fig.)

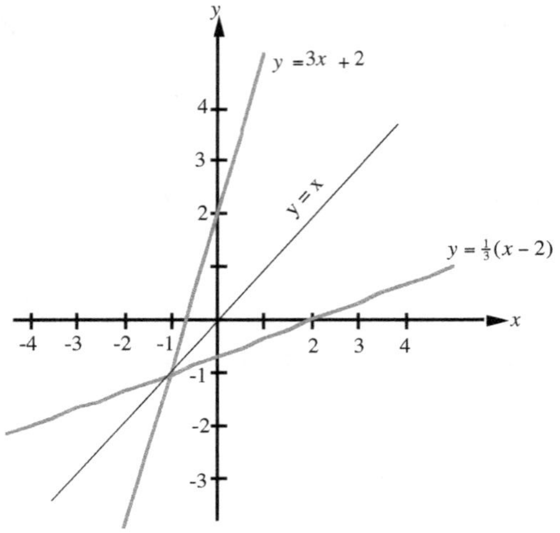

Figure 4: The graphs of $y = 3x + 2$ and $y = \frac{1}{3}(x - 2)$

Note above that, we could also find the equation of the graph, find its inverse equation and then use this equation to sketch the inverse graph.

Determining if two functions are inverses of each other

Case 1: Given the equations of the functions

Example Are the functions $y = 4x + 2$ and $y = \frac{x-2}{4}$ inverses of each other?

Solution

Method 1 We will use the principle of composition of functions which states that:

Two functions f_1 and f_2 are inverses of each other if

1. $f_1[f_2(x)] = x$ and **2.** $f_2[f_1(x)] = x$ (Note that inverse functions reverse the action of each other.)

Let $f_1 = 4x + 2$ and $f_2 = \frac{x-2}{4}$

$$f_1[f_2(x)] = 4[\frac{x-2}{4}] + 2$$

$$= \not{4}[\frac{x-2}{\not{4}}] + 2$$

$$= x - 2 + 2$$

$$= x$$

$$f_2[f_1(x)] = [\frac{(4x+2)-2}{4}]$$

$$= \frac{4x+2-2}{4}$$

$$= \frac{4x}{4}$$

$$= x$$

Since $f_1[f_2(x)] = f_2[f_1(x)] = x$,

$y = 4x + 2$ and $y = \frac{x-2}{4}$ are inverses of each other.

Method 2 Find the inverse of one of the functions and compare this inverse with the other function.

Step 1: We find the inverse of $y = 4x + 2$ (by interchanging x and y and solving for y)

$$x = 4y + 2$$

$$y = \frac{x-2}{4} \quad \text{(solving for } y\text{)}$$

Step 2: Clearly, this inverse is identical with the other function.

Therefore, $y = 4x + 2$ and $y = \frac{x-2}{4}$ are inverses of each other.

Case 2: Given the graphs of the functions

Procedure: Fold the page along the line $y = x$ and if the two graphs coincide, then the functions are inverses of each other. See Figures 2 and 3 (page 29, 30)

Lesson 5 Exercises

A Find the inverse relation in each of the following and state if the inverse relation is a function:

1. $A = \{(2,3),\ (4,5),\ (6,7)\}$ **2.** $B = \{(0,1),\ (2,5),\ (4,10)\}$ **3.** $C = \{(b,a),\ (c,d),\ (a,b)\}$

4. $\{(1,3),(2,3),(4,5)\}$

5. Given a table for a function, determine if the inverse relation represents an inverse function.

x	2	4	8
y	5	6	10

Determine if the inverse relation in each of the following graphs represents an inverse function:

6 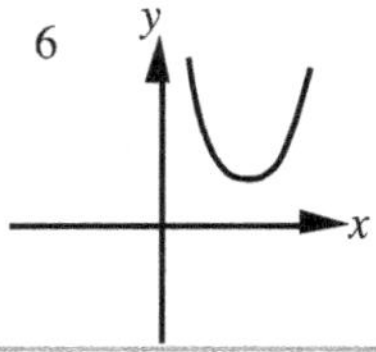

7. 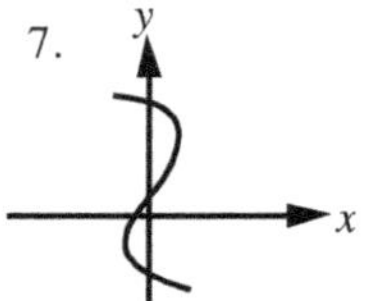

Answers: 1. $\{(3,2),(5,4),(7,6)\}$. It is a function; **2.** $\{(1,0),(5,2),(10,4)\}$ .It is a function.
3. $\{(a,b),(d,c),(b,a)\}$ .It is a function; **4.** $\{(3,\ 1),(3,\ 2),(5,\ 4)\}$ .It is not a function.
5. The inverse relation is a function; **6.** The inverse relation is **not** a function;
7. The inverse relation is a function; even though the given relation is not a function.

B Find the inverse relation of the following and indicate which of the inverse relations are functions.

1. $y = -4x + 1$; **2.** $y = x^2 - 3$ **3.** $y - 5 = x^2$; **4.** $y = x^3$; **5.** $x - 2y = 6$

Determine which of the following pairs are inverses of each other.

6. $y = \dfrac{1}{x+2} - 3$ and $y = \dfrac{1}{x+1} - 4$; **7.** $y = 5x + 2$ and $y = \dfrac{x-2}{5}$; **8.** $y = x^3$ and $y = x^{\frac{1}{3}}$

Answers: **1.** $y = -\frac{x}{4} + \frac{1}{4}$ (a function); **2.** $y = \pm\sqrt{x+3}$ (**not** a function);
3. $y = \pm\sqrt{x-5}$ (**not** a function); **4.** $y = x^{\frac{1}{3}}$ (a function); **5.** $y = 2x + 6$ (a function);
6. No; **7.** Yes; **8.** Yes.

C. In each of the following graphs, sketch its inverse by reflecting the graph in the line $y = x$:

1. 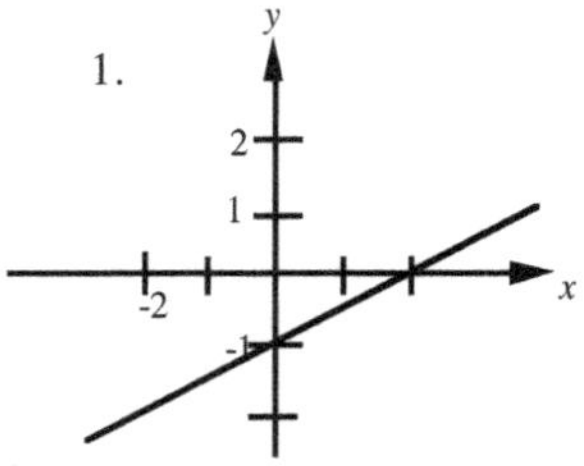

2. 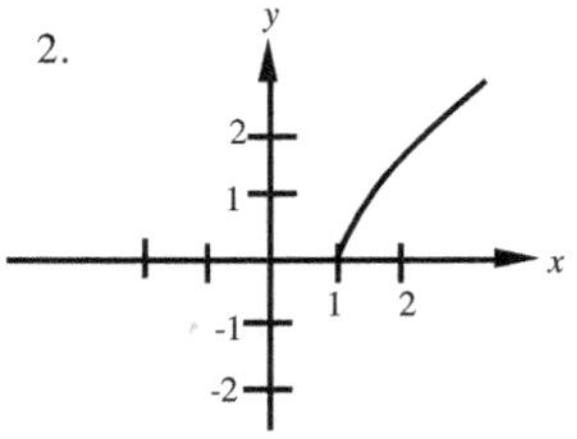

Answers:

1. 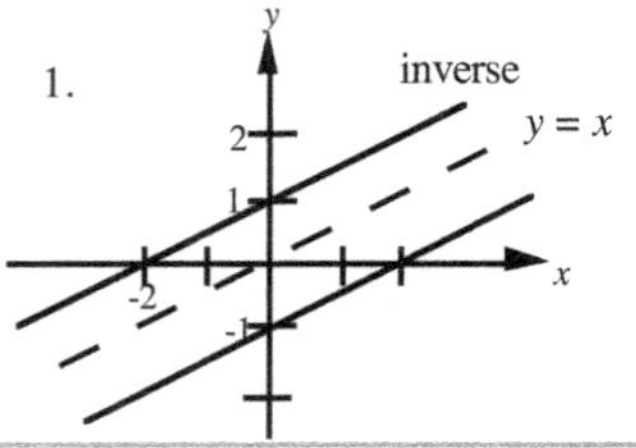

2. 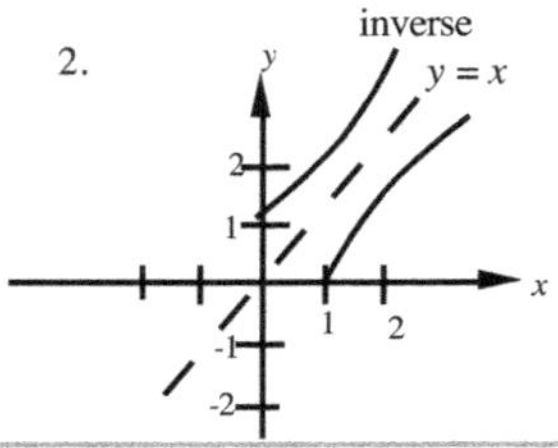

CHAPTER 3
Basic Review of Functions II

Lesson 6: **Continuous and Discontinuous Functions**
Lesson 7: **Asymptotes**
Lesson 8: **Graphs of Rational Functions**

Lesson 6
Continuous and Discontinuous Functions

We use the concepts of continuous and discontinuous functions in sketching the graphs of functions. The graphs of polynomial, sine, and cosine functions are continuous functions. The graphs of rational functions, the tangent ,the cosecant and cotangent functions as well as the graphs of the hyperbolas are generally discontinuous functions.

The meanings given below to continuous and discontinuous functions will be sufficient for our conceptual and qualitative understanding of college algebra and trigonometry. In calculus, we shall learn more formal and complete definitions of continuous and discontinuous functions.

Continuous Functions

Graphically, a function is continuous at a point or on a interval if the graph (curve or line) has no breaks, "jumps" or "holes" in it at that point or on that interval. We can also view a continuous function as one whose line or curve can be drawn without lifting the pencil from the paper.

Discontinuous Functions

Graphically, a function may be discontinuous at a point or on an interval if it has any of the "defective" properties of breaks, jumps, or holes in its curve or line. The breaks represent the excluded values in the domain or range of the function.

There are two main types of discontinuities, namely finite discontinuity and infinite discontinuity.

Finite Discontinuity, Holes

A **finite discontinuity** is a discontinuity which can be removed by redefining the function. Sometimes, we call a finite discontinuity removable or temporary discontinuity. (By redefining the function, in this case, we " bypass or go around the hole".)

Hole: A hole is a circle representing a break in a line or curve at a finite discontinuity. (Figure)

Infinite (Essential) Discontinuity

An **infinite discontinuity** is a discontinuity (a break in a curve) which cannot be removed by redefining the function.

In calculations, we sometimes meet ratios in which certain values when substituted for a variable in the denominator make the denominator zero, and consequently the value of each such ratio (fraction) is undefined.

Examples are **1.** $\frac{3x}{2x+1}$ which is undefined at $x = -\frac{1}{2}$; **2.** $\frac{3x^2 - 7x + 2}{x - 1}$ which is undefined at $x = 1$.

We call such ratios of polynomials **rational expressions**. A rational expression is the ratio of two polynomials.

However, we must note that there are rational expressions which are defined for all real values of the variable in the denominator. For example, $\frac{8}{x^2+4}$ is defined for all real values of the variable in the denominator The denominator x^2+4 is positive for all real values of x and never zero, since the square of any nonzero real number is always positive.
A rational expression may be proper or improper.

Proper Rational Expression

In a proper rational expression, the degree of the numerator polynomial is lower than the degree of the denominator polynomial.

Example $\frac{x+1}{x^2-4}$ (Degree of the numerator polynomial is 1; degree of the denominator polynomial is 2.)

Improper Rational Expression

In an improper fraction, the degree of the numerator polynomial is greater than or equal to the degree of the denominator polynomial.

Examples 1. $\frac{x^2-9}{x-2}$ (Degree of the numerator polynomial is 2; degree of the denominator polynomial is 1.)

2. $\frac{x}{x+2}$ (Degree of the numerator polynomial is 1; degree of the denominator polynomial is 1.)

In working with rational expressions, we shall exclude those values (called **excluded values**) of the variable in the denominator which make the denominator zero. At the excluded values, the rational expressions are undefined.

With polynomial functions, the graph is one smooth unbroken (continuous curve). With rational functions, there may be broken (discontinuous) lines or curves. The curves are broken at the excluded values mentioned above.

Excluded Values and Graphing of Rational Functions

In sketching the graph of a rational function, we can use either holes or asymptotes (see next section) to indicate the places where there are discontinuities in the curve or line. For finite discontinuities, we shall use holes, but for infinite (essential) discontinuities, we shall use asymptotes.

In sketching the curves at or near the infinite discontinuities we shall use the asymptotes as "guiding straight lines". In the next section, we cover the definition of an asymptote and then learn how to find equations of asymptotes, as well as sketch selected curves with asymptotes. We shall also justify why so much effort has been devoted to study asymptotes, something that most current text books do not do at this level.

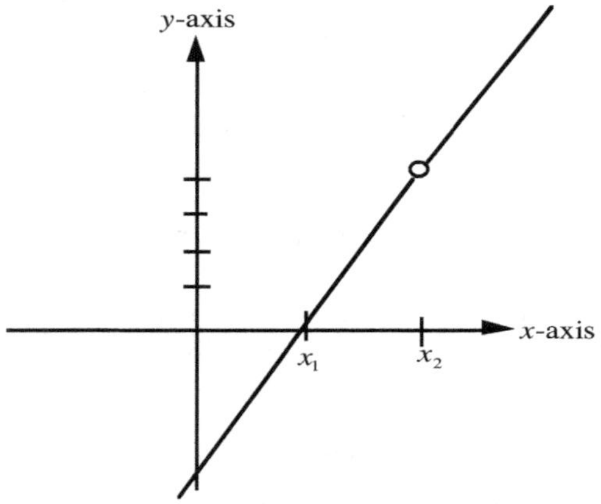

Figure: Function with a finite discontinuity at x_2.

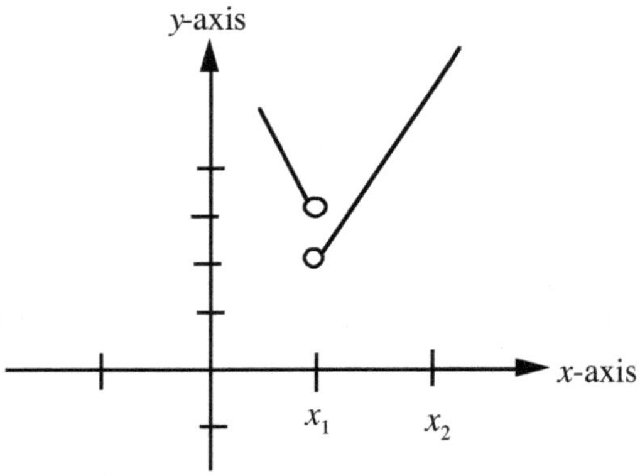

Figure: Function with a hole at x_1

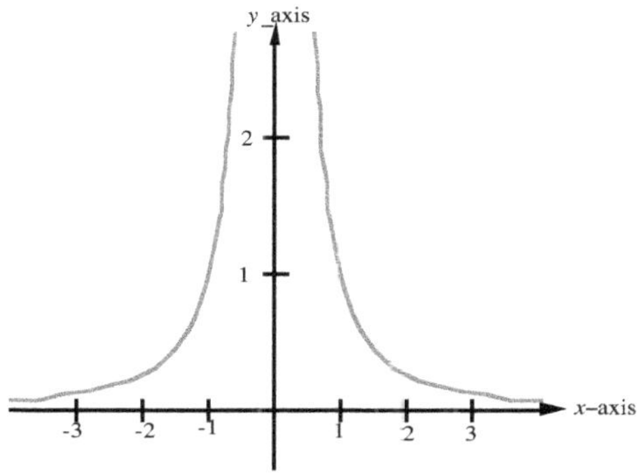

Figure Graph of a function with infinite discontinuity at $x = 0$.

Lesson 6 Exercises

A **1.** Define or explain the following from memory

(a). Continuous functions, **(b)** Discontinuous functions; **(c)** Finite discontinuity, **(d)** Infinite discontinuity; **(e)** Proper rational Expression; (**f)** Improper rational expression **(g)** Excluded values

2. How do we indicate graphically the discontinuities in a curve or line?

3. Qualitatively, distinguish between the graph of a continuous function and the graph of a discontinuous function. Give an example of each.

4. What is the significance of (a) a hole, (b) an asymptote in sketching the graphs of functions?

Answers: See text.

B State whether the discontinuity in each of the following is finite or infinite,

1

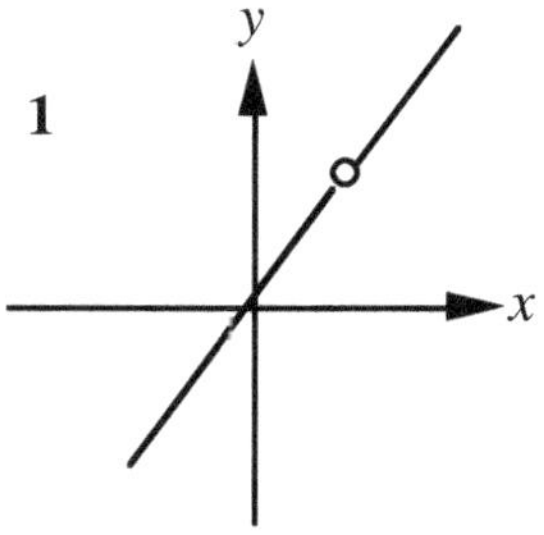

2

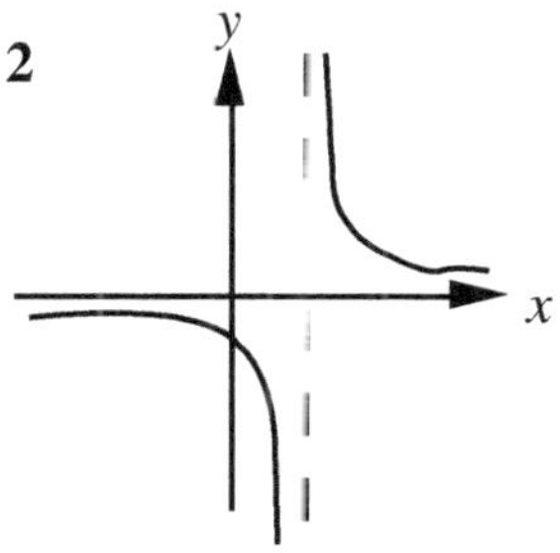

3

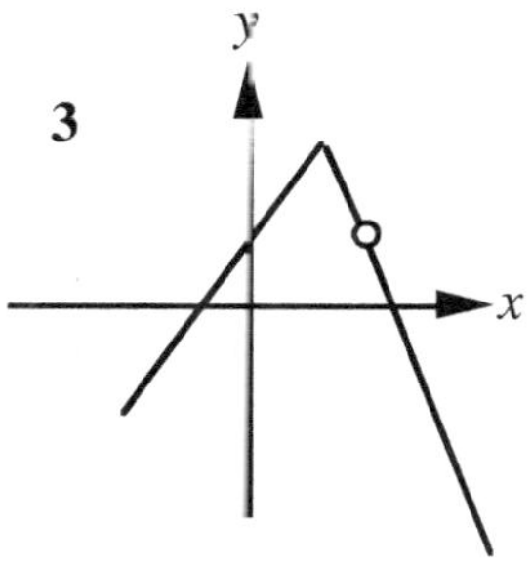

4

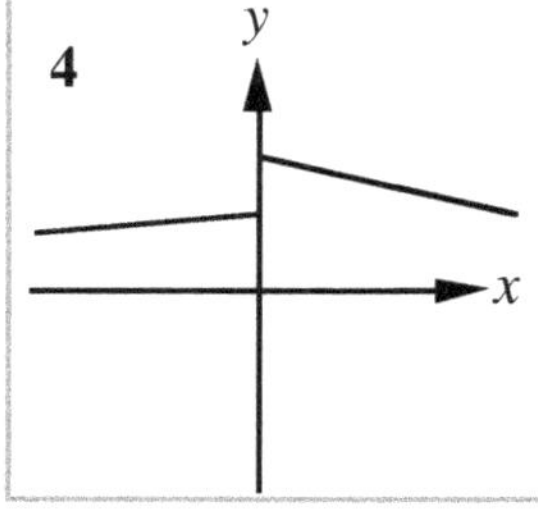

5

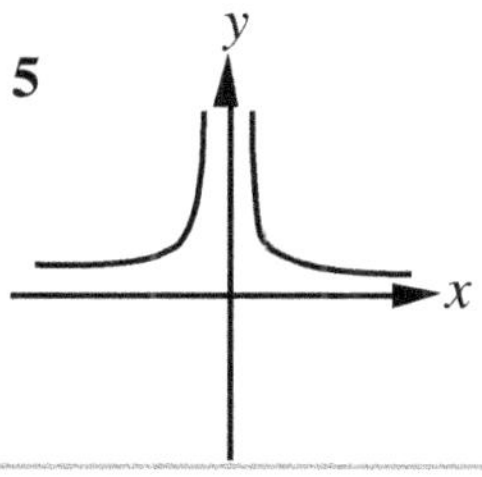

6

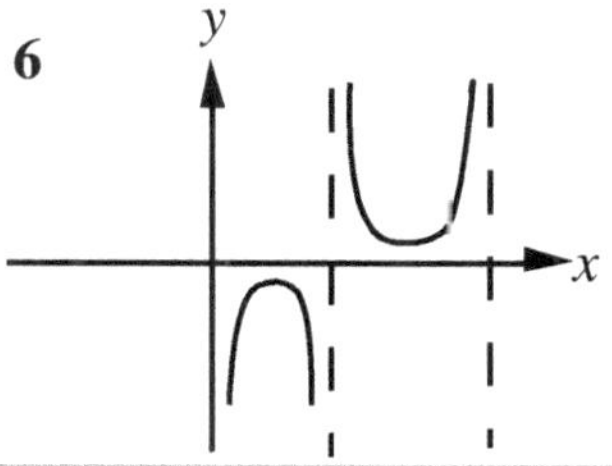

Answers: **1.** Finite; **2.** infinite; **3.** finite; **4.** finite;? **5.** infinite; **6.** infinite.

Lesson 7

Asymptotes

The word asymptote comes from the Greek word " asymptotos" (Latin: asymptota) which means "not meeting".

Definition: An **asymptote** to a given curve is a straight line which the curve approaches (nearer and nearer) as the curve is produced or extended.

Since an asymptote is a straight line, an asymptote possesses all the properties of straight lines that we learned in the past. An asymptote may be a vertical line (vertical asymptote) with equation of the form $x = a$ (Example: $x = 3$). It may be a horizontal line (horizontal asymptote) with equation of form $y = b$ (Example: $y = 2$); or it may be an oblique line (slant line) with equation of the form $y = mx + b$ (Examples: $y = 3x + 2$, $y = 5x$) where $m \neq 0$.

Vertical Asymptote

A **vertical asymptot**e to a curve is the vertical line $x = a$ (where a is the x–intercept) which the equation of the curve approximates as x approaches a and y increases or decreases without bound .Geometrically, the curve smoothly approaches the vertical asymptote nearer and nearer as y increases or decreases without bound.

Horizontal Asymptote

A **horizontal asymptote** to a curve is the horizontal line $y = b$ (where b is the y–intercept) which the equation of the curve approximates as y approaches b and x increases or decreases without bound. Geometrically, the curve smoothly approaches the horizontal asymptote nearer and nearer as x increases or decreases without bound.

Determining the equations of the vertical asymptotes, given the equation of the function

Step 1: Solve the given equation for y in terms of x (if it has not been solved for).

Step 2: Set the denominator from Step 1 to zero and solve for x.

Step 3: The real values of x obtained in Step 2 give the equations of the vertical asymptotes. If the values are **not** real, then there are no vertical asymptotes.

Example 1 Find the equations of the vertical asymptotes of the rational function whose equation is

$$y = \frac{4x^2 + 3}{x^2 - 1}$$

Solution

Step 1: Fortunately, the given equation has been solved for y. We therefore go on to the next step.

Step 2: Set the denominator to zero and solve for x.

Then $x^2 - 1 = 0$

Solving, $x = 1$, or $x = -1$

The equations of the vertical asymptotes are the lines $x = 1$, and $x = -1$.

Example 2 Find the equations of the vertical asymptotes of the rational function whose equation is $y(x^2 - 1) = 4x^2 + 3$

Solution Step 1: Solve for y

Then $y = \dfrac{4x^2 + 3}{x^2 - 1}$

For the rest of the steps, see Example 1, above.

Example 3 Find the equations of the vertical asymptotes of the rational function whose equation is given by : $y(x^2 + 1) = 4x^2 + 3$

Solution

Step 1: Solve for y

Then $y = \dfrac{4x^2 + 3}{x^2 + 1}$

Step 2: Set the denominator to zero and solve for x.

$x^2 + 1 = 0$

Solving, $x = \sqrt{-1}$, which is not real, There are **no** vertical asymptotes.

Note that the denominator $x^2 + 1$ is positive for all real values of x and never zero, since the square of any nonzero real number is always positive.

Determining the equations of the horizontal asymptotes, given the equation of the function

There are two main methods. One method is similar to the method used in fincing the vertical asymptotes. The other method considers the use of limits (by considering large values of x).

Method 1

Step 1: Solve the given equation for x in terms of y.

Step 2: Set the denominator obtained in Step 1 to zero and solve for y. The real values of y obtained give the equations of the horizontal asymptotes. If there are no real values, then there are no horizontal asymptotes.

Method 2

Step 1: Solve the given equation for y.

Step 2: Divide every term in both the numerator and the denominator by the highest power of x in in the **denominator.**

Step 3: Consider large values of x. Note for instance that, for large values of x, $\dfrac{1}{x} \approx 0$, $\dfrac{1}{x^2} \approx 0$.

Step 4: The values of y obtained from Step 3 give the equations of the horizontal asymptotes.

Example 4 Determine the horizontal asymptotes of the curve given by

$$y = \frac{4x^2 + 3}{x^2 - 1}$$

Solution

Method 1

Step 1: Solve the given equation for x in terms of y.

Cross multiplying, $y(x^2 - 1) = 4x^2 + 3$

$$x^2y - y = 4x^2 + 3$$

$$x^2y - 4x^2 = y + 3$$

$$x^2(y - 4) = y + 3 \Longleftarrow$$

$$x^2 = \frac{y+3}{y-4}$$

$$x = \pm\frac{\sqrt{y+3}}{\sqrt{y-4}}$$

(factoring in order to obtain a single term for x^2. If you have never met this technique before, try to master it. This factoring (monomial factoring) becomes necessary, sometimes, when solving literal equations for one of the variables)

Step 2: Set the denominator to zero and solve for y.

$$\sqrt{y-4} = 0$$

$$\left(\sqrt{y-4}\right)^2 = 0^2$$

$$y - 4 = 0$$

$$y = 4$$

Therefore, the equation of the horizontal asymptote is $y = 4$.

Method 2

Example Find the horizontal asymptotes of $y = \dfrac{4x^2 + 3}{x^2 - 1}$

Solution

Step 1:

$$y = \frac{\frac{4x^2}{x^2} + \frac{3}{x^2}}{\frac{x^2}{x^2} - \frac{1}{x^2}}$$

$$y = \frac{4 + \frac{3}{x^2}}{1 - \frac{1}{x^2}}$$

Step 2: For large values of x, $\frac{3}{x^2} \approx 0$, $\frac{1}{x^2} \approx 0$, and

$$y \approx \frac{4+0}{1-0}$$

$$y \approx \frac{4}{1}$$

$$y \approx 4$$

The equation of the horizontal asymptote is the line $y = 4$. (We obtain the same result as by the first method.)

Example 5 Find the horizontal asymptotes of

$$y = \frac{\sqrt{x+6}}{x-1}$$

Solution We use Method 2

Step 1: Divide every term by x (That is, divide both the numerator and the denominator by x)

$$y = \frac{\frac{1}{x}\sqrt{x+6}}{\frac{1}{x}(x-1)}$$ (Multiplying by $\frac{1}{x}$ is equivalent to dividing by x)

$$= \frac{\sqrt{\frac{1}{x^2}(x+6)}}{\frac{1}{x}(x-1)}$$ (**Note:** We square $\frac{1}{x}$ before writing it as a factor of the radicand)

$$= \frac{\sqrt{\left(\frac{1}{x}+\frac{6}{x^2}\right)}}{1-\frac{1}{x}}$$

Step 2 For large values of x, $\frac{1}{x} \approx 0$, $\frac{6}{x^2} \approx 0$, and

$$y = \frac{\sqrt{0+0}}{1-0}$$

$$y = 0$$

The equation of the horizontal asymptote is $y = 0$. (The x-axis)

How to determine the equations of oblique (slant) asymptotes

The equation of an oblique asymptote is of the form $y = mx + b$, where m is not zero, and b is any real number. There are a number of methods for finding these equations. However, we shall cover only one method which involves long division.

Example Find the equation of the oblique asymptote of the curve given by

$$y(x-1) = x(x+1)$$

Solution

Step 1: Solve for y.

$$y(x-1) = x(x+1)$$

$$y = \frac{x(x+1)}{x-1}$$

$$y = \frac{x^2+x}{x-1}$$

Step 2: On the right-hand side, divide the numerator by the denominator, using long division to obtain $y = x + 2 + \dfrac{2}{x-1}$

The polynomial partial quotient $x + 2$ is the equation of the oblique asymptote.
Therefore, the equation of the oblique asymptote to the given curve is the line $y = x + 2$.

Non-existence of Vertical and Horizontal Asymptotes

If on solving an equation for x and y to determine vertical or horizontal asymptotes, we obtain non-real or undefined values, then there are no vertical or horizontal asymptotes .

The following generalization about horizontal and oblique asymptotes will be useful: for checking the methods covered

Case 1. If the degree of the numerator polynomial **equals** the degree of the denominator polynomial then the equation of the horizontal asymptote is

$$y = \frac{\text{coefficient of the leading term in the numerator}}{\text{coefficient of the leading term in the denominator}}$$

Case 2. If the degree of the numerator polynomial is **less than** the degree of the denominator polynomial, then the equation of the horizontal asymptote is the line $y = 0$.

Case 3. If the degree of the numerator polynomial **is greater** than the degree of the denominator polynomial then there are no horizontal asymptotes.

Case 4: If the degree of the numerator polynomial **equals** 1 plus the degree of the denominator polynomial, then there are no horizontal asymptotes; but there are **oblique asymptotes**.

Case 5. If the degree of the numerator polynomial is greater than 1 plus the degree of the denominator polynomial, then there are no horizontal asymptotes, and no oblique asymptotes.

Asymptotic Formula for a Given Equation or Expression

The asymptotic formula is the formula which approximates the exact formula when the independent variable (argument) becomes very large (increases indefinitely).

Example 1 Given the exact formula $y = x^2 + x + \frac{5}{x}$. Find the asymptotic formula for y.

Solution For large values of x, $\frac{5}{x} \approx 0$, and the equation becomes

$$y = x^2 + x + 0$$
$$y = x^2 + x$$

The asymptotic formula is $y = x^2 + x$

Example 2 Find the asymptotic expansion of the expression given by $(2x + \frac{1}{x})^2$

Solution Let $y = (2x + \frac{1}{x})^2$

Then $y = 4x^2 + 4 + \frac{1}{x^2}$

For large values of x, $\frac{1}{x^2} \approx 0$, and

$$y = 4x^2 + 4 + 0$$
$$y = 4x^2 + 4 + 0$$

The asymptotic expansion of $(2x + \frac{1}{x})^2$ is $4x^2 + 4$

How to Draw a Curve and its Asymptote

Example 1 Let us consider the graph of a function (Figure 1) with the following properties:

1. The point A (-7, 0) is an x-intercept of the curve. This point is the nearest known x-intercept on this side of the vertical asymptote, x = -2.
2. The function is positive on the interval from x = -7 to x = -2. (The curve is above the x-axis) We draw this branch curve so that the curve smoothly approaches its asymptote (Figure)

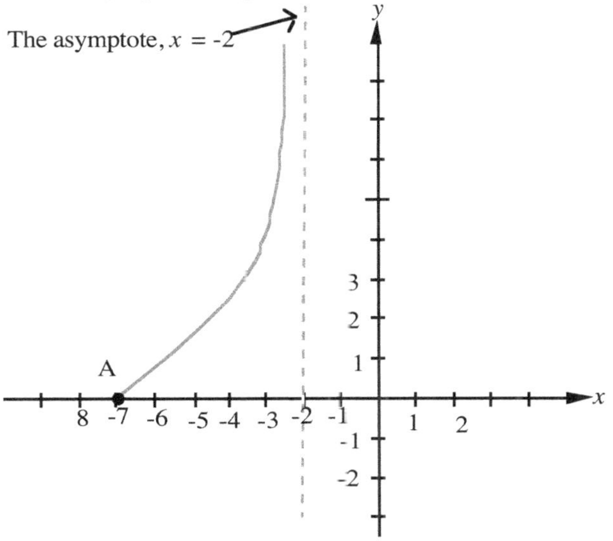

Figure 1: The function is positive between -7 and -2 and has a vertical asymptote at x = -2.

Example 2 In Example 1, all the properties remain unchanged except that the curve is below the x-axis (i.e., the function is negative) on the interval from -7 to 2. The graph is shown below.

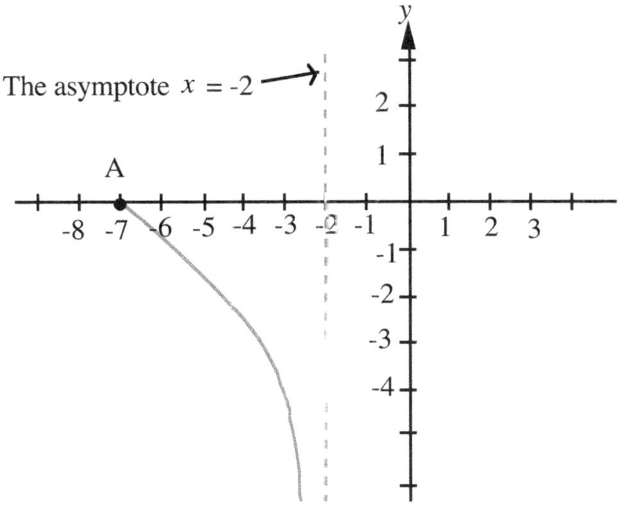

Figure: The function is negative between -7 and -2.

Example 3 Suppose that it is known that A is a point on a curve (either given or calculated); and that between *D* and *C* (Figure) the function is positive; between *C* and *B*, the function is negative. Also, suppose that the curve has a vertical asymptote at $x = -2$, and an x-intercept at *C*.

Noting that we are trying to draw the sketch as continuously as permissible, we connect the points and draw the curve so as to approach its asymptote gradually (Figure)

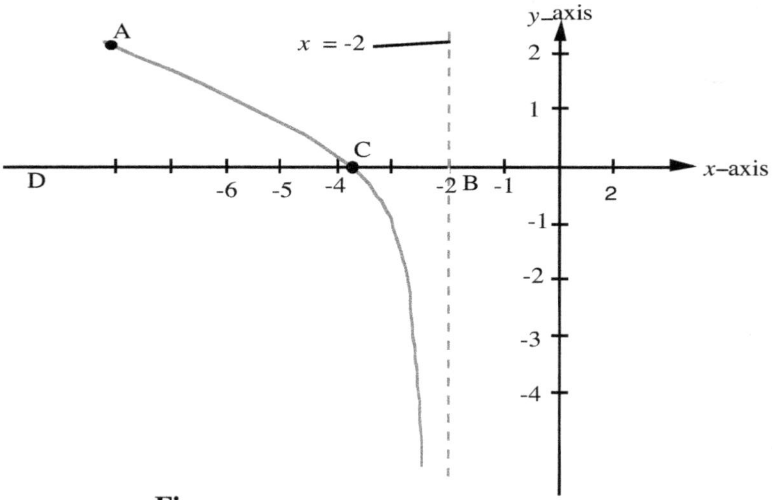

Figure

Example 4 : Everything remains the same as in Example 3 except that there is no x-intercept between A_1 and *B*; and that the curve is above the x-axis (i.e., the function is positive) between *D* and *B*. The sketch is shown in Figure below.

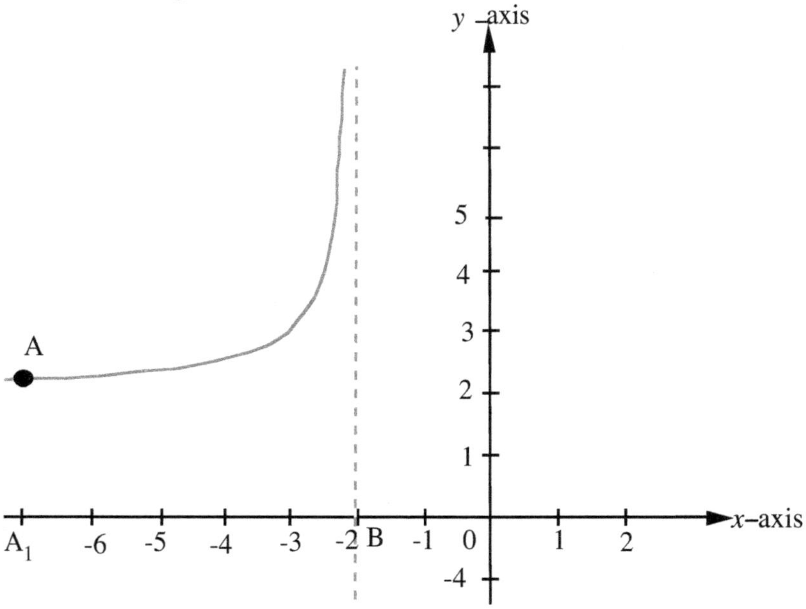

Figure:

Example 5 A curve has the following properties: x-intercept at $x = 7$; and no other x-intercept between $x = 3$ and $x = 7$; and a vertical asymptote at $x = 3$. We sketch the curve as shown in Figure below.

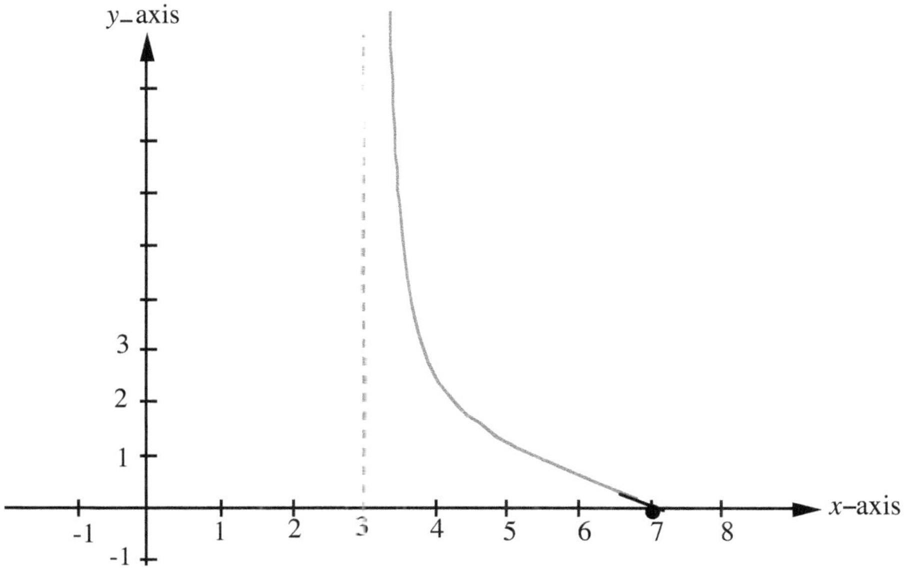

Example 6 The graph of a function has the following properties:
x-intercepts at $x = -3, x = 0, x = 4$; vertical asymptotes at $x = -1, x = 2$. Curve is above the x-axis between $x = -3$ and $x = -1$; and between $x = 0$ and $x = 2$. Curve is below the x-axis between $x = -1$ and $x = 0$; and between $x = 2$ and $x = 3$.

Step 1: Plot the x-intercepts.

Step 2: Draw the vertical asymptotes using dotted or broke lines.

Step 3: Using the experience gained in the previous examples, we draw the various branches of the curve, noting that the curves are to be drawn through the intercepts and also that the curves are to be asymptotic to the asymptotes on the respective intervals (Figure below).

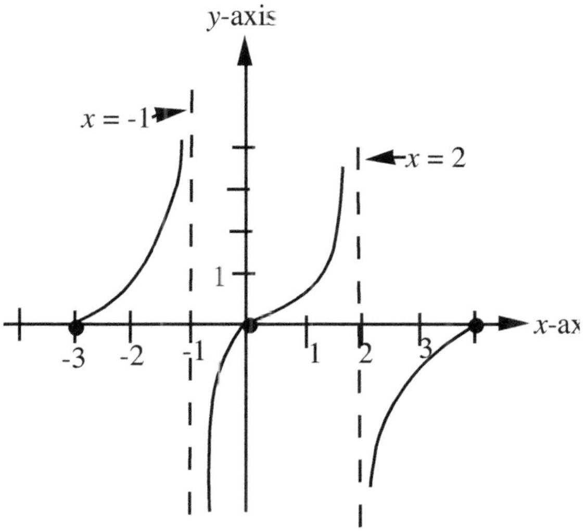

Example 7 The graph of a function has the following properties:

1. x-intercepts at $x = 0, x = 5$; **2**. Vertical asymptote at $x = 3$; **3**. Horizontal asymptote at $y = 4$

4. The function is positive between $-\infty$ and 0; and between $x = 5$ and $+\infty$.

5. The function is negative between $x = 0$ and $x = 3$; and between $x = 3$ and $x = 5$.

6. The function is increasing to the right of $x = 3$; **7**. The function is decreasing to the left of $x = 3$.

Solution

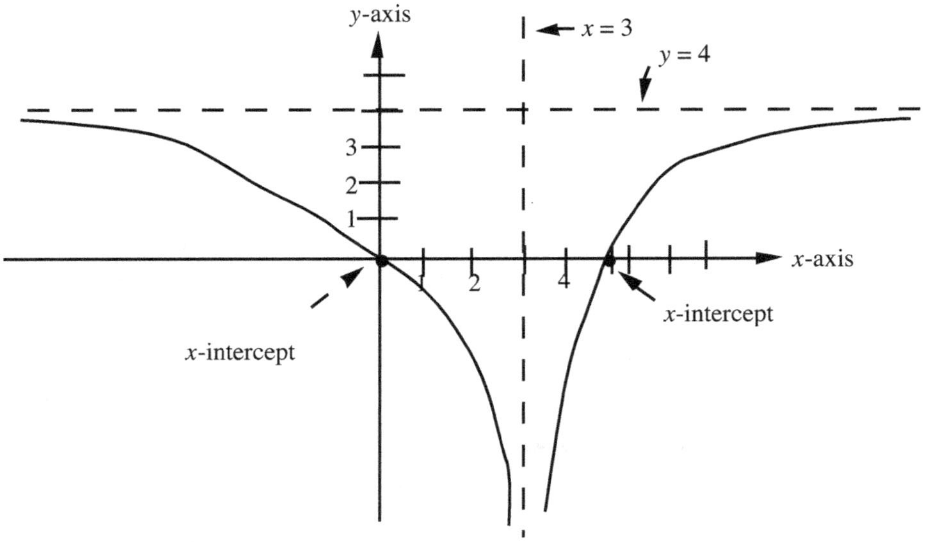

Importance of Devoting Time to Study Asymptotes

Let us justify why we have devoted so much effort and time to study asymptotes.

We do this by presenting examples of the different types of functions or relations that are covered in this book and all of which involve asymptotes.

A **Rational Functions** (Chapter 3)

1. $f(x) = \frac{1}{x}$; **2**. $f(x) = \frac{3x-1}{x-4}$; **3**. $f(x) = \frac{5x^2}{(x-1)(x+2)(x+3)}$; 4. $f(x) = \frac{1}{x^2}$.

B **Exponential Functions**

1. $f(x) = e^x$; **2**. $f(x) = 2^x$; **3**. $f(x) = e^x + 3$

C **Logarithmic Functions**

1 **1**. $f(x) = \log x$; **2**. $f(x) = \log(x+2)$

D **Trigonometric Functions** (Chapter 11)

1. $f(x) = \sec x$; **2**. $f(x) = \csc x$; **3**. $f(x) = \tan x$; **4**. $f(x) = \cot x$

E **Others** which are not functions but relations are the hyperbolas.

1. $\frac{x^2}{a^2} - \frac{y^2}{b^2} = 1$; **2**. $\frac{y^2}{a^2} - \frac{x^2}{b^2} = 1$.

We should note that, in the main, it is the asymptotic behavior of these functions which make them different, graphically, from the polynomial and continuous trigonometric functions..

Lesson 7 Exercises

A What is an asymptote to a given curve?

B Find the equations of the vertical and horizontal asymptotes (if any) for each of the following:

1. $y = \frac{3x+1}{x+2}$; **2**. $y = \frac{x+1}{4-x}$; **3**. $y = \frac{5}{x+1}$; **4**. $y = \frac{3}{x^2+2}$; **5**. $y = \frac{4x^2+5}{x^2-9}$

6. $y = \frac{5x^2}{x^2-6x}$; **7**. $y = \frac{2+x^2}{x^2-8x+15}$

8. Find the equation of the oblique asymptote to $y(x-2) = x(x+2)$

9. Given that $A = x^3 + \frac{1}{x^2+x+12}$, find the asymptotic formula for A.

10. Find the asymptotic expansion for $(x - \frac{1}{x})^2$

Answers: B **1**. Vertical asymptote, $x = -2$, horizontal asymptote, $y = 3$.

2. Vertical asymptote, $x = 4$; horizontal asymptote, $y = -1$;

3. Vertical asymptote, $x = -1$; horizontal asymptote, $y = 0$.

4. No Vertical asymptotes, horizontal asymptote, $y = 0$.

5. Vertical asymptotes, $x = -3, x = 3$; horizontal asymptote, $y = 4$.

6. Vertical asymptotes, $x = 0, x = 6$; horizontal asymptote, $y = 5$.

7. Vertical asymptotes, $x = 3, x = 5$; horizontal asymptote, $y = 1$.

8. $y = x + 4$; **9.** $A = x^3$; **10.** $x^2 - 2$

Lesson 8

From the Graphs of Polynomial Functions to the Graphs of Rational Functions

A **rational function** is a function which is the ratio of two polynomial functions.

Examples

1. $f(x) = \frac{1}{x}$; **2**. $f(x) = \frac{6}{x-2}$; **3**. $f(x) = \frac{3x}{x^2-9}$;

4. $f(x) = \frac{1}{(x-2)(x+1)(x+3)}$; **5**. $f(x) = \frac{x^3+5x^2+3x-1}{x^2-2x-1}$

We have already learned how to sketch the graphs of polynomial functions. We have also learned how to find equations of asymptotes as well sketch graphs with asymptotes.
We shall now learn how to sketch the graphs of rational functions.

Graphs of $f(x) = x^n$ and $f(x) = x^{-n}$

The graphs of $f(x) = x^n$ and $f(x) = x^{-n}$ are in the same quadrants, except that the graphs of $f(x) = x^{-n}$ are hyperbolic in shape or in character.

Example: The graphs of $f(x) = x$ and $f(x) = x^{-1} = \frac{1}{x}$ are in the same quadrants except that the graph of $f(x) = x^{-1}$ is hyperbolic in shape (Figure 2, below).

Graphs of the Reciprocals of Some Simple Continuous Functions

The author has suggested the following descriptions for relationships between some simple continuous functions and their reciprocals.

For the function $f(x) = \frac{1}{x}$ (the reciprocal of $f(x) = x$)

Whenever the originally continuous line, $y = x$ (Figures) becomes infinitely discontinuous at a point, the line breaks up into two pieces (at this point); a vertical asymptote and a horizontal asymptote are formed at this point; and each piece bends and orientates itself such that one end smoothly becomes asymptotic to the x–axis (the horizontal asymptote) and the other end becomes asymptotic to the y–axis (the vertical asymptote). Similar behavior can be described for

$f(x) = \frac{1}{mx+b}$ (the reciprocal of $f(x) = mx + b$)

When $n \geq 2$ for $f(x) = \frac{1}{x^n}$ (e.g., $f(x) = \frac{1}{x^2}$) **and reciprocals of trigonometric functions**

Whenever a given curve becomes infinitely discontinuous at a point (Figures), the given curve b reaks up into two pieces at this point; a vertical asymptote and a horizontal asymptote are formed at this point, and each piece reverses its concavity and orientates itself such that the end of each piece smoothly becomes asymptotic to the asymptotes so formed.

The above described behavior can be applied to the reciprocals of some continuous functions such as the reciprocals of polynomial functions, the reciprocals of some trigonometric functions such as the reciprocals of the sine, cosine and tangent functions. Generally, with the reciprocals of trigonometric functions, only vertical asymptotes are formed at the discontinuous points

Lesson 8: From the graphs of polynomial functions to the graphs of rational functions

Given the graph of a typical polynomial function, we can readily and by inspection sketch the graph of its reciprocal. Examples of the above behavior are presented below. (Figures)

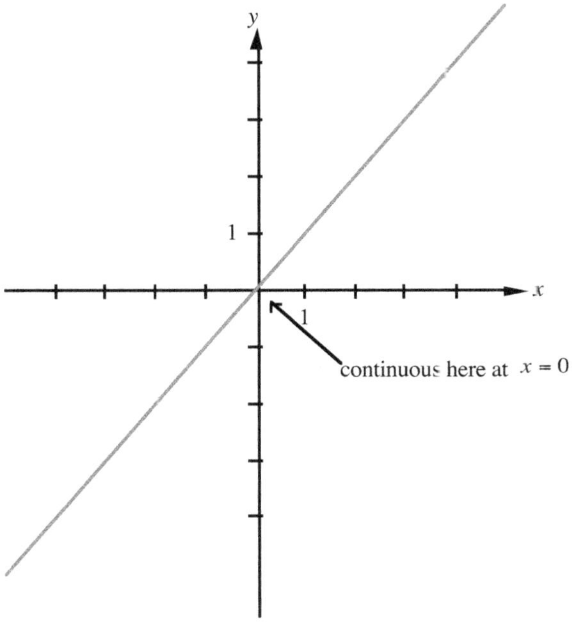

(a) **Figure 1:** The graph of $y = x$ (Polynomial function)

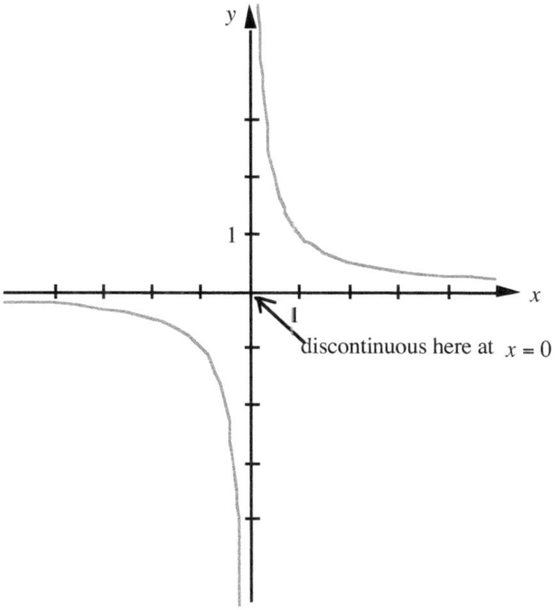

(b) **Figure 2: The** graph of $y = \frac{1}{x}$ (Rational function)

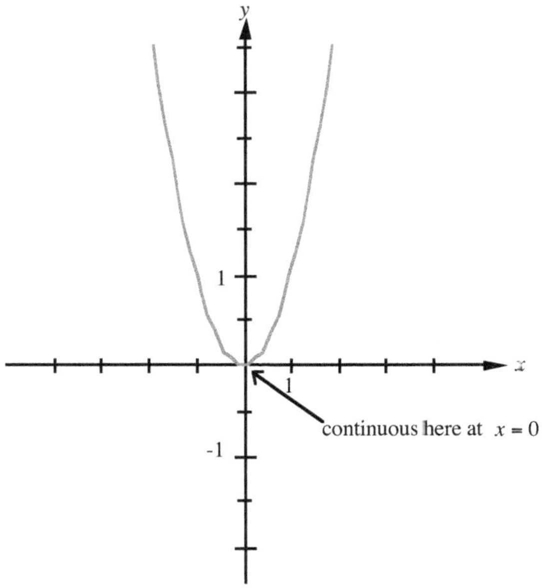

(c) The graph of $y = x^2$ (Polynomial function)

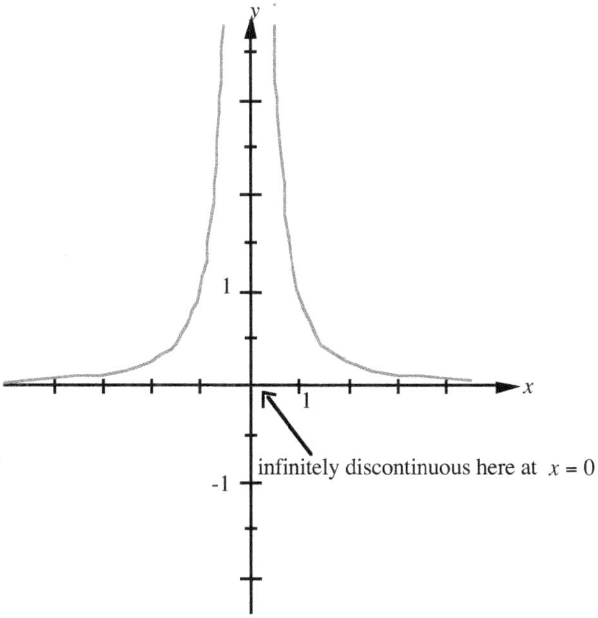

(d) The graph of $y = \frac{1}{x^2}$ (Rational function)

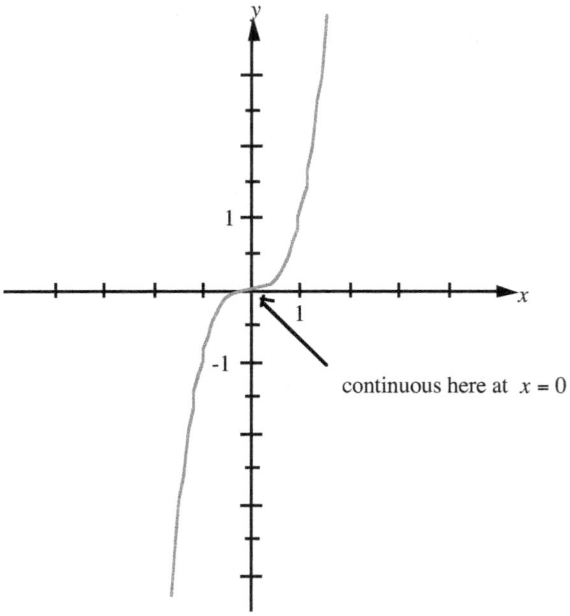

(e) The graph of $y = x^3$ (Polynomial function) .

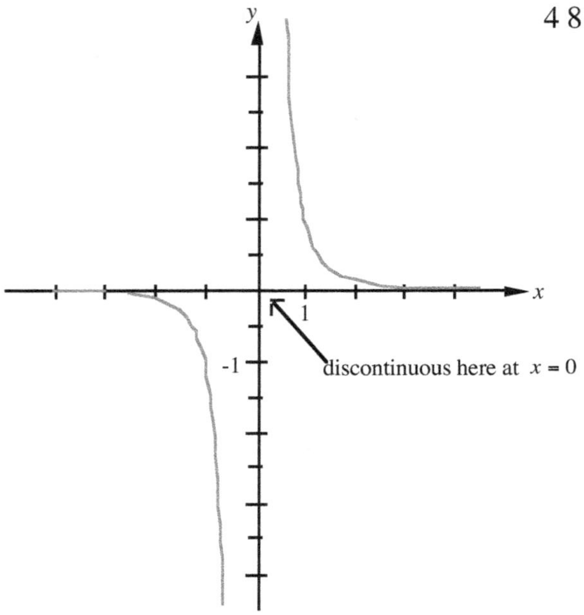

(f) The graph of $y = \frac{1}{x^3}$ (Rational function)

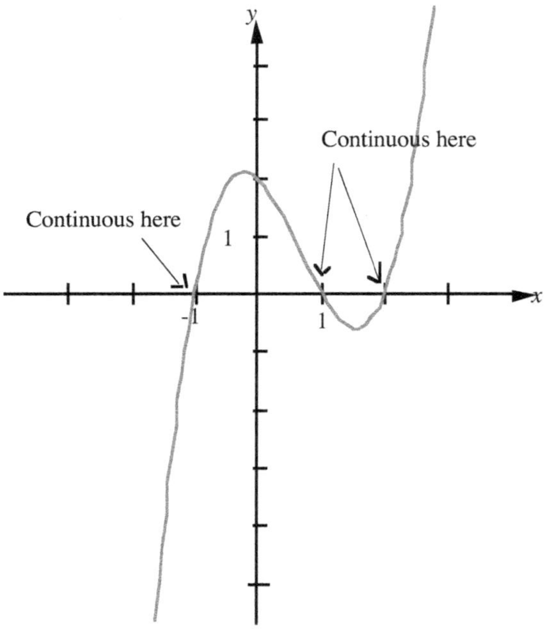

(g) The graph of $y = (x-1)(x-2)(x+1)$.
(Polynomial function)

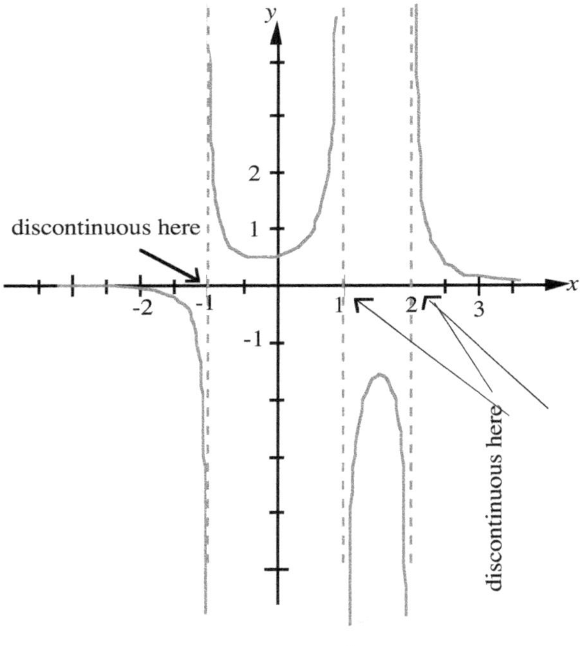

(h) The graph of $y = \frac{1}{(x-1)(x-2)(x+1)}$
(Rational function)

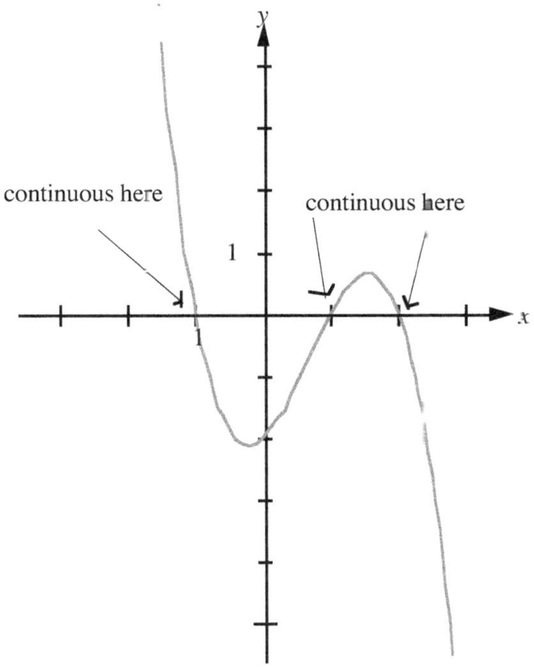

(i) The graph of $y = -(x-1)(x-2)(x+1)$.
(Polynomial function)

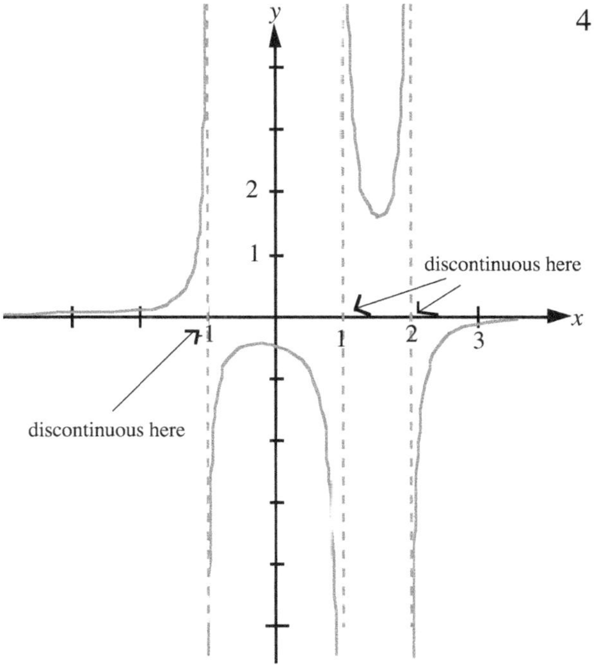

(j) The graph of $y = -\dfrac{1}{(x-1)(x-2)(x+1)}$
(Rational function)

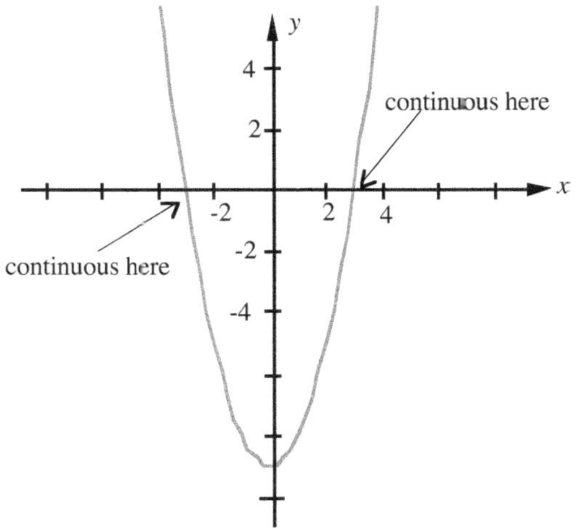

(k) The graph of $y = (x+3)(x-3)$.
Polynomial function)

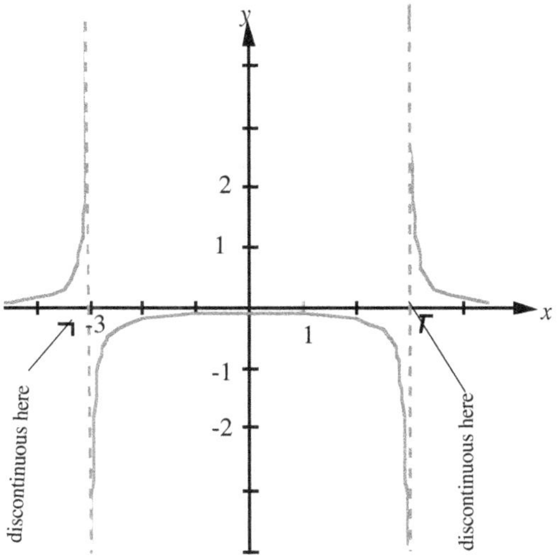

(l) The graph of $y = \dfrac{1}{(x+3)(x-3)}$
(Rational function)

The following two examples from trigonometry are **not** polynomial or rational functions but exhibit similar behavior. See Chapter 3).

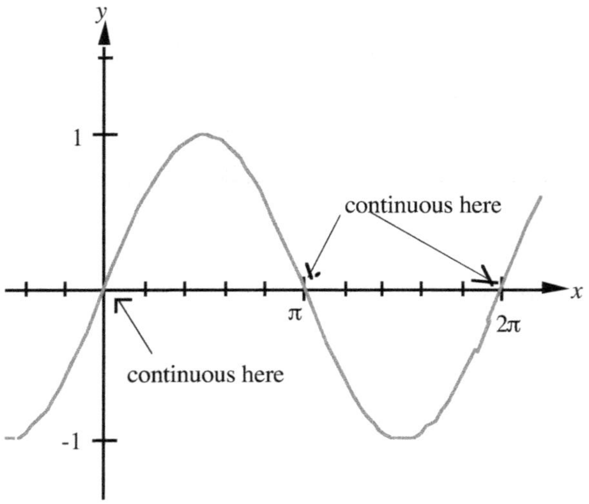

(m) The graph of $y = \sin x$.

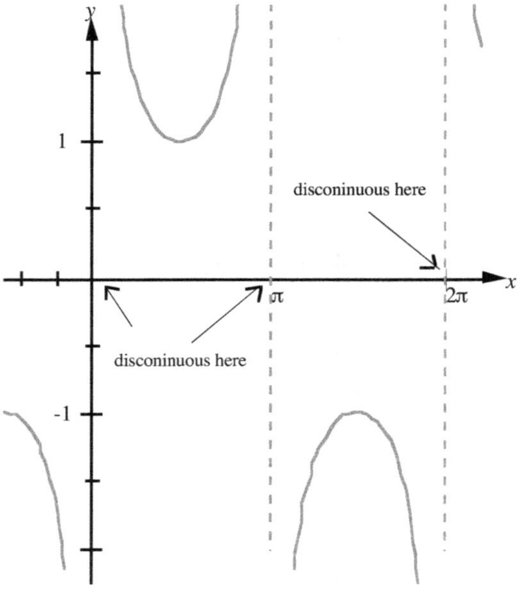

(n) The graph of $y = \frac{1}{\sin x} = \csc x$

Sketching the Graphs of $f(x) = \frac{1}{x}$ and $f(x) = \frac{1}{x-h}$

The function given by $f(x) = \frac{1}{x}$ is not defined when $x = 0$.

Let $y = f(x)$. Then, $y = \frac{1}{x}$ and $x = \frac{1}{y}$. Similarly, $x = \frac{1}{y}$ is not defined when $y = 0$.
Thus, the lines $x = 0$ (the y-axis) and $y = 0$ (the x-axis) are vertical and horizontal asymptotes to $y = \frac{1}{x}$ and $x = \frac{1}{y}$, respectively (Figures)

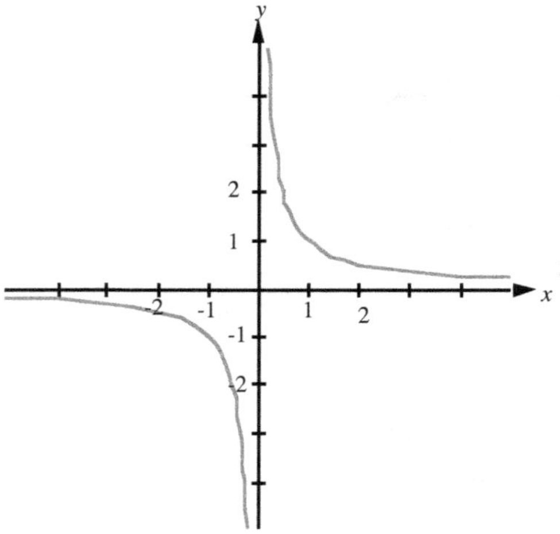

Figure : The graph of $y = \frac{1}{x}$

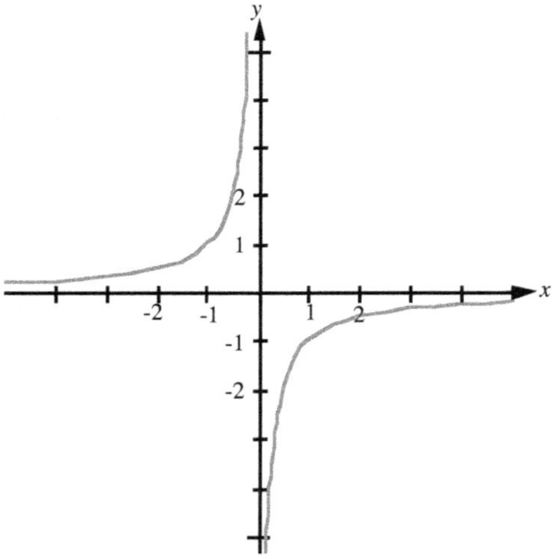

Figure: Graph of $y = -\frac{1}{x}$

Lesson 8: From the graphs of polynomial functions to the graphs of rational functions

Example 1 Sketch the graphs of $y = -\frac{1}{x}, y = \frac{2}{x}$; $y = \frac{1}{x-3}$; and $\frac{2x}{x-3}$ from the graph of $y = \frac{1}{x}$

Solution Reflect the graph of $y = \frac{1}{x}$ about the x-axis. (Figure)

Similarly, the graphs of $y = \frac{2}{x}$, $y = \frac{1}{x-3}$, and $\frac{2x}{x-3}$ are shown in Figure.

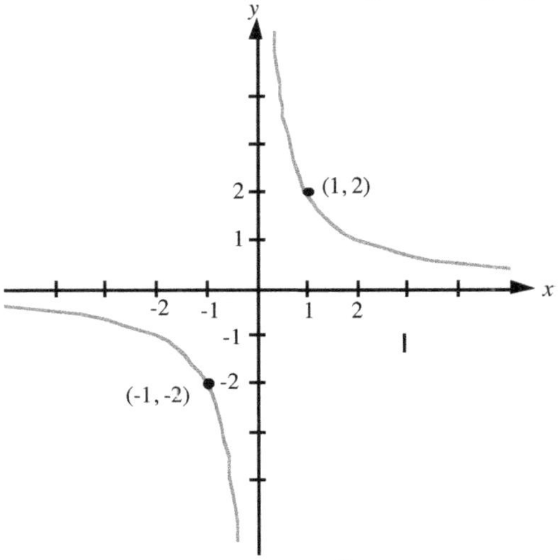

Figure The graph of $y = \frac{2}{x}$

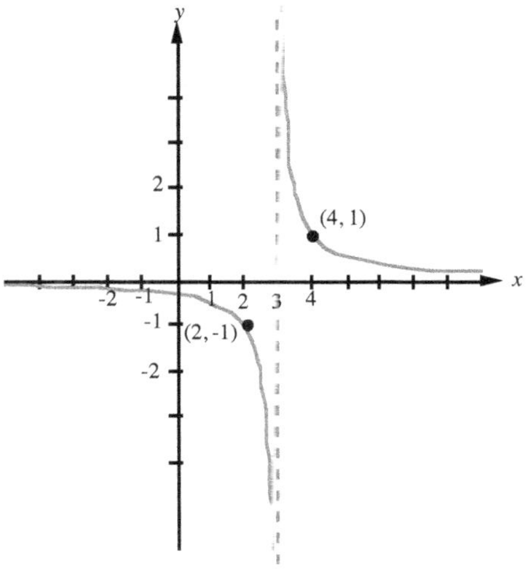

Figure : The graph of $y = \frac{1}{x-3}$

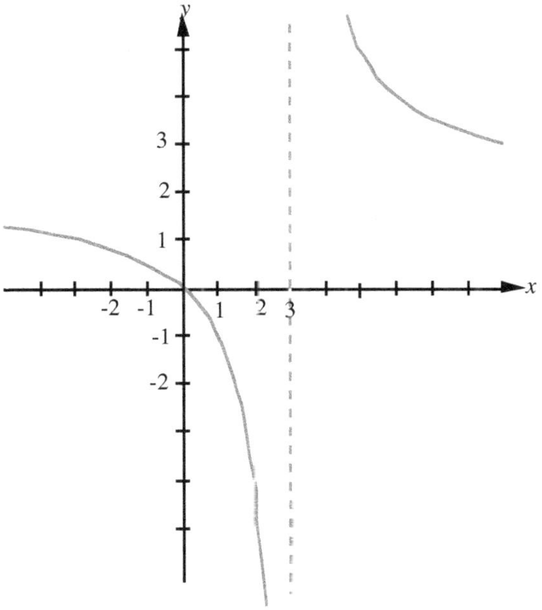

Figure :The graph of $\frac{2x}{x-3}$

General Method of Sketching the Graphs of Rational Functions
(Another method)

Before we learn the general method of sketching the graphs of rational functions, we will discuss two types of rational functions with the corresponding discontinuities.

Reducible and Irreducible Rational Functions

Consider the reducible rational function given by

$$f(x) = \frac{(x+1)(x+2)}{x+1}.$$

If we reduce immediately without examining the denominator to note that $x \neq -1$, we will obtain

$$f(x) = x + 2.$$

Now, if we sketch the graph of $f(x) = x + 2$, we might forget to indicate the discontinuity at $x = -1$ (Figures). Therefore, given a reducible rational function, we must always examine, note and record the excluded values (if any) before proceeding to reduce the fraction, otherwise , we may "lose track" of the excluded values.

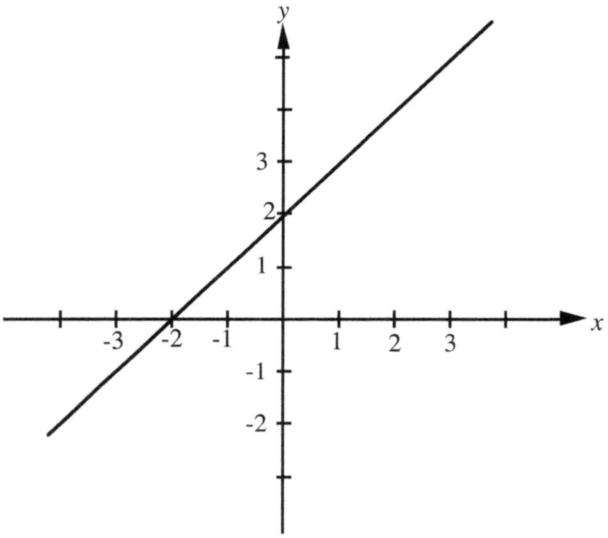

Figure: Graph of $y = x + 2$

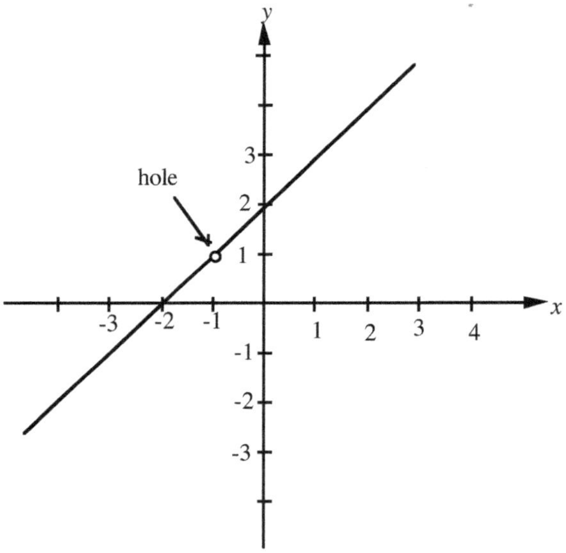

Figure: Graph of $y = \frac{(x+1)(x+2)}{x+1}$ (a hole at $x = -1$)

Now, we shall cover properties of examples on irreducible and reducible functions without graphing.

Example 1 $f(x) = \frac{1}{x-2}$.

This fraction is proper and irreducible. The graph has an infinite discontinuity and a vertical asymptote at $x = -2$.

Example 2 $f(x) = \frac{1}{(x+3(x-5)}$.

This fraction is proper and irreducible. The graph has infinite discontinuities and vertical asymptotes at $x = -3, x = 5$.

Example 3 $f(x) = \frac{x+1}{(x+1)(x-2)}$ (reducible)

This fraction is proper but reducible. The common factor is $(x + 1)$. The graph has a finite discontinuity and a hole at $x = -1$. It also has an infinite discontinuity and a vertical asymptote at $x = 2$; and a horizontal asymptote at $y = 0$.

Example 4 $f(x) = \frac{x^2+5}{x+1}$

This fraction is improper and irreducible. It has an infinite discontinuity and a vertical asymptote at $x = -1$. By the long division process the function can be expressed as the sum of a polynomial quotient and a proper, irreducible fraction. In this case, the polynomial quotient is the equation of the oblique asymptote.

$\frac{x^2+5}{x+1} = x - 1 + \frac{6}{x+1}$. The equation of the oblique asymptote is $y = x - 1$.

Example 5 $f(x) = \frac{(x+3)(x-5)}{(x-5)}$

This fraction is improper and reducible. The common factor $(x - 5)$ yields a finite discontinuity (a hole) at $x = 5$. The graph is a straight line ($y = x + 3$) with a hole at $x = 5$. There is **no** vertical asymptote.

Example 6 $f(x) = \frac{x}{x}$

The fraction is improper and reducible. The graph is that of the line $y = 1$, with a finite discontinuity (a hole) at $x = 0$. (Note: On reducing, $\frac{x}{x} = 1$)

Example 7 $f(x) = \frac{x+2}{x+2}$

The fraction is improper and reducible. The graph is that of the line $y = 1$ with a finite discontinuity (a hole at $x = -2$) . (Note: On reducing, $\frac{x+2}{x+2} = 1$)

We should note that **not** all rational functions have discontinuities and consequently not all rational functions have vertical asymptotes or holes in their graphs.

For example, the function $f(x) = \frac{x}{x^2+1}$ is defined for all real values of, since the square of any (non-zero) real number is positive. Moreover, if we attempt to find vertical asymptotes by setting the denominator to zero, we shall obtain non-real solutions. However, the graph of this function has a horizontal asymptote given by the line $y = 0$.

How to sketch the Graphs of Proper Irreducible Rational Functions

This is a direct method.

Step 1: Find the x-intercepts by setting $y = 0$, or by setting the numerator to zero and solving for x. If there are no real solutions, then there are no x-intercepts.

Step 2: Find the y-intercept by setting $x = 0$ and solving for y. If the value of y obtained is undefined or non-real, then there is no y-intercept.

Step 3: Find the vertical asymptotes by setting the denominator polynomial to zero and solving for x. If there are no real solutions, then there are no vertical asymptotes.

Step 4: Find the horizontal asymptotes. (See page 36) If there are no real and defined values, then there are no horizontal asymptotes.

Step 5: Using broken or dotted lines, draw the vertical and the horizontal asymptotes as determined in Steps 3 and 4. Plot the x- and y-intercepts from Steps 1 and 2.

Step 6: Using a sign diagram (See page), determine the signs of the function on each side of the vertical asymptotes and also the signs of the function between the x-intercepts (or zeros) as was done for the factorable polynomials (page). (That is, determine whether the curve is above or below the x-axis on each side of the vertical asymptotes and between the intervals created by the x-intercepts.) Instead of using sign diagrams, we can use convenient points as was done for factorable polynomials (see page).

Step 7: Additional or convenient points may be plotted by choosing convenient x-values and calculating the corresponding y-values, to obtain ordered pairs.

Step 8: Connect the points within each interval by a smooth solid curve noting that the curve does not meet its asymptote but rather approaches it gradually and smoothly as the curve and the asymptote are extended indefinitely (Review examples on page)

Example Sketch the graph of $y = \frac{3x}{x^2 - 9}$

Step 1: To find the x-intercepts, we set the numerator (or y) to zero.
`Then $3x = 0$ and from which $x = 0$.
The x-intercept $= 0$

Step 2: To find the y-intercept, we set $x = 0$.

Then $y = \frac{3(0)}{0 - 9}$

$y = 0$

Therefore, the y-intercept $= 0$

Step 3: To find the vertical asymptotes we set the denominator polynomial to zero and solve for x.

$x^2 - 9 = 0$

$(x + 3)(x - 3) = 0$

Solving, $x = -3$, $x = 3$. (We could use the quadratic formula if the factors are not easily recognizable)

There are vertical asymptotes at $x = -3$ and at $x = 3$.

Step 4: To find the horizontal asymptotes, we can use the method of limits (page 37).

$$y = \frac{\frac{3x}{x^2}}{\frac{x^2}{x^2} - \frac{9}{x^2}}$$

$$y = \frac{\frac{3}{x}}{1 - \frac{9}{x^2}}$$

For large values of x, $y \approx \frac{0}{1-0} \approx \frac{0}{1}$ (Noting that $\frac{3}{x} \approx 0, \frac{1}{x^2} \approx 0$)

$$y \approx 0$$

Therefore the line $y = 0$ (the x-axis) is a horizontal asymptote.

Step 5: We shall consider the behavior of the function on the intervals from $-\infty$ to $x = -3$; from $x = -3$ to $x = 0$; from $x = 0$ to $x = 3$; and from $x = 3$ to $+\infty$. We shall use sign a diagram (Figure) to determine the signs of the function on the intervals (See Sign diagram). We can alternatively choose convenient x-values, calculate y-values to determine if the graph is above or below he x-axis on each interval.

Step 6: With the help of the sign diagram, we draw curves (Figure) within each interval taking note of the hyperbolic nature of the curve. We can plot additional convenient points, especially around the turning points.

Number line: $-\infty$ ... -3 ... 0 ... 3 ... $+\infty$

	Factor	Column 1	Column 2	Column 3	Column 4
		Signs of the intervals			
Row 1	$3x$	$-$	$-$	$+$	$+$
Row 2	$x - 3$	$-$	$-$	$-$	$+$
Row 3	$x + 3$	$-$	$+$	$+$	$+$
Row 4	$\frac{3x}{(x-3)(x+3)}$	$-$	$+$	$-$	$+$

Sign diagram for $y = \frac{3x}{x^2 - 9}$.

Note the following in sketching the curve.

Between $-\infty$ and -3, the function is negative. (The curve is below the x-axis; Row 4 Column 1

Between -3 and 0, the function is positive. (The curve is above the x-axis; Row 4 Column 2

Between 0 and 3, the function is negative. (The curve is below the x-axis; Row 4 Column 3

Between 3 and $+\infty$, the function is positive. (The curve is above the x-axis; Row 4 Column 4

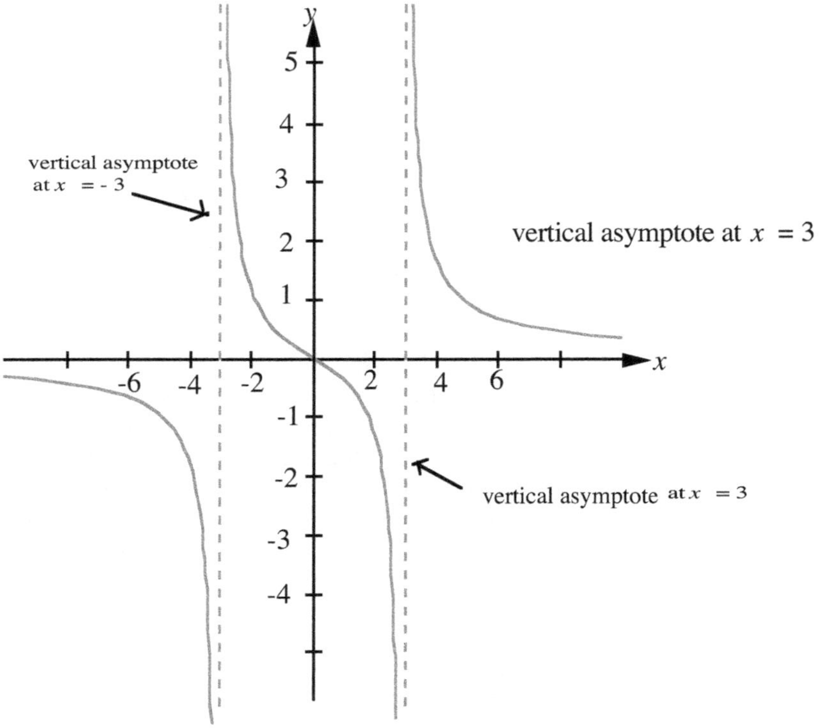

Figure : Graph of $y = \dfrac{3x}{x^2 - 9}$

Note:
Although we have not mentioned (or stressed) anything about the use of oblique asymptotes in sketching the graphs of rational functions, wherever the existence of the oblique asymptote is obvious from the equation, we should not hesitate to determine it and use it to sketch a more accurate graph. An equation has an oblique asymptote if the degree of the numerator polynomial is 1 degree more than the degree of the denominator polynomial.

For example, $f(x) = \dfrac{x^2 + 3}{x + 1}$ has an oblique asymptote which can easily be found by long division (See also page 39)

Lesson 8 Exercises

A Given the graphs of the following, rapidly sketch the graphs of their reciprocals
1. $y = (x+1)(x-3)$; **2**. $y = (x+3)(x-1)(x-2)$; **3**. $f(t) = t$

B With examples, discuss the similarities and differences between each the following:
(a) an improper rational function and an irreducible function.
(b) a proper rational function and an irreducible function; (c) a reducible function and an improper function.

C Sketch the graphs of the following functions:

1. $f(x) = \dfrac{3}{x-4}$; **2**. $f(x) = \dfrac{3x+1}{x-2}$; **3**. $f(x) = \dfrac{x+1}{4-x}$; **4**. $f(x) = \dfrac{5}{x+1}$; **5**. $f(x) = \dfrac{3}{x^2+2}$

6. $f(x) = \dfrac{4x^2+3}{x^2-9}$; **7**. $f(x)\dfrac{2x^2}{x^2-6x}$; **8**. $f(x) = \dfrac{2(x-1)(x-2)}{x-2}$;

9. $f(x) = \dfrac{(x+3)(x-1)(x+2)(x-2)}{(x+3)(x-1)(x+2)}$.

Answers: C

1.

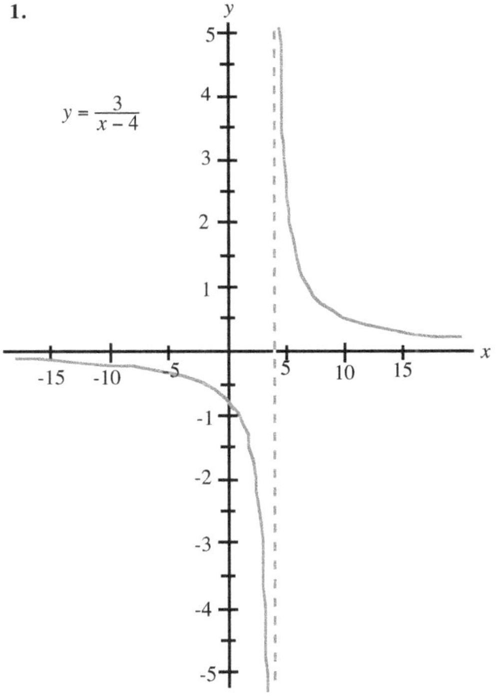

2.

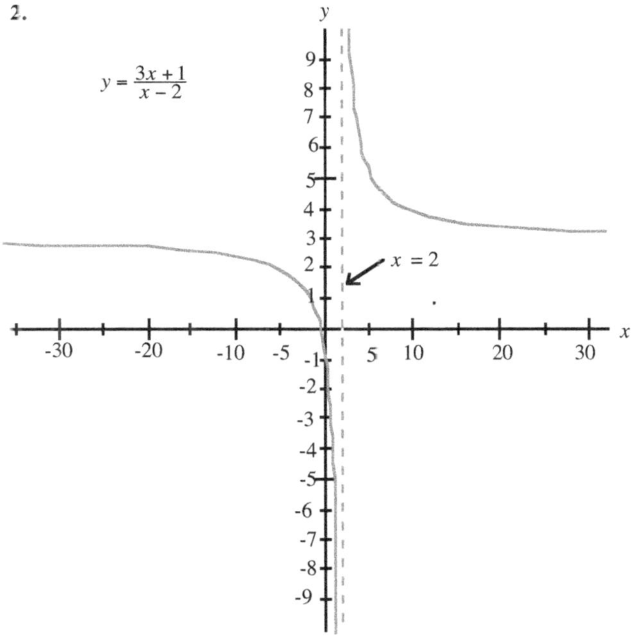

3.

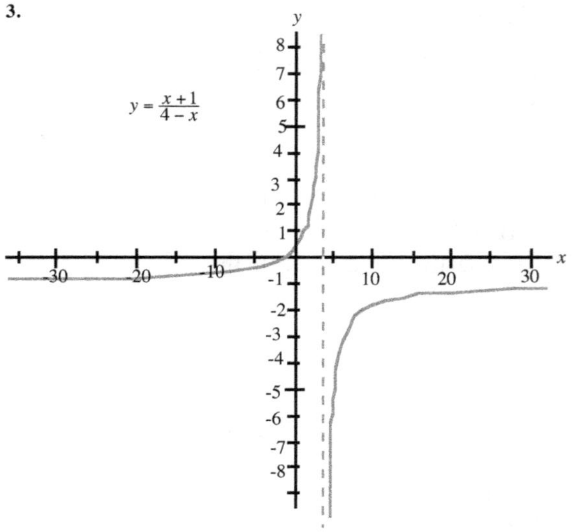

$y = \frac{x+1}{4-x}$
y
x
-30
-20
-10
10
20
30
8
7
6
5
4
3
2
1
-1
-2
-3
-4
-5
-6
-7
-8

4.

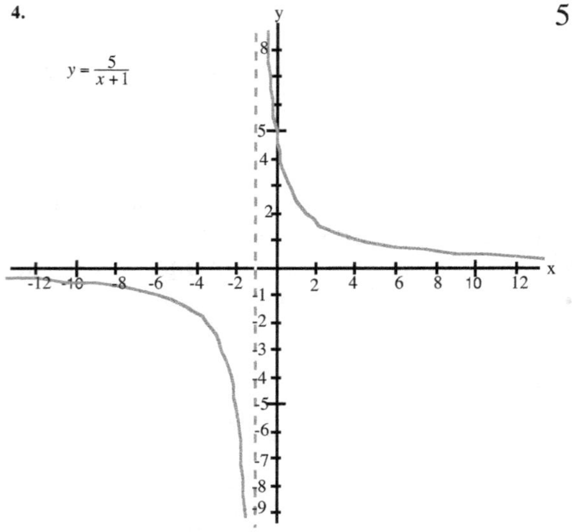

$y = \frac{5}{x+1}$
y
x
-12
-10
-8
-6
-4
-2
2
4
6
8
10
12
8
5
4
2
1
2
3
4
5
6
7
8
9

5.

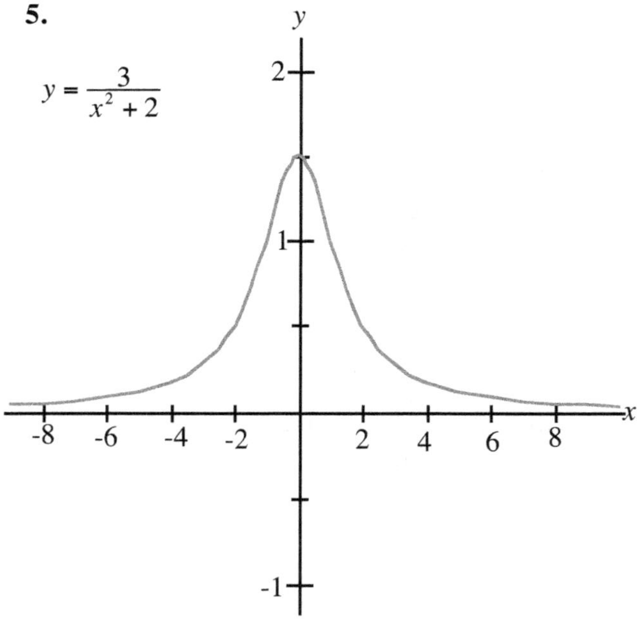

$y = \frac{3}{x^2+2}$
y
x
2
1
-1
-8
-6
-4
-2
2
4
6
8

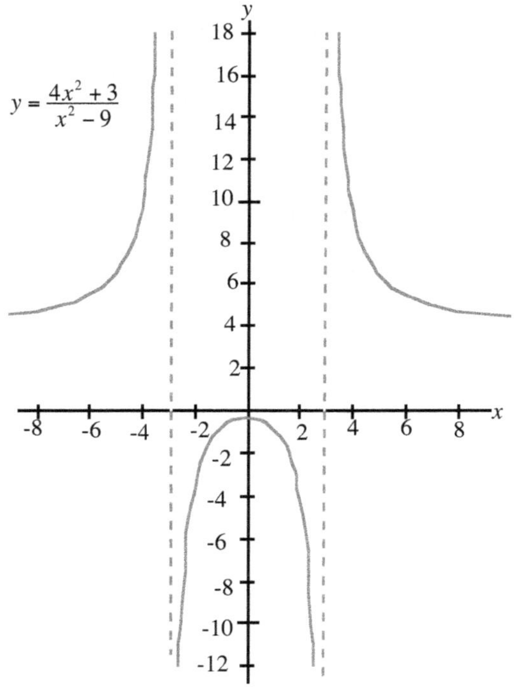

$y = \frac{4x^2+3}{x^2-9}$
y
x
18
16
14
12
10
8
6
4
2
-2
-4
-6
-8
-10
-12
-8
-6
-4
-2
2
4
6
8

7.

$y = \dfrac{2x^2}{x^2 - 6x}$

8.

$y = \dfrac{2(x-1)(x-2)}{x-2}$

9.

$y = \dfrac{(x+3)(x-1)(x+2)(x-2)}{(x+3)(x-1)(x+2)}$

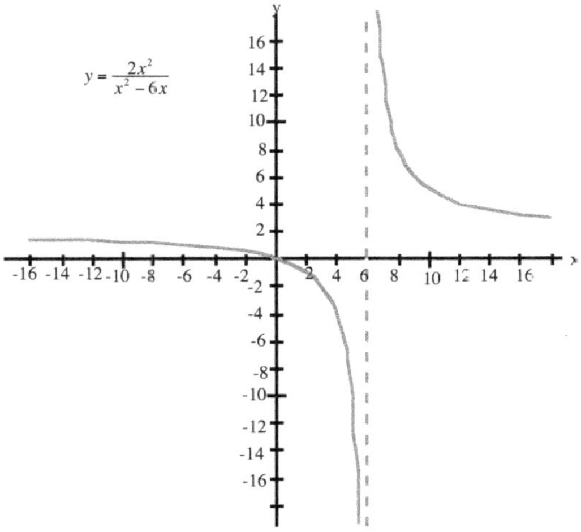

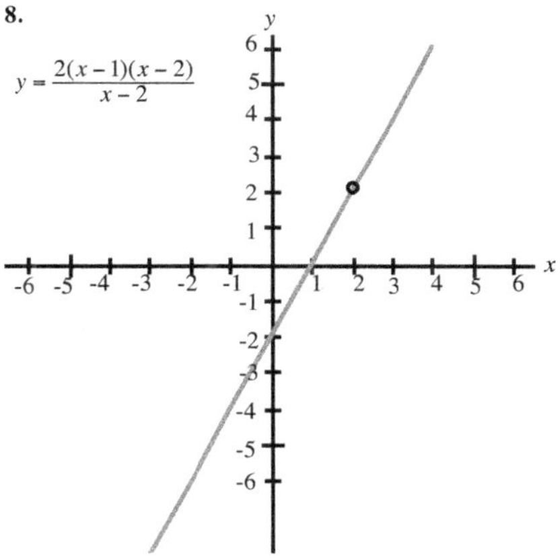

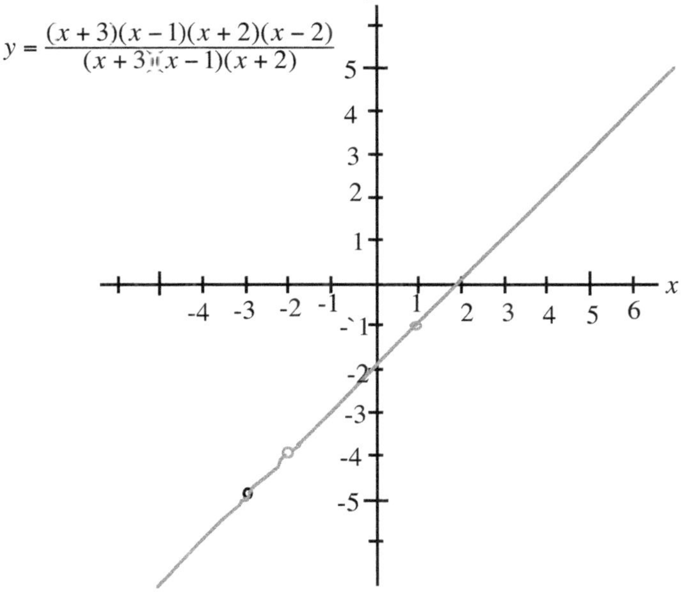

CHAPTER 4

Basic Review of Functions III

Lesson 9: **Positive, Negative, Increasing, and Decreasing Functions**
Lesson 10: **Concavities of Curves; Critical Points**
Lesson 11: **Reflection of Points, Lines and Curves**
Lesson 12: **Introduction to Transformation of Functions and Relations**
Lesson 13: **Vertical Stretching and Shrinking of graphs**
Horizontal Stretching and Shrinking
Lesson 14: **Symmetry; Even and Odd Functions**

Lesson 9

Positive, Negative, Increasing, and Decreasing Functions

Introduction

We will need the above terms in describing the behavior of functions. A conceptual understanding of these terms will facilitate a "unified " or " lumped" approach in covering functions. We will make use of the concept of a positive or a negative function in sketching the graphs of polynomial functions, rational functions, exponential functions, logarithmic functions, and trigonometric functions. For example, in order to draw the correct connection (graph) between, say, any two zeros (roots) of a polynomial function, a knowledge of whether the function is positive or negative between the zeros will be invaluable.

Positive and Negative Functions

When we say that a function is **positive** on a certain interval, say , the interval from $x = a$ to $x = b$, we mean that, algebraically, all the y-coordinate values on this interval from a to b are **positive**, and graphically, the entire curve lies above the x-axis, on this interval (Figure).

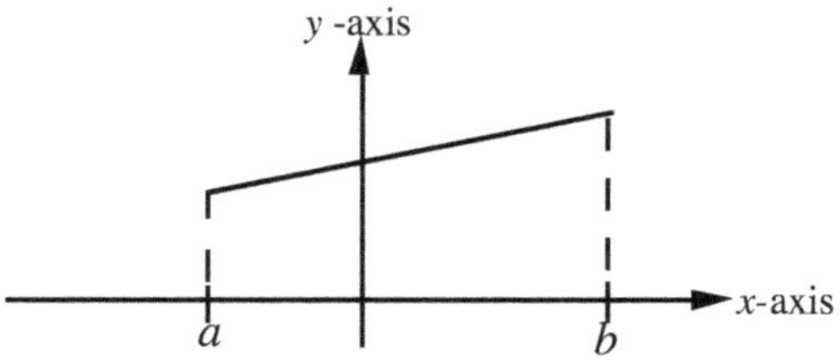

Figure

Similarly, when we say that a function is **negative** on the interval from $x = a$ to $x = b$, we mean that all the y-coordinate values are negative (have minus signs) on this interval from a to b, and that , graphically, the entire curve lies below the x-axis (Figure)

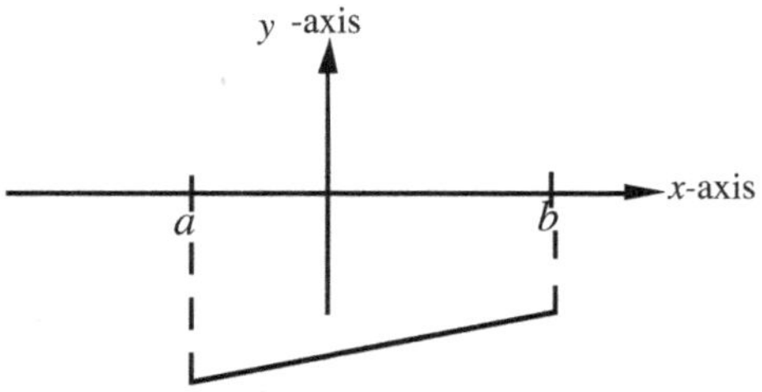

Figure

Decreasing and Increasing Functions

We use the above terms in describing critical points, namely, minimum and maximum points, and points of inflection. We will use these terms in describing the behavior of functions as well as in sketching their graphs. For example, we will use the term " concavity " (e.g. concave up or down) in defining critical points (turning points); in describing the behavior of polynomial, rational, trigonometric, exponential and logarithmic functions as well as conic relations, namely the hyperbola and the ellipse.

Increasing Function (strictly increasing function)

Geometrically, a strictly increasing function is one whose graph (curve or line) rises from left to right, as one reads from left to right (or simply, the curve leans to the right). That is, as we move our eyes horizontally to the right in the coordinate plane, we encounter higher and higher values of y. Thus, any point on the curve is higher than any other point (on the curve) to its left. (Recall from elementary math that a straight line leaning to the right in the page has a positive slope).

More formally, a **function** f is increasing (or strictly increasing) on an interval containing the numbers x_1, x_2 if whenever $x_2 > x_1$, $f(x_2) > f(x_1)$ (See Figure)

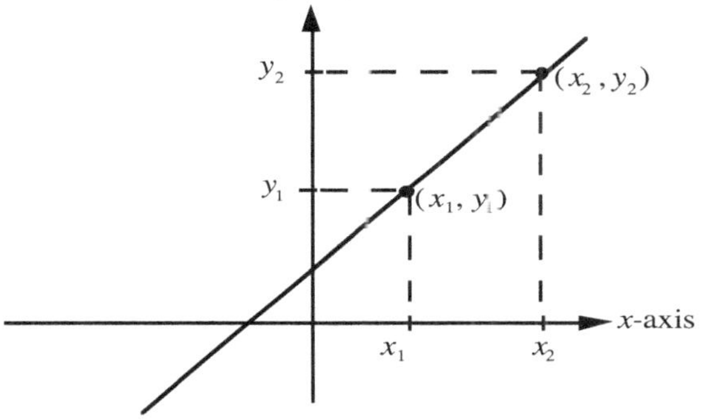

Figure: Graph of a increasing function

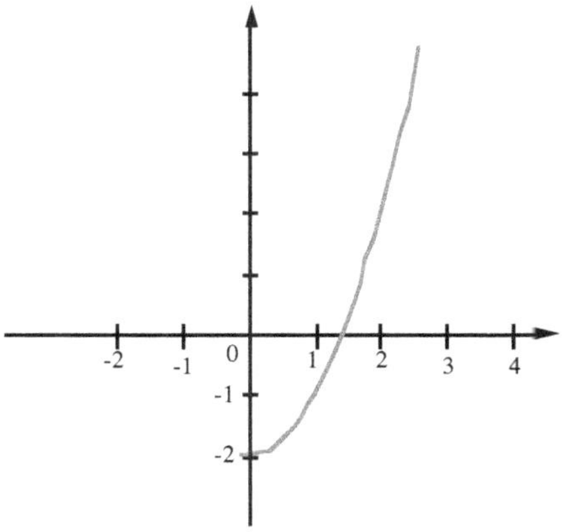

Figure : The graph of an increasing function

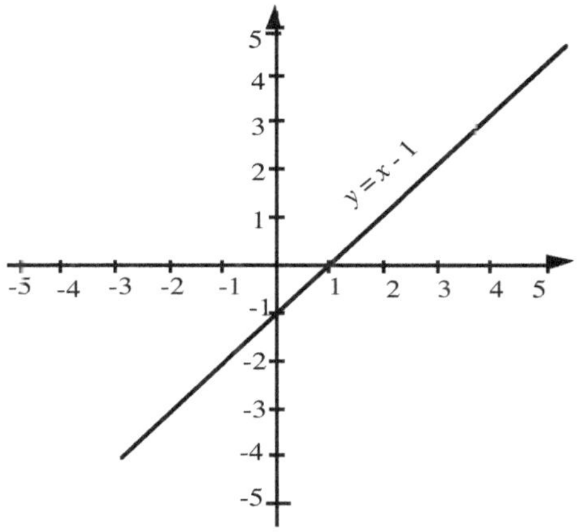

Figure : Graph of $y = x - 1$ (an increasing function)

Example The function given by $f(x) = x - 1$ is an increasing function. (Fig.)

Let us test for $x_1 = 2, x_2 = 5$ (that is $x_2 > x_1$) in $f(x) = x - 1$

Then $f(x_1) = f(2) = 2 - 1 = 1$; and $f(x_2) = f(5) = 5 - 1 = 4$

Clearly, $5 > 2$ implies $f(5) > f(2)$ (since $f(5) = 4$ and $f(2) = 1$

Note: The above is **no** proof, but is only an illustration

Decreasing Function (strictly decreasing function)

Geometrically, a strictly **decreasing function** is one whose graph (curve or line) falls (drops) from left to right as one reads from the left to the right,
(or simply, the curve leans to the left). That is, as we move our eyes horizontally from the left to the right in the coordinate plane, we encounter lower and lower values of y. Thus, any point on the curve is lower than any other point (on the curve) to its left (Figure). The curve leans to the left in the page.

More formally, a **function** f is decreasing (or strictly decreasing) on an interval containing the numbers x_1, x_2 if whenever $x_2 > x_1$, $f(x_2) < f(x_1)$
(That is, as the x-values increase, the y-values decrease) (See Figure)

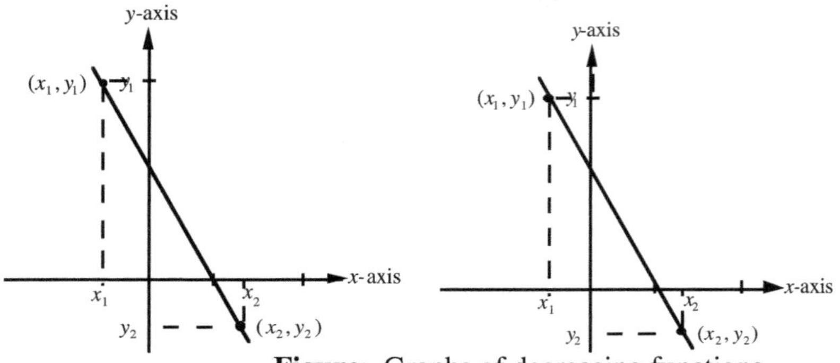

Figure: Graphs of decreasing functions

Example

The function given by $f(x) = -x - 1$ is decreasing on the interval $[-5, -2]$

Let us test for $x_1 = -5, x_2 = -2$ (that is, $x_2 > x_1$) in $f(x) = -x - 1$

Then $f(x_1) = f(-5) = -(-5) - 1 = 5 - 1 = 4$

$f(x_2) = f(-2) = -(-2) - 1 = 2 - 1 = 1$. (We could also use -4 or -3 for testing)

Clearly, $-2 > -5$ implies $f(-2) < f(-5)$ (since $f(-2) = 1$ and $f(-5) = 4$)

Note: The above is **no** proof, but is only an illustration

Some functions are neither increasing nor decreasing. A function which is neither increasing nor decreasing is called a **constant function.** More formally, a function is **constant** on an interval containing the numbers x_1, x_2 if whenever $x_2 > x_1$, $f(x_2) = f(x_1)$. Example; see below.

Figure: The graph of a constant function:
$f(x) = 5$

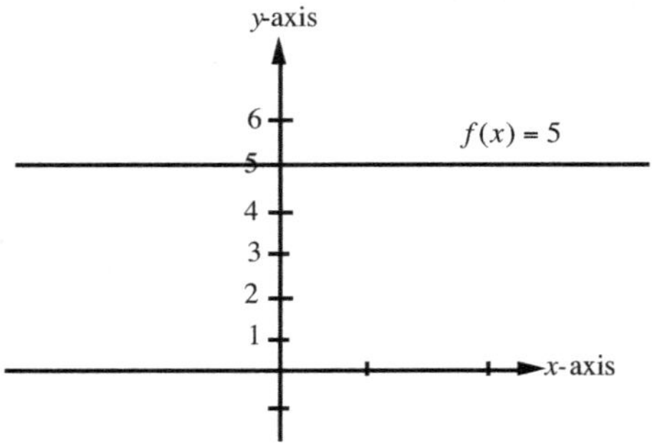

Lesson 9 Exercises

Give both graphical and algebraic answers (with examples) to the following:

1. What is meant by saying that a function is positive on an interval?
2. What is meant by saying that a function is negative on an interval?
3. What is meant by saying that a function is increasing on an interval?
4. What is meant by saying that a function is decreasing on an interval.
5. What is a constant function?

Answers: See text.

About endpoints of increasing intervals and decreasing intervals from graphs

In writing the intervals of increasing or decreasing functions from graphs, we will include th e endpoints whenever the endpoints are defined. For example, with reference to the diagram below, $f(x)$ is increasing on the intervals $[-2, -1)$ and $[1, 2]$,, and decreasing on the intervals $(-1, 1]$, and $[2, 3]$.

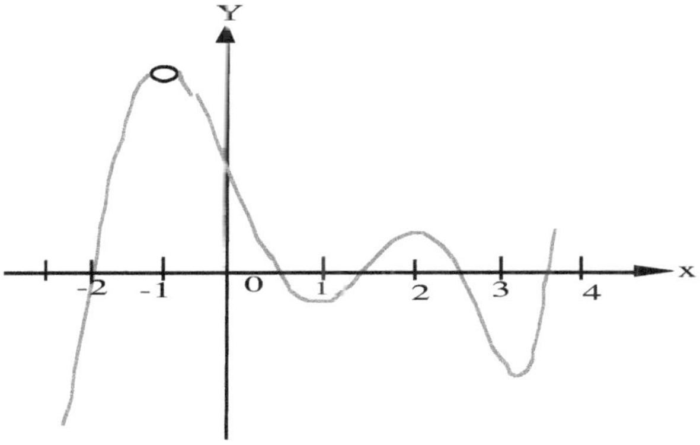

Other Terminology

Nonincreasing function (monotone decreasing function):

A function f is nonincreasing on an interval containing the numbers x_1, x_2 if whenever $x_2 > x_1$, $f(x_2) \leq f(x_1)$

(Note that nonincreasing implies "does not increase" and therefore, it is either constant or decreasing on that interval) Note: Using nonincreasing is consistent with terms such as a "nonpositive" integer which is either zero or a negative integer.

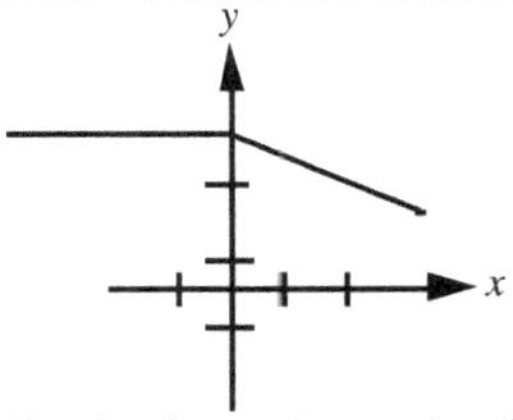

Graph of a nonincreasing function

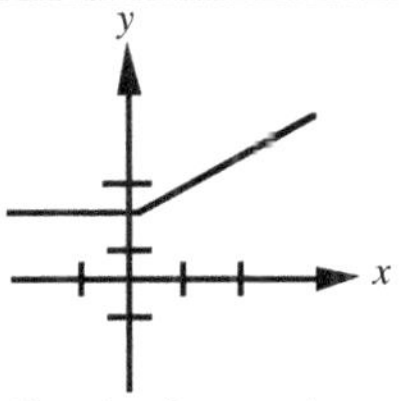

Graph of a nondecreasing function

Nondecreasing function (monotone increasing function):

A function f is nondecreasing on an interval containing the numbers x_1, x_2 if whenever $x_2 > x_1$, $f(x_2) \geq f(x_1)$ (Note that nondecreasing implies "does not decrease" and therefore, it is either constant or increasing on that interval)

Lesson 10

Concavities of Curves; Critical Points

Concavities of Curves

The following meanings given to concavities will be useful in describing the turning behavior of curves at certain points or locations. (In calculus, we will learn more formal definitions of concavities) . In particular, we will illustrate the "sense of concavity " (direction of the concavity).

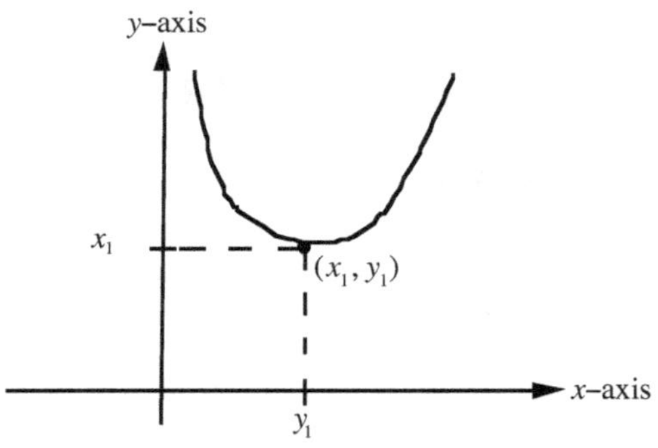

Figure 1: Curve is concave up (opens upwards) at (x_1, y_1)

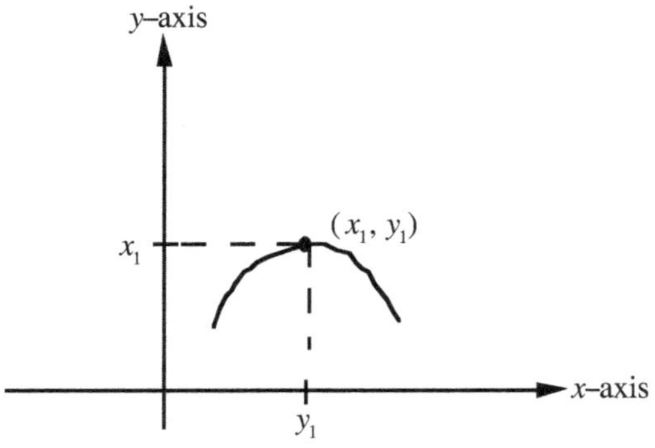

Figure 2: Curve is concave down (opens downwards) at (x_1, y_1)

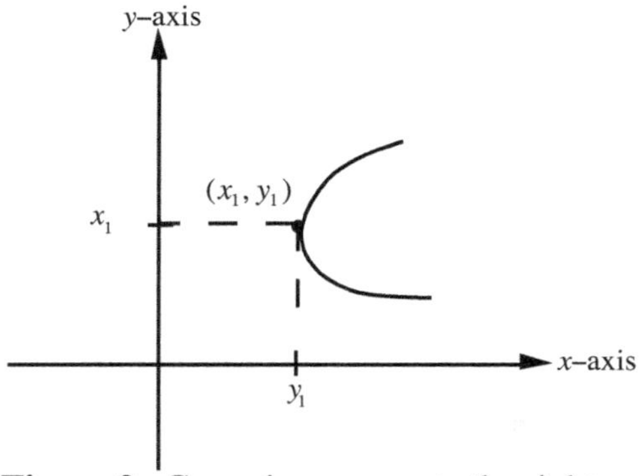

Figure 3: Curve is concave to the right (opens to the right) at (x_1, y_1))

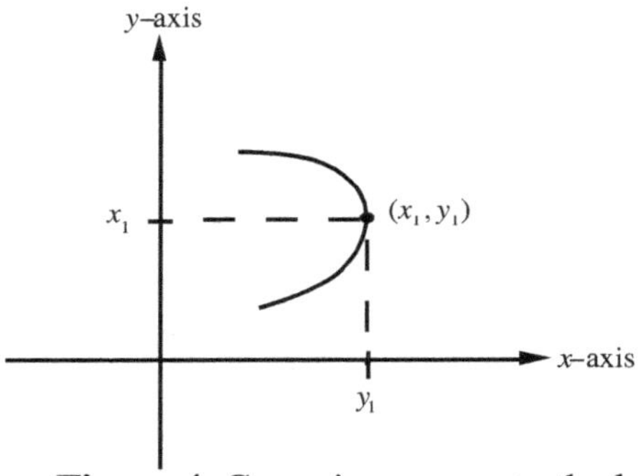

Figure 4: Curve is concave to the left (opens to the left) at (x_1, y_1))

Note above that Figures 1 and 2 are graphs of functions; but Figures 3 and 4 are **not** functions by the vertical line rule.

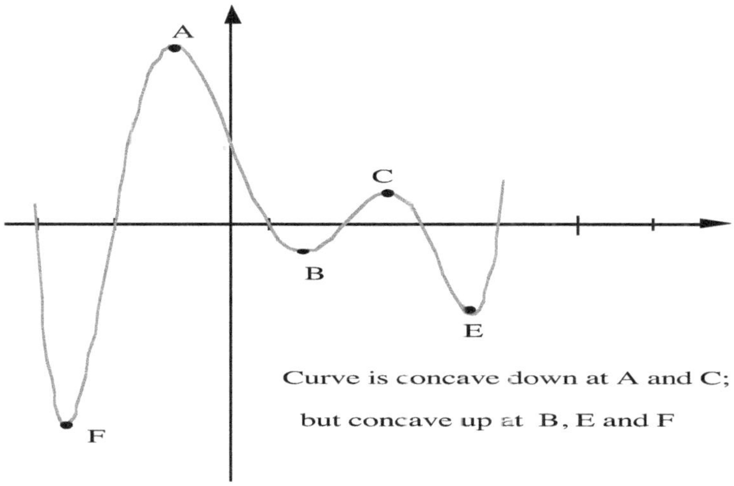

Extra:

The author also proposes the following new system:

1. Concave north (concave up)

2. Concave south (concave down)

3. Concave east (concave to the right).

4. Concave west (concave to the left)

5. Concave North-East (N-E);

6. Concave North-West (N-W);

7. Concave South-West (S-W)

8. Concave South-East (S-E)

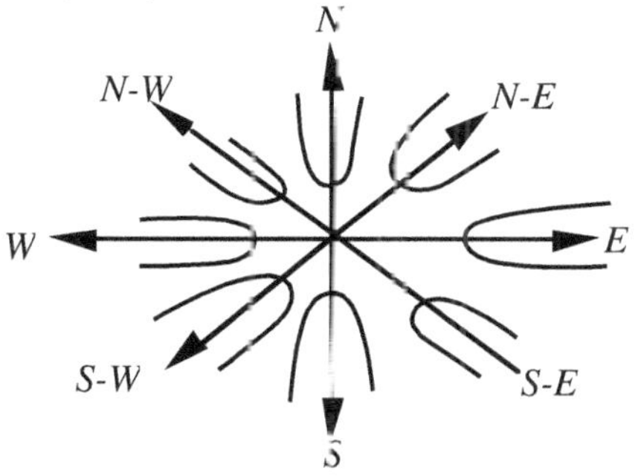

Figure: Concavity directions based on the North-South Line

Note above that only concave north and concave south are graphs of functions.

Critical Points

Turning Points, Minimum Points, Maximum Points, Inflection Points

Maximum Point: A given point (turning point) on a curve is a maximum point of the curve if the given point is higher than any other point on the curve in the immediate vicinity (on both sides of the point). The y-coordinate of this point is called a maximum value of the curve. At a maximum point, the curve is concave downwards. Sometimes, a maximum point of a curve may not be the highest point on the curve. When this happens, this maximum point is higher than only points sufficiently near it and there are other points on the curve which are higher than this point. In this case, we call this maximum point a relative maximum. However, if this maximum point is the highest point on the curve (irrespective of the domain of the function), then we say that this point is the absolute maximum.

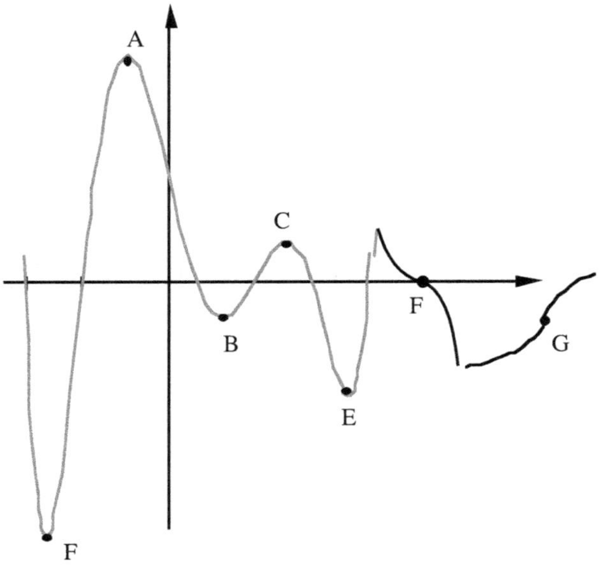

Minimum Point: A given point (turning point) on a curve is a minimum point of the curve if the given point is lower than any other point on the curve in the immediate vicinity (on both sides of the point). At a minimum point, the curve is concave upwards. The y-coordinate of this point is called a minimum value of the curve. Sometimes, a minimum point of a curve may not be the lowest point on the curve. When this happens, this minimum point is lower than only points sufficiently n ear it and there are other points on the curve which are lower than this point. In this case, we call this minimum point a relative minimum. However, if this minimum point is the lowest point on the curve (irrespective of the domain of the function), then we say that this point is the absolute minimum.

Generally, for a polynomial of degree n, there are at most $n-1$ relative minima or maxima.

For example, $y = x^2$ has only one minimum point (and that point is an absolute minimum).

Point of Inflection: A given point (turning point) on a curve is a point of inflection if the curve at this point, changes from being concave up to being concave down or vice versa. At a point of inflection, the curve is either concave down to the left of the point and concave up to the right of the point, or it is concave up to the left of the point and concave down to the right of the point. Thus, the sense of concavity to the right of the point is opposite to the sense of concavity to the left of the point. Note that a point of inflection is neither a minimum point nor a maximum point.

More formally, for a function f, $f(x_0)$ is a relative maximum of f on an open interval containing x_0 if $f(x_0) > f(x)$ for all $x \neq x_0$ on the interval. Similarly, for a function f, $f(x_0)$ is a relative minimum of f on an open interval containing x_0 if $f(x_0) < f(x)$ for all $x \neq x_0$ on the interval..

Lesson 10 Exercises

1. Identify the concavities of the following graphs:

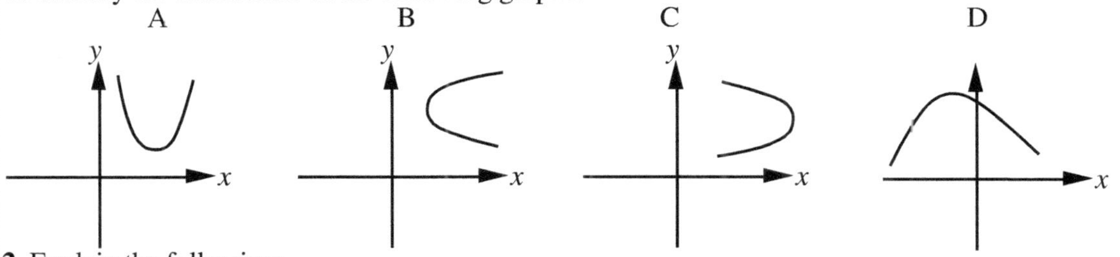

2. Explain the following:

(a) Turning Point, **(b)** Minimum Point, **(c),** Maximum Point, **(d)** Inflection Point

Answers: **1.** Concave upwards; **2.** Concave to the right; **3.** Concave to the left; **4.** Concave downwards.

Lesson 11

Reflection of Points, Lines and Curves

We will cover reflections about the y-axis, the x-axis, the origin and the line $y = x$.

Let us discuss a few examples of the usefulness of being able to rapidly reflect points, lines and curves. If we know that a curve whose equation we are to find is the reflection of, say, the curve $y = x^2$ about the x-axis, then we could immediately write the equation of the reflected curve as $y = -x^2$ Similarly, the curves $y = e^x$ and $y = e^{-x}$ are reflections of each other about the y-axis.

Given the graph of a function or a relation, we can find the graph of the inverse of the function or relation by reflecting the given graph about the line $y = x$.

Reflection of a point about the y-axis

Let $P(x,y)$ be a point in an x-y coordinate plane. Then the coordinates of the **point** of reflection of the given point about the y-axis is obtained by replacing the x-coordinate by $-x$, keeping the y-coordinate unchanged. The point of reflection is thus given by $R(-x,y)$.

Example Find the point of reflection of the point (4, 2)about the y-axis.

Solution Change the sign of the x-coordinate and keep the y-coordinate unchanged. The reflected point is $P(-4, 2)$. (Figure). If the page were folded along the y-axis, the two points (4, 2) and (-4, 2) would coincide (occupy the same space).

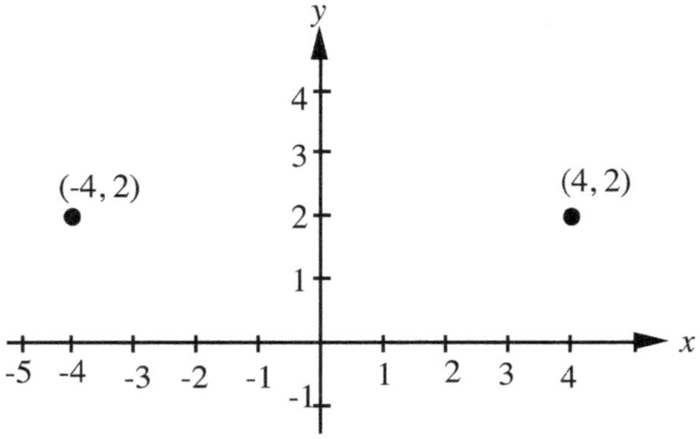

Figure: The graph of the reflection of (4, 2) about the y-axis.

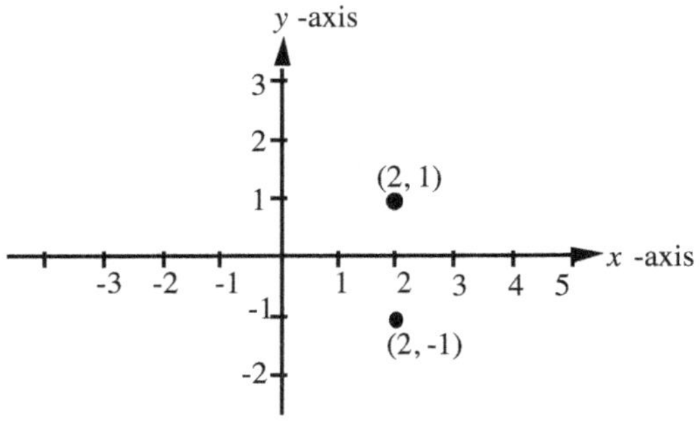

Figure: The graph of the reflection of (2, 1) about the x-axis

Reflection of a point about the x-axis

Let $P_1(x,y)$ be a point in an x-y coordinate plane. Then the coordinates of the point of reflection of the given point about the x-axis is obtained by replacing the y-coordinate by $-y$, keeping the x-coordinate unchanged. The point of reflection is thus given by $P_2(x,-y)$.

Example Find the point of reflection of the point (2, 1) with respect to the x-axis.

Solution Change the sign of the y-coordinate and keep the x-coordinate unchanged. The point of reflection is (2,-1). (Figure)

If the page were folded along the x-axis the two points (2 , 1) and (2 -1) would coincide.

Reflection of a point about the origin

The point of reflection of the point $P(x,y)$ about the origin is $P(-x,-y)$, obtained by replacing x by $-x$ and y by $-y$, That is, change the signs of both the x- and y-coordinates.

Example Find the point of reflection of (2, 4) about the origin.

Solution The point of reflection is given by $P(-x,-y)$ and in the present case it is Q(-2, -4). Practically, if the page were folded first along the positive y-axis (1st quadrant on to 2nd quadrant), and hen along the x-axis (1st and 2nd quadrant together on to the 3rd quadrant), the two points (2 4), and (-2 -4), would coincide (Figure); or simply, we would fold the 1st quadrant on to the 3rd quadrant so that the negative x-axis coincides with the positive x-axis; and that the positive y-axis is coincident with the negative y-axis.

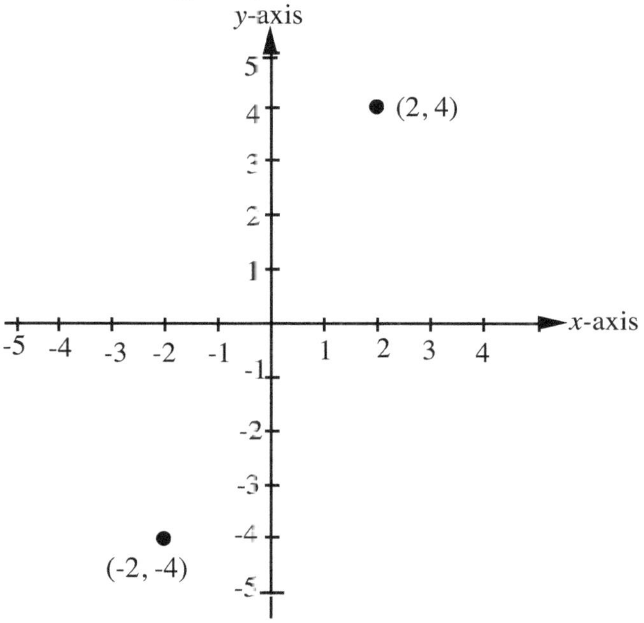

Summary for reflection

1. To reflect about the y-axis, change the sign of the x-coordinate, keeping the y-coordinate unchanged.

2. To reflect about the x-axis, change the sign of the y-coordinate, keeping the x-coordinate unchanged.

3. To reflect about (or through) the origin, change the signs of both the x- and y-coordinates.

Note that changing the sign of the x-coordinate, or the y-coordinate is equivalent to multiplying or dividing he coordinate by -1.

We should note above that the given point and the reflected point are reflections of each other.

Reflection of a point about a vertical line

Geometrically, two points *A* and *B* are reflections of each other about a given line if and only if the given line is the perpendicular bisector of *AB* (i.e., the line divides the segment *AB* into two equal parts at right angles).

Also: **Reflection of a point about the vertical line** $x = x_m$

Let the point to be reflected be (x_1, y_1).

Let the reflected point be (x_2, y_1)

(The *y*-coordinate is constant since the reflection is about a vertical line)

Let the midpoint of the line connecting (x_1, y_1) and (x_2, y_1) be (x_m, y_1).

Then $x_m = \dfrac{x_1 + x_2}{2}$ (Note that the equation of the line about which (x_1, y_1) is reflected is $x = x_m$)

Note: For the reflection about the *y*-axis, $x_m = 0$ (The *y*-axis is the line $x = 0$.)

Example Find the point of reflection of the point *A* (4, 6) about
(a) the *y*-axis ; (*b*) the line $x = 2$.

Solution

Method 1

(a)

Step 1: Plot the point (4, 6). See Figure.

Step 2: Draw a horizontal line from (4, 6) to intersect the *y*-axis (the line $x = 0$) at C.
Since the point C is mid-way between the point (4, 6) and the reflected point, count 4 units horizontally to the left, stop and place a dot here. This point is the point of reflection.

The point of reflection is *B*(-4, 6).

(*b*) Since the point D is 2 units from the point (4, 6), count 2 units horizontally to the left, stop and place a dot here. This is the point of reflection about the line $x = 2$.

The point of reflection is *C*(0, 6).

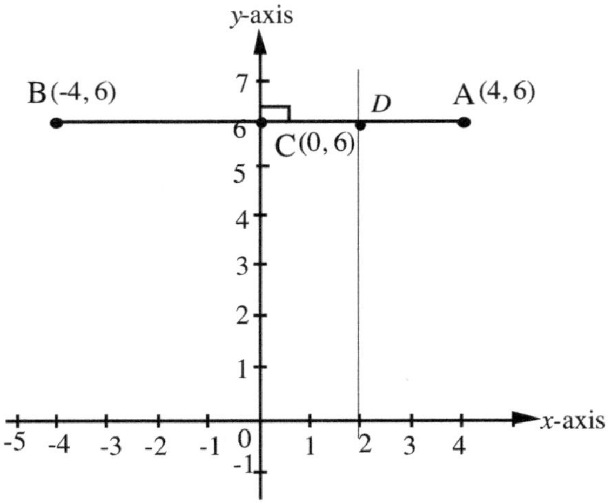

Figure: Graph of the reflection of the point *A* (4, 6) about the *y*-axis and the line $x = 2$.

Note above that in (a) *BC* = *CA* and *BA* intersects the *y*-axis at right angles at *C*; and in (b) *CD* = *DA* and *CA* intersects the line $x = 2$ at right angles at *D*. Note that *D* is equidistant from *A* and *C*.

Method 2

(a)

Apply $x_m = \frac{x_1 + x_2}{2}$ (Note that the equation of the line about which (x_1, y_1) is reflected is $x = x_m$.)

$x_1 = 4$, $x_m = 0$ (on the y-axis, $x = 0$). Substituting,

$$0 = \frac{4 + x_2}{2}$$

$$0 = 4 + x_2$$

$$-4 = x_2$$

The x-coordinate of the reflected point -4

Therefore, the point of reflection about the y-axis is B(-4, 6). (the y-coordinate does not change)

(b)

Apply $x_m = \frac{x_1 + x_2}{2}$ (Note that the equation of the line about which (x_1, y_1) is reflected is $x = x_m$.)

$x_1 = 4$, $x_m = 2$ (on the line $x = 2, x = 2$). Substituting,

$$2 = \frac{4 + x_2}{2}$$

$$4 = 4 + x_2$$

$$0 = x_2$$

The x-coordinate of the reflected point is 0.

Therefore, the point of reflection about the line $x = 2$ is $C(0, 6)$. (the y-coordinate does not change)

Reflection of a point about the horizontal line $y = y_m$

Let the point to be reflected about the line $y = y_m$ be (x_1, y_1)

Let the reflected point be (x_1, y_2),

(The x-coordinate is constant since the reflection is about a horizontal line)

Let the midpoint of the line connecting (x_1, y_1) and (x_1, y_2) be (x_1, y_{m1}).

$$y_m = \frac{y_1 + y_2}{2}$$ (Note that the equation of the line about which (x_1, y_1) is reflected is $y = y_m$)

Note: For the reflection about the x-axis, $y_m = 0$ (The x-axis is the line $y = 0$.)

Example Find the point of reflection of (2, 1) about the x-axis.

Solution $y_1 = 1$, $y_m = 0$ (On the x-axis, $y = 0$), Substituting in $y_m = \frac{y_1 + y_2}{2}$, we obtain

$$0 = \frac{1 + y_2}{2}$$

$$0 = 1 + y_2$$

$$-1 = y_2$$

The y-coordinate of the reflected point is -1.

Therefore, the point of reflection about the x-axis is (2, -1). This problem was solved previously by merely changing the sign of the y-coordinate, keeping the x-coordinate unchanged.

Reflection of a given curve (or line) about the *y*-axis

Step 1: Sketch the given curve using broken or dotted lines.

Step 2: Reflect critical points or important points on the curve about the *y*-axis .(see p.68)

Step 3: Connect the points of reflections from Step l by a smooth solid curve (if it is a curve) or by a straight line (if it is a straight line).

Reflection of a given curve (or line) about the *x*-axis

Method: Repeat steps 1, 2 and 3 above but about the *x*-axis.

Step 1: Sketch the given curve using broken or dotted lines.

Step 2: Reflect critical points or important points on the curve about the *x*-axis .(see p.68)

Step 3: Connect the points of reflections from Step l by a smooth solid curve (if it is a curve) or by a straight line (if it is a straight line).

Reflection of a given curve about (or through) the origin

Step 1: Using broken lines rapidly sketch the graph of the given curve.

Step 2 :Reflect critical or important points about the origin (as explained on page 69. .

Step 3. Connect the points of reflections by a solid curve.

Example 1. Sketch the reflection of the line whose equation is $y = 3x + 2$ about (*a*) the *y*-axis; (*b*) the *x*-axis.

(a) Step 1. Sketch the graph of $y = 3x + 2$ using broken lines.

Step 2: Reflect two points on $y = 3x + 2$ about the *y*-axis and connect the new points by a solid line. (Fig.1)

(b) Step 1: Sketch the graph of $y = 3x + 2$ using broken lines.

Step 2: Reflect two points on $y = 3x + 2$ about the *x*-axis and connect the new points by a solid line. (Fig. 2)

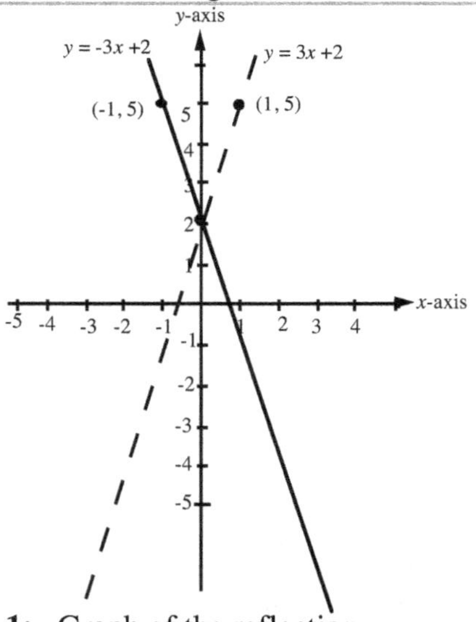

Fig. 1: Graph of the reflection of the line y = 3x + 2 about the *y*-axis

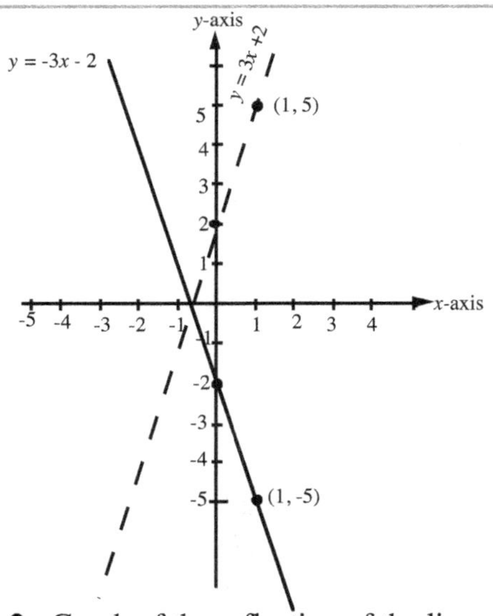

Fig. 2: Graph of the reflection of the line $y = 3x + 2$ about the *x*-axis

Note above: You may also use the slope-intercept method to draw the graph of a straight line.

Reflection of a curve or a line about the line $y = x$

Finding the reflection about the line $y = x$ graphically, is synonymous with finding the inverse of the given graph.

Step 1: Reflect important (critical) points on the given curve about the line $y = x$.

To accomplish the above, **interchange** the ordered pairs (x, y) on the given curve or line and plot the results in a rectangular coordinate system.

Step 2: Connect the points (from Step 1) by a solid curve or line accordingly.

Practically, if the page were folded along the line $y = x$, the given curve (or line) and the reflected curve would coincide.

Example 2 Sketch the reflection of the line whose equation is $y = 3x + 2$ about the line $y = x$

Solution

Step 1. Sketch the graph of $y = 3x + 2$ using broken lines.

Step 2: Reflect two points on $y = 3x + 2$ about the line $y = x$, and connect the new points by a solid line.

Extra: New Theorem by the author
If two lines are inverses of each other, their slopes are reciprocals of each other. If the slope of a line is m, the slope of the inverse of this line is $\frac{1}{m}$.

(see also p.80)

Proof: Let the equation of a line with slope m and y-intercept b be given by $y = mx + b$.

The inverse of this line is

$x = my + b$ or $y = \frac{1}{m}x - \frac{b}{m}$

Clearly, the slope of the inverse is $\frac{1}{m}$. Q.E.D

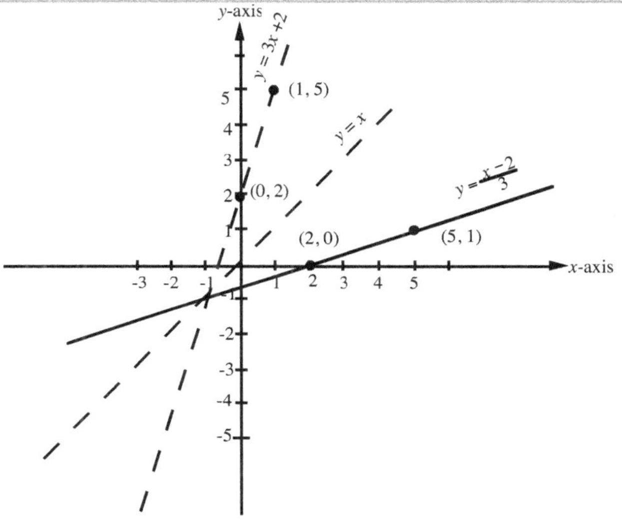

Fig. 3 Graph of the reflection of $y = 3x + 2$ about the line $y = x$.

Equation of the reflection of a curve, given the equation of the curve

(a) **About the *y*-axis**

Replace x by $-x$ in the given equation.

$y = (-x)^3 + 2(-x)^2 + 3(-x) + 1$

$y = -x^3 + 2x^2 - 3x + 1.$

(b) **About the *x*-axis.**

Replace y by $-y$ and solve for y in the given equation.

$-y = x^3 + 2x^2 + 3x + 1$

$y = -(x^3 + 2x^2 + 3x + 1) = -x^3 - 2x^2 - 3x - 1$

Example 4 Find the "reflected equation" of $y = 3x + 2$ about the line $y = x$.

The solution is the same as finding the inverse of $y = 3x + 2$. (We repeat a previous solution)

Step 1 Interchange x and y in the given equation.

Then, we obtain $x = 3y + 2$. (2)

By tradition, we want to keep the x-axis horizontal and express y as a function of x.

Step 2: Solving for y, $y = \frac{x-2}{3}$

The reflected equation is $y = \frac{x-2}{3}$.

Lesson 11 Exercises

Find the point of reflection about the y-axis for each of the following:

1. $(3, 2)$; **2.** $(-4, 3)$; **3.** $(-2, -5)$; **4.** $(4, -6)$

Exercises 5 – 8: Repeat Exercises 1 – 4 but with respect to the x–axis.

Exercises 9 – 12. Repeat Exercises 1 – 4 but with respect to the origin.

13. Repeat Exercise 1 – 4 with respect to the line $x = 2$.

Find an equation of the reflection with respect to (a) the x–axis, (b) the y–axis, (c) the origin, (d) the line $y = x$ for the following:

14. $y = 4x + 3$; 15. $y = x^2 - 4x + 3$; **16.** $y = x^2 - 3x - 2$; **17.** $-y = x^2 - 3x - 2$.

18. Sketch the graph of $y = 2x + 3$ and reflect the graph with respect to (a) the x–axis, (b) the y–axis.

19. Sketch the graph of $y = -3x + 2$ and reflect the graph with respect to the line $y = x$.

Answers:

1. $(-3,\ 2)$; **2**. $(4,\ 3)$; **3**. $(2,\ -5)$; **4**. $(-4,\ -6)$; **5**. $(3,\ -2)$; **6**. $(-4,\ -3)$; **7**. $(-2,\ 5)$; **8**. $(4,\ 6)$; **9**. $(-3,\ -2)$; **10**. $(4,\ -3)$ **11**. $(2,\ 5)$; **12**. $(-4,\ 6)$; **13**. (a) $(1,\ 2)$; (b) $(8,\ 3)$; (c) $(6,\ -5)$; (d) $(0,\ -6)$; **14**. (a) $y = -4x - 3$; (b) $y = -4x + 3$; (c) $y = 4x - 3$; (d) $x = 4y + 3$ or $y = \frac{1}{4}x - \frac{3}{4}$; **15**. ($a$) $y = -x^2 + 4x - 3$; (b) $y = x^2 + 4x + 3$; (c) $y = -x^2 - 4x - 3$; (d) $x = y^2 - 4y + 3$; **16**. (a) $y = -x^2 + 3x + 2$; (b) $y = x^2 + 3x - 2$; (c) $y = -x^2 - 3x + 2$; (d) $x = y^2 - 3y - 2$; **17**. (a) $y = x^2 - 3x + 2$; (b) $y = -x^2 - 3x + 2$; (c) $y = x^2 + 3x - 2$; (d) $x = -y^2 + 3y + 2$;

18.

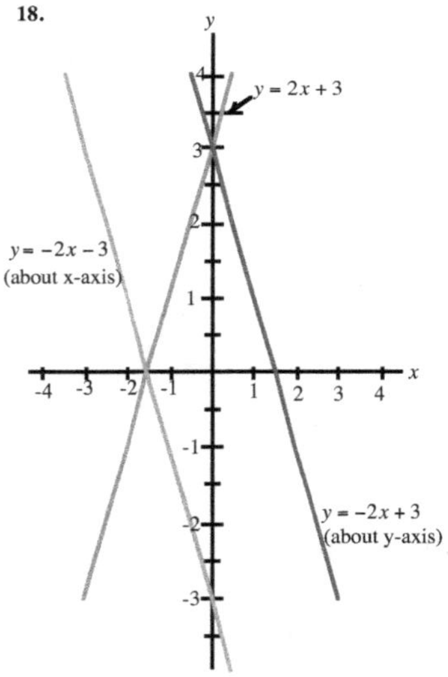

19.

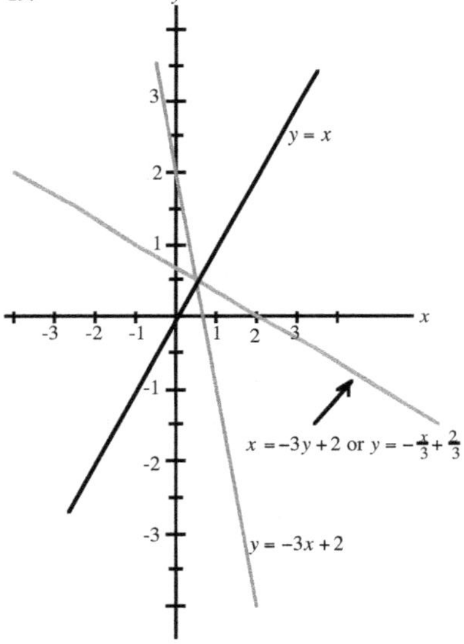

Introduction to Transformation of Functions and Relations
Translation of Points, Relations, Functions and Axes

There are three main methods for sketching the graphs of functions and relations in an x-y coordinate system of axes, namely, the general method, transformation method, and specific methods for specific types of functions. For example, the slope-intercept method is for linear functions involving two variables, and the "vertex-axis of symmetry" method is for quadratic functions. Each method has its relative merits. In the general method,, we choose convenient values for one of the variables and calculate the corresponding values of the other variable to obtain ordered pairs which are then plotted and the points connected. In **transformations**, we are given points (ordered pairs), graphs or equations of known basic functions, and we are required to sketch new graphs using the given nformation. To sketch the graph of a function or a relation, what we need is to obtain importan t points on the graph, plot these points and connect them accordingly. Quickly construct a table of values for the given or known points and using the formulas in the box below, determine the new points, plot them and connect them using the appropriate curves or lines. Let $y = f(x)$ Then the general transformation equation is given by

$$y - k = af[b(x - h)]$$

x–modifiers

y–modifiers x–modifiers

(A)   OR $$y = af[b(x - h)] + k$$ (B)

y–modifiers

The parameters k and a are y–modifiers (change the y-coordinates); b and h are x–modifiers (change the x-coordinates).

Some books write the k on the right-hand side of the equation as in (B) above. The form with k on the left-hand side (as in (A) above) makes the signs of k and h consistently easier to recall.

The parameters k and h are for translation (shifting): k is the vertical shift, h is the horizontal shift ; a and b are for stretching and shrinking. In constructing tables for the x- and y-coordinates, k and h are addends to the known or given y-coordinates and x-coordinates respectively, while a is a multiplier of the y-coordinates, b is a divisor of the x-coordinates. Note also that the parameters, k, a, b ($b \neq 0$), and h are real.

Let x_0 = original or known x-coordinate, x_n = new x-coordinate. Then $x_n = \dfrac{x_0}{b} + h$	Let y_0 = original or known y-coordinate, y_n = new y-coordinate. Then $y_n = y_0(a) + k$

Note the following examples:
1. In $y - 2 = 3f[4(x - 5)]$
   $k = 2, h = 5$, (set $y - 2 = 0, x - 5 = 0$, and solve).
   $a = 3, b = 4$ (by comparison with (A) above)
2. In $y + 2 = -3f[-4(x + 5)]$
   $k = -2, h = -5$, ($y + 2 = 0, x + 5 = 0$, and solve),
   $a = -3, b = -4$.

3. For $y = \sqrt{-x + 2}$ rewrite as
   $y = \sqrt{-(x - 2)}$ (basic function: $y = \sqrt{x}$)
   Here, $k = 0, a = 1; h = 2, b = -1$
4. For $y = \sqrt{-2x + 3}$ rewrite as
   $y = \sqrt{-2(x - \frac{3}{2})}$ (basic function: $y = \sqrt{x}$)
   Here, $k = 0, a = 1, h = \frac{3}{2}, b = -2$

To use the usual algebraic **order of operations**, write the equation in the above standard forms (A) or (B). See also Examples 3 and 4 above. **Note:** To divide or multiply by -1, just change the sign. We will first graph a most general case, and later, we will graph less general cases. Two main steps would be involved, namely, **algebraic** action for which we perform the needed calculations; and **geometric** action for which we plot the points and connect them. The **graphical** result, may be a translation (shifting), a stretching, a shrinking, an expansion or a contraction or compression of the given or basic graphs.

Example 1 Given $y = x^2$, draw the graph of $y + 4 = 3([-2x + 2]^2)$

Rewrite $y + 4 = 3([-2x + 2]^2)$ as $y + 4 = 3([-2(x - 1)]^2)$. Here, $k = -4$, $h = 1$, $a = 3$, $b = -2$

$y = x^2$		$y + 4 = 3([-2(x-1)]^2)$		
x_0	y_0	$x_n = \frac{x_0}{b} + h$	$y_n = y_0(a) + k$	(x_n, y_n)
0	0	$\frac{0}{-2} + 1 = 0 + 1 = 1$	$0(3) - 4 = 0 - 4 = -4$	$(1, -4)$
1	1	$\frac{1}{-2} + 1 = -\frac{1}{2} + 1 = \frac{1}{2}$	$1(3) - 4 = 3 - 4 = -1$	$(\frac{1}{2}, -1)$
2	4	$\frac{2}{-2} + 1 = -1 + 1 = 0$	$4(3) - 4 = 12 - 4 = 8$	$(0, 8)$
3	9	$\frac{3}{-2} + 1 = -\frac{3}{2} + 1 = -\frac{1}{2}$	$9(3) - 4 = 27 - 4 = 23$	$(-\frac{1}{2}, 23)$
4	16	$\frac{4}{-2} + 1 = -2 + 1 = -1$	$16(3) - 4 = 48 - 4 = 44$	$(-1, 44)$
-1	1	$\frac{-1}{-2} + 1 = \frac{1}{2} + 1 = \frac{3}{2}$	$1(3) - 4 = 3 - 4 = -1$	$(\frac{3}{2}, -1)$
-2	4	$\frac{-2}{-2} + 1 = 1 + 1 = 2$	$4(3) - 4 = 12 - 4 = 8$	$(2, 8)$
-3	9	$\frac{-3}{-2} + 1 = \frac{3}{2} + 1 = \frac{5}{2}$	$9(3) - 4 = 27 - 4 = 23$	$(\frac{5}{2}, 23)$
-4	16	$\frac{-4}{-2} + 1 = 2 + 1 = 3$	$16(3) - 4 = 48 - 4 = 44$	$(3, 44)$

Step 1: Quickly prepare a a table of values for x and y, using the basic parabola equation and convenient x-values.

Step 2: Using the table of values from Step 1, determine the new coordinates x_n and y_n.

Step 3: Plot the new points and connect them by a solid curve (Fig. 1 below)

Students: Try to sketch the same graph using the general method, and compare the two methods.

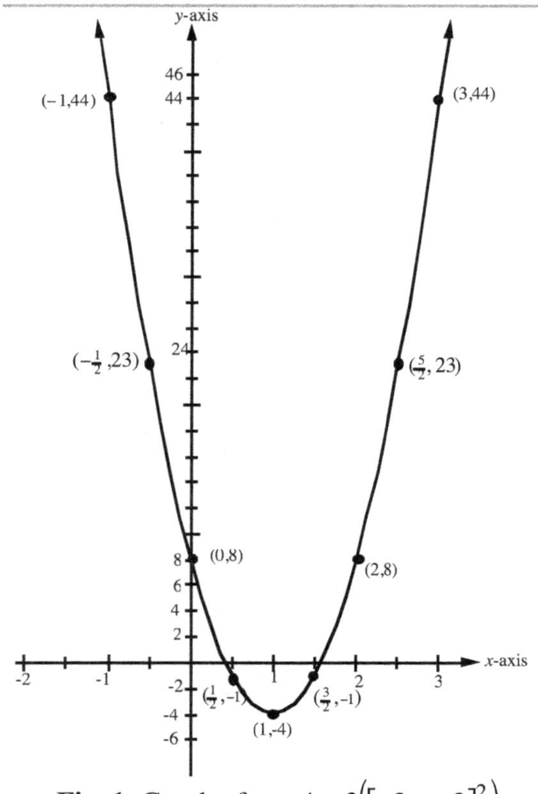

Fig. 1: Graph of $y + 4 = 3([-2x + 2]^2)$

Less General Examples

In Example 1: $y + 4 = 3([-2(x - 1)]^2)$ (A)

$k = -4$ $h = 1$, $a = 3$. $b = -2$

Now, let us observe some **less general examples**.

Case 1: when $k = 0$ (A) becomes

$y = 3([-2(x - 1)]^2)$

Case 2: when $k = 0$, $h = 0$ (A) becomes

$y = 3([-2(x)]^2)$

Case 3: when $k = 0$, $h = 0$, $a = 1$ (A) becomes $y = [-2(x)]^2$

Case 4: when $k = 0$, $h = 0$, $a = 1$, $b = 1$, (A) becomes $y = x^2$

Case 5: when $b = 1$, $a = 1$ (A) becomes

$y + 4 = (x - 1)^2$

Study the above example very well since you can easily apply the experience gained to trigonometric functions such as $y - 1 = 3\sin(4x + \pi)$. (Lessons 30, 33, 35, 36))

Note: Apart from the transformation method, if given the equation of the function, you can use the general method where you choose x, and calculate y to obtain ordered pairs for graphing.

Finding a Transformation Equation,

Knowing the Graphs of a Function and the Transformed Function

We can write down a specific transformation equation if we know the numerical values of the parameters k, a, b and h of the general transformation equation $y - k = af[b(x - h)]$.

Example 2 We do Example 1, p.76 backwards. Find the transformation equation, given or knowing the following information in the table below, where
x_0 = original or known x-coordinate, x_n = new x-coordinate,
y_0 = original or known y-coordinate, y_n = new y-coordinate.

$y = f(x)$		
x_0	y_0	(x_n, y_n)
0	0	$(1,-4)$
1	1	$(\frac{1}{2},-1)$
2	4	$(0,8)$

Solution:

Step 1: Find b and h.

We will use two x_0-values and two x_n-values together with the equation $x_n = \frac{x_0}{b} + h$ to set up a system of two equations in which the unknowns are b and h.

Substituting $x_0 = 0$, $x_n = 1$ (1st row of table) in $x_n = \frac{x_0}{b} + h$, we obtain $1 = \frac{0}{b} + h$ (A)

Similarly, using $x_0 = 2$, $x_n = 0$ (3rd row of table) in $x_n = \frac{x_0}{b} + h$, we obtain $0 = \frac{2}{b} + h$ (B)

We now solve (A) and (B) simultaneously.

From (A) $1 = 0 + h$ and $h = 1$

Substituting $h = 1$ in (B)

$$0 = \frac{2}{b} + 1$$

$$0 = 2 + b \quad (h = 1)$$

$$b = -2$$

Step 2: Find k and a.

We will use two y_0-values and two y_n-values, together with the equation $y_n = y_0(a) + k$ to set up a system of two equations in which the unknowns are k and a.

Substituting $y_0 = 0$, $y_n = -4$ in $y_n = y_0(a) + k$, we obtain $-4 = 0(a) + k$ (C)

Similarly, using $y_0 = 4$, $y_n = 8$ (3rd row of table) in $y_n = y_0(a) + k$, we obtain

$$8 = 4a + k \quad \text{(D)}$$

We now solve (C) and (D) simultaneously.

From (C) $-4 = k$ or $k = -4$

Substituting $k = -4$ in (D)

$$8 = 4a - 4 \quad (k = -4)$$

$$12 = 4a$$

$$a = 3$$

Step 3: Now we substitute $k = -4$, $a = 3$, $b = -2$, and $h = 1$ in $y - k = af[b(x - h)]$, to obtain

$$y - (-4) = 3f[-2(x - 1)]$$

$y + 4 = 3f[-2(x - 1)]$ or $y + 4 = 3f[-2x + 2]$

If $f(x)$ is quadratic, we obtain $y + 4 = 3[(-2x + 2)^2]$

Note above that given the graphs of a function and the transformed function, you could read the points $P_0(x_0, y_0)$ and $P_n(x_n, y_n)$ from the graph and construct a table of values as above.

You need two P_0 points and two P_n points.. Match the P_0's and P_n's.

Translating (Shifting) the Graphs of Functions, and Relations

Given the graph of $y = f(x)$, we can sketch the graph of $y - k = f(x - h)$ by moving (shifting) each point, on $f(x)$, h units horizontally and k units vertically simultaneously.
We agree that if the function is of the form $y - k = f(x - h)$, then the shift is h units to the right and k units up. In the above form, h is positive and k is positive. We agree also that, if the equation is of the form $y + k = f(x + h)$, then the h-shift is to the left and the k-shift is down. In this form, h is negative k is negative.
From the general equation $y - k = af[b(x - h)]$, with $a = 1$, $b = 1$, we obtain $y - k = f(x - h)$.

Example 3 Given $y = x^2$, draw the graph of $y + 4 = (x - 1)^2$ (review p.226)

In $y + 4 = (x - 1)^2$, $k = -4$, $h = 1$, $a = 1$, $b = 1$

$y = x^2$		$y + 4 = (x - 1)^2$		
x_0	y_0	$x_n = x_0 + h$	$y_n = y_0 + k$	(x_n, y_n)
0	0	$0 + 1 = 1$	$0 - 4 = -4$	$(1, -4)$
1	1	$1 + 1 = 2$	$1 - 4 = -3$	$(2, -3)$
2	4	$2 + 1 = 3$	$4 - 4 = 0$	$(3, 0)$
3	9	$3 + 1 = 4$	$9 - 4 = 5$	$(4, 5)$
4	16	$4 + 1 = 5$	$16 - 4 = 12$	$(5, 12)$
-1	1	$-1 + 1 = 0$	$1 - 4 = -3$	$(0, -3)$
-2	4	$-2 + 1 = -1$	$4 - 4 = 0$	$(-1, 0)$
-3	9	$-3 + 1 = -2$	$9 - 4 = 5$	$(-2, 5)$
-4	16	$-4 + 1 = -3$	$16 - 4 = 12$	$(-3, 12)$

Procedure:

Step 1: Quickly prepare a table of values for x and y using the basic parabola equation and convenient x-values.
Step 2: Using the table of values from Step 1, determine the new coordinates x_n and y_n.
Step 3: Plot the new points and connect them by a solid curve., (Fig. 1 below)

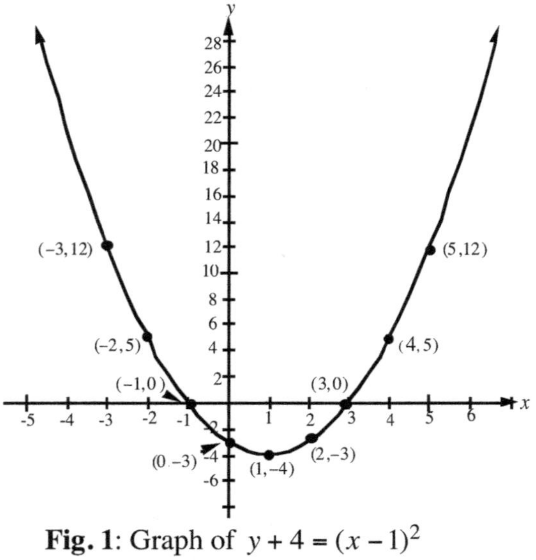

Fig. 1: Graph of $y + 4 = (x - 1)^2$

About Translation of Functions and Relations

Example 2 Given the graph of $y = f(x)$), describe the shifts to be made to obtain the graph of $y - 3 = f(x - 4)$

Solution Every point on the original graph of $y = f(x)$ is shifted 3 units up and 4 units to the right simultaneously.

Example 3 Given the graph of $y = x^2$,describe the shifts to be made to obtain the graph of $y - 3 = (x - 4)^2$

Solution Same as in Example 2.. Every point on $y = x^2$ is shifted 3 units up and 4 units to the right simultaneously.

Finally, even though descriptions such as "Every point on $y = x^2$ is shifted some units up or down", the practical approach to bring about these translations is to follow the method of Example 1 above in which we construct a table of values for x and y, plot and connect. This approach is also good for "bookkeeping" purposes as well as if $a \neq 1$, $b \neq 1$, in $y - k = af[b(x - h)]$
Study the introduction to transformations of functions (page 75).

Example 4 Given the graph of $y = x^2$, describe the shifts to be made to transform it to the graph of $y = (x-3)^2$

In standard form, $y = (x-3)^2$ becomes

$$y - 0 = (x-3)^2, \text{ where, } k = 0, h = 3$$

Each point on the graph is shifted 3 units to the right, keeping the y-values unchanged $(k = 0)$.

Translation of Axes

Consider a point P (Figure) with the original coordinates x_0, y_0 and with the coordinate reference axes O_0X_0, O_0Y_0, where x_0 and y_0 are measured from O_0Y_0 and O_0X_0, respectively.

Suppose that the origin, $O_0(0,0)$, is moved to a new point $O_n(h,k)$, giving rise to a new set of reference coordinate axes O_nX_n and O_nY_n. Let the new coordinates of the point P be x_n, and y_n where x_n and y_n are measured with reference to the new axes, O_nX_n and O_nY_n, respectively.

We must note that the point P does not move but only the origin moves, and that (x_0, y_0) and (x_n, y_n) represent the same point. We also assume that the axes, O_0X_0 and O_nX_n, have the same scalar unit and that O_0X_0 and O_nY_n have the same scalar units. From geometric considerations, we obtain the following relationships:

$$x_0 = x_n + h \qquad (1) \qquad \text{(old coordinate = new coordinate + shift)}$$

$$y_0 = y_n + k \qquad (2)$$

From equations (1) and (2)

$$x_n = x_0 - h \qquad (3)$$

$$y_n = y_0 - k \qquad (4)$$

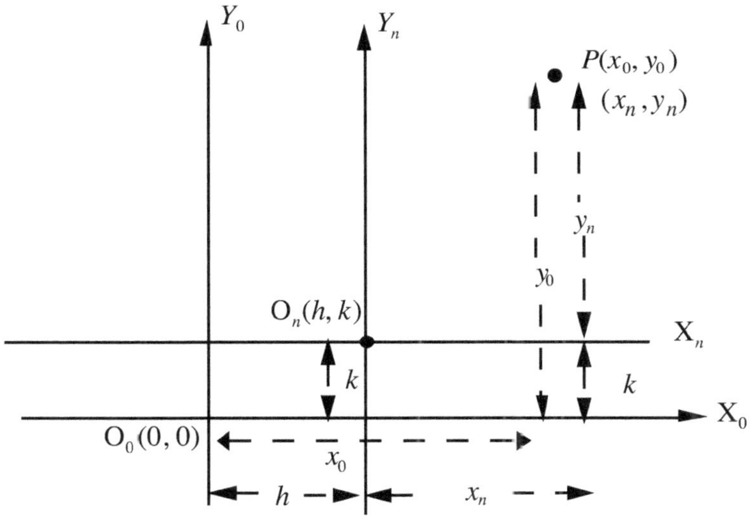

Figure: Graph of translation of axes

Application of Translation of Axes

Example 5 Suppose a point having original coordinates $x = 2, y = -3$ has its origin moved to the new origin $O_n(-1, 4)$. Find the new coordinates with reference to the new origin.

Solution: In this example, $x_0 = 2, y_0 = -3, \ h = -1, \ k = 4$.

Applying equation (3) and substituting,

$$x_n = x_0 - h$$

(new coordinate = old coordinate - horizontal shift)

$$x_n = 2 - (-1)$$

$$x_n = 3$$

Similarly, substituting in equation (4),

$$y_n = y_0 - k$$

(new coordinate = old coordinate - vertical shift)

$$= -3 - 4$$

$$y_n = -7$$

The new coordinates are $x = 3, y = -7$.

Let us say, your old house address was $(2,-3)$ according to the old numbering system. According to the new system, the address of the same house is now $(3,-7)$.

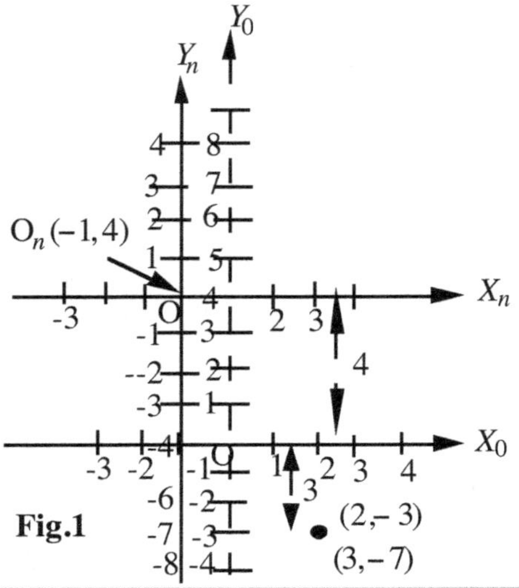

Fig.1

Example 6 Given that the equation of an ellipse is $4x^2 + y^2 + 24x - 2y + 21 = 0$, assuming that the origin of this ellipse is moved to the point (-2, 3), write this equation with reference to the new origin.

Solution: **Approach 1**

We will apply equations (1) and (2) from previous page.

Let $x_0 = x, y_0 = y, \ h = -2, \ k = 3$

Then $x = x_n + h$

$= x_n + (-2)$

$x = x_n - 2$ (A)

Similarly, $y = y_n + 3$ (B)

We now substitute equations (A) and (B) in the given equation. Then we obtain

$$4(x_n - 2)^2 + (y_n + 3)^2 + 24(x_n - 2) - 2(y_n + 3) + 21 = 0.$$

Simplifying , we obtain

$$4x_n^2 + 8x_n + y_n^2 + 4y_n - 8 = 0$$

If we complete the square, we obtain

$4(x_n + 1)^2 + (y_n + 2)^2 = 16$, and if we drop the subscript n, we obtain

$4(x + 1)^2 + (y + 2)^2 = 16$. <---This is the new equation with reference to the new origin.)

Approach 2

We complete the square first and then substitute $x_n - 2$ and $y_n + 3$ or add -2 and 3 respectively to the x-term and the y-term within the parentheses.

Step 1: $4x^2 + y^2 + 24x - 2y + 21 = 0,$

$4(x + 3)^2 + (y - 1)^2 = 16$ (by completing the square)

Step 2: $4(x + 3 - 2)^2 + (y - 1 + 3)^2 = 16$ (adding -2 and +3 to adjust for the new origin)

$4(x + 1)^2 + (y + 2)^2 = 16.$

Motivation for Learning Transformation of Functions and Relations

Transformation can be used to sketch new graphs from other/basic graphs as exemplified below.

Polynomials

From $y = x^2$, we can sketch the graphs of:

$(a)\ \ y = x^2 + 4$; $(b)\ \ y - 5 = 3(x-2)^2$;

$(c)\ \ y = x^2 + 6x + 12$. (hint : complete the square)

From $y = |x|$, we can sketch $y = |x - 2|$

Rational Functions From $y = \frac{1}{x}$, we can sketch

$(a)\ y = \frac{1}{x} + 2$; $(b)\ \ y = \frac{1}{x-2}$; $(c)\ \ y = \frac{1}{x+4}$

Logarithmic Functions

From $y = \log x$, we can sketch

$(a)\ \ y = 2\log x + 5$; $(b)\ \ y = \log(x-2)$.

Exponential Functions

From $y = e^x$, we can sketch

$(a)\ \ y = e^x - 3$; $(b)\ \ y = e^{x+5}$

Trigonometric Functions

From $y = \sin x$, we can sketch $(a)\ y = \sin x + 2$;

$(b)\ \ y = \sin x - 1$; $(c)\ \ y = 4\sin(x+3) + 1$;

Conic Relations

From $x^2 + y^2 = r^2$, $\frac{x^2}{a^2} + \frac{y^2}{b^2} = 1$, $\frac{x^2}{a^2} - \frac{y^2}{b^2} = 1$,

we can sketch the graphs of say,

$(x-h)^2 + (y-k)^2 = r^2$, $\frac{(x-h)^2}{a^2} + \frac{(y-k)^2}{b^2} = 1$,

.

Lesson 12 Exercises

A Given the graph of $y = f(x)$, describe the shifts to be made to obtain the following:

1. $y - 2 = f(x-1)$; **2.** $y + 2 = f(x+1)$; **3.** $y = f(x-3) + 2$; **4.** $y = f(x+2) - 4$

5. Sketch the graphs of $y = 2x - 1$ and $y + 4 = 2(x-3) - 1$ on the same set of rectangular coordinate system of axes.

Answers: **1.** Move each point on the graph 2 units up and one unit up simultaneously.

2. Move each point on the graph 2 units down and one unit to the left simultaneously.

3. Move each point on the graph 2 units up and 3 units to the right simultaneously.

4. Move each point on the graph 4 units down and 2 units to the left simultaneously.

B **1.** Find the new coordinates if each of the following points is translated 3 units to the right: (a) (2, 0); (b) (-5, 0); (c) (4, 1)

2. The point (- 2, 3) is translated 1 unit up and 4 units to the left. Find the coordinates of the new point.

3. The point (4 -1) is translated 2 units down and 3 units to the right. Find the coordinates of the new point.

4. The point (3 , -2) has its origin translated to a new origin (-4,1). Find the new coordinates with reference to the new origin.

5. The origin of the line $y = 3x - 2$ is moved to a new origin (-3, 2). Find an equation of this line with reference to the new origin..

6. The equation of an ellipse is $4x^2 + 9y^2 - 16x + 18y - 11 = 0$. Assuming that the origin of this equation is moved (translated) to the new point (--3, 2), transform the above equation in terms of the new origin..

7. From the graph of $y = x^2$, sketch the graph of $y + 4 = (x-1)^2$

Answers: **1.** (a) (5, 0); (b) (−2, 0); (c) (7, 1); **2.** (−1, −1); **3.** (7, −3); **4.** (7, −3);

5. $y_n = 3x_n - 13$, where x_n and y_n are the new coordinates; **6.** $4(x_n - 5)^2 + 9(y_n + 3)^2 = 36$

7. See Example 1, p. 226

Lesson 13
Vertical Stretching and Shrinking

Given the sketch of $y = x^2$ and knowing how to stretch and shrink the graphs of functions, we can rapidly sketch the graphs of $y = 3x^2$ and $y = \frac{1}{3}x^2$.

Similarly, from $y = e^x$, we can sketch $y = 4e^x$ and $y = \frac{1}{2}e^x$; from $y = \log x$, we can sketch $y = 5\log x$ and $y = \frac{1}{3}\log x$; from $y = \frac{1}{x}$ we can sketch $y = \frac{2}{x}$ and $y = \frac{1}{2x}$.

Vertical Stretching or Stretching in the y-direction

Action: Keeping the x-coordinates the same while increasing the y-coordinates.
Effect : The curve or line is pulled towards the y-axis.

Example 1 From the graph of $y = x^2$, sketch the graph of $y = 2x^2$.

Solution (Note here that $a = 2$, $k = 0$, $h = 0$, $b = 1$ in $y - k = af[b(x - h)]$

Step 1: Using broken lines, rapidly sketch the graph of $y = x^2$.

Step 2: Construct a table of values for $(x, 2x^2)$. from a table for (x, x^2) as was done on page 76, (Example 1). Each point on $y = x^2$, is shifted to a new point $(x, 2x^2)$.

Step 3: Using a smooth solid curve, connect the new points (**Figure 1**) to obtain the characteristic parabolic curve. Pictorially, each half of this curve will be between the curve, $y = x^2$ and the y-axis. The characteristic. U-shape is narrower than that of $y = x^2$. The curve is contracted in the x-direction. The curve is pulled towards the y-axis.

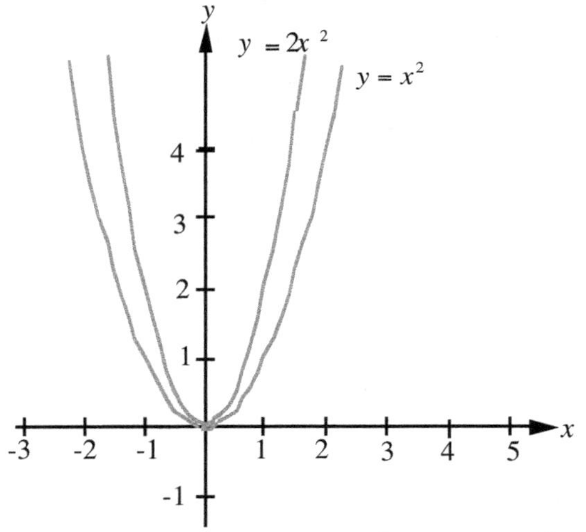

Figure 1: The graphs of $y = x^2$ and $y = 2x^2$.

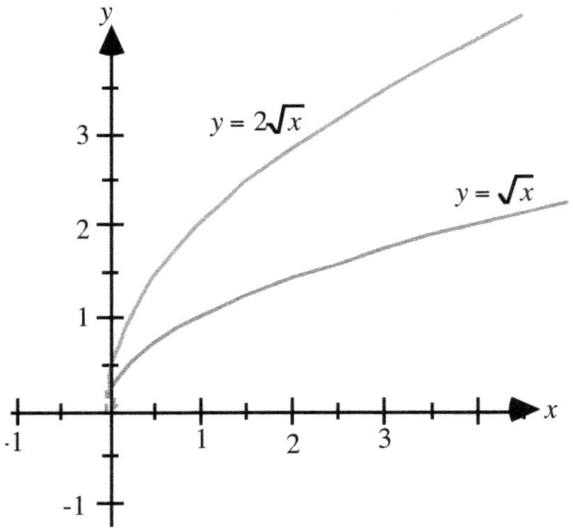

Figure 2: The graphs of $y = \sqrt{x}$, $y = 2\sqrt{x}$.

Example 2 From the graph of $y = \sqrt{x}$ sketch the graph of $y = 2\sqrt{x}$

Solution (Note here that $a = 2$, (stretch factor) $k = 0$, $h = 0$, $b = 1$ in $y - k = af[b(x-h)]$

Step 1: Using broken lines, rapidly sketch the graph of $y = \sqrt{x}$.

Step 2. Construct a table of values for $(x, 2\sqrt{x})$. from a table for $(x, \sqrt{x})$ as was done on page 76, (Example 1).

Step 3: On the same coordinate axes as was used for $y = \sqrt{x}$, plot the new points and connect them (**Figure** 2 above) to obtain the characteristic curve. Note that each point on $y = \sqrt{x}$, is moved to a new point $(x, 2\sqrt{x})$. Pictorially, the curve $y = 2\sqrt{x}$ will be between the curve, $y = \sqrt{x}$ and the y-axis. The curve is pulled towards the y-axis.

Vertical Shrinking or Shrinking in the y-direction

Action: Keeping the x-coordinates the same while decreasing the y-coordinates.
Effect : Curve or line is pushed away from the y-axis and pulled towards the x-axis .

Example 3 From the graph of $y = x^2$, sketch the graph of $y = \frac{1}{3}x^2$,

Solution (Note here that $a = \frac{1}{3}$ (shrink factor); $k = 0$, $h = 0$, $b = 1$ in $y - k = af[b(x-h)]$)

Step 1: Using broken lines, sketch the graph of $y = x^2$.

Step 2: Construct a table of values for $(x, \frac{1}{3}x^2)$. from a table for (x, x^2) as was done on page 76, (Example 1).

Step 3: On the same coordinate axes as was used for $y = x^2$, connect the new points (**Figure 3**) to obtain the characteristic curve. Pictorially, this curve will be between the curve, $y = x^2$ and the x-axis. The characteristic U-shape will be more widely open than that of $y = x^2$. In effect, the curve is pushed away from the y-axis and pulled towards the x-axis. The curve is also expanded in the x-direction. For $x > 0$, $y = \frac{1}{3}x^2$ strictly increases more slowly than $y = x^2$. For $x < 0$, $y = \frac{1}{3}x^2$ strictly decreases more slowly than $y = x^2$.

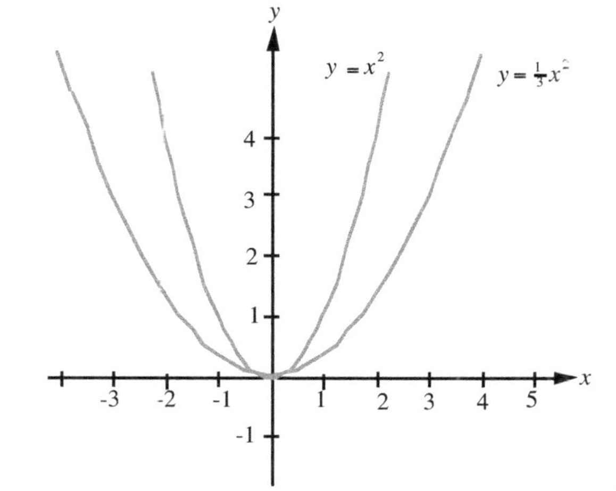

Figure 3 : Graphs of $y = x^2$ and $y = \frac{1}{3}x^2$.

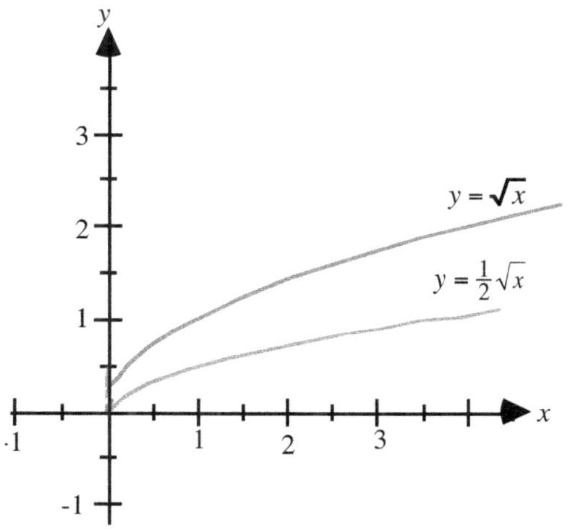

Figure 4: The graphs of $y = \sqrt{x}$, and $y = \frac{1}{2}\sqrt{x}$.

Example 4: From the graph of $y = \sqrt{x}$ sketch the graph of $y = \frac{1}{2}\sqrt{x}$.

Procedure (Note here that $a = \frac{1}{2}$ (shrink factor), $k = 0$, $h = 0$, $b = 1$

Step 1: Using broken lines, rapidly sketch the graph of $y = \sqrt{x}$

Step 2: Construct a table of values for $(x, \frac{1}{2}\sqrt{x})$. from a table for $(x, \sqrt{x})$ as was done on page 76, (Example 1). On the same coordinate axes, plot the new points.

Step 3: Using a solid curve, connect the new points together to obtain the characteristic curve (**Figure 4** above)

Horizontal Stretching and Shrinking

(Keep the y-coordinates the same and change the x-coordinates)

Example 5 Given $y = x^2$, draw the graph of $y = (-2x)^2$. Note that $y_0 = y_n$

Here, $k = 0, h = 0, a = 1, b = -2$; $y = (-2x)^2 = y - 0 = 1([-2(x-0)]^2)$

$y = x^2$		$y = (-2x)^2$		
x_0	y_0	$x_n = \frac{x_0}{b}$	y_n	(x_n, y_n)
0	0	$\frac{0}{-2} = 0$	0	(0, 0)
1	1	$\frac{1}{-2} = -\frac{1}{2}$	1	$(-\frac{1}{2}, 1)$
2	4	$\frac{2}{-2} = -1$	4	(−1, 4)
3	9	$\frac{3}{-2} = -\frac{3}{2}$	9	$(-\frac{3}{2}, 9)$
4	16	$\frac{4}{-2} = -2$	16	(−2, 16)
-1	1	$\frac{-1}{-2} = \frac{1}{2}$	1	$(\frac{1}{2}, 1)$
-2	4	$\frac{-2}{-2} = 1$	4	(1, 4)
-3	9	$\frac{-3}{-2} = \frac{3}{2}$	9	$(\frac{3}{2}, 9)$
-4	16	$\frac{-4}{-2} = 2$	16	(2, 16)

Horizontal Shrinking

Step 1: Quickly prepare a table of values for x and y using the basic equation $y = x^2$ and convenient x-values and calculating corresponding y-values

Step 2: Using the table of values from Step 1, determine the new coordinates x_n and y_n

Note in this example that y_0 from Step 1 is the same as in Step 2, since $k = 0$, and $a = 1$.

Step 3: Plot the points and connect them by a solid curve. **(Fig. 5** below)

Note above that $y = (-2x)^2 = 4x^2$.

We could therefore graph $y = 4x^2$, with $a = 4$ (stretch factor) as in Example 1 (p. 232) to obtain the same graph.

Observe the pairs (x_n, y_n): (0, 0), (−1, 4), (−2, 16), (1, 4), (2, 16).

Example 6 Given $y = \sqrt{x}$, draw the graph of $y = \sqrt{2x}$ .Note that $y_0 = y_n$

Here, $k = 0, h = 0, a = 1, b = 2$; $(y = \sqrt{2x} = y - 0 = 1\sqrt{2(x-0)})$

$y = \sqrt{x}$		$y = \sqrt{2x}$		
x_0	y_0	$x_n = \frac{x_0}{b}$	y_n	(x_n, y_n)
0	0	$\frac{0}{2} = 0$	0	(0, 0)
1	1	$\frac{1}{2} = \frac{1}{2}$	1	$(\frac{1}{2}, 1)$
4	2	$\frac{4}{2} = 2$	2	(2, 2)
9	3	$\frac{9}{2} = \frac{9}{2}$	3	$(\frac{9}{2}, 3)$

Step 1: Quickly prepare a table of values for x and y using the basic parabola equation and convenient x-values.

Step 2: Using the table of values from Step 1, determine the new coordinates x_n and y_n.

Step 3: Plot the points and connect them by a solid curve. **(Fig. 6** below)

Note: If we are to graph only $y = \sqrt{2x}$ we can choose convenient x-values such as 0, 2, 8, 32; calculate y and plot.

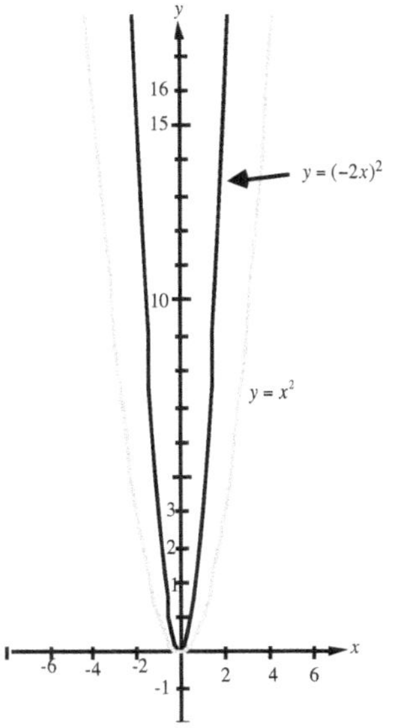

Fig. 5 : Graph of $y = (-2x)^2$

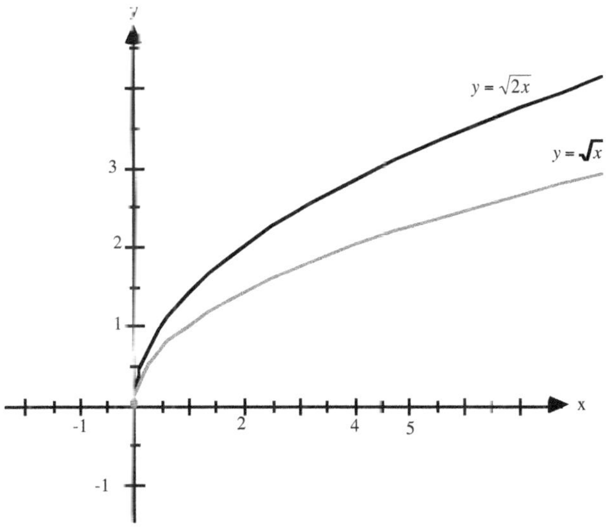

Fig. 6 Graph of $y = \sqrt{2x}$

So far, we have learned how to transform basic graphs to new graphs. In the future we will cover more direct ways of sketching the graphs of some functions such as the quadratic functions.

Remember that it is good practice to construct a table of values for x and y to obtain ordered pairs (points) which are then plotted and connected.

Lesson 13 Exercises

Sketch the graph of $y = x^2$, and on the set of axes, sketch the graphs of the following:

1. $y = \frac{1}{2}x^2$; **2.** $y = \frac{x^2}{3}$; **3**. $y = 3x^2$; **4.** $y = 12x^2$

From the graph of $y = \pm\sqrt{x}$ sketch the graphs of the following:

5. $x = \frac{1}{2}y^2$; **6.** $x = \frac{1}{3}y^2$; **7.** $x = 3y^2$; **8.** $x = 12y^2$.

Answers:

1.- 4.

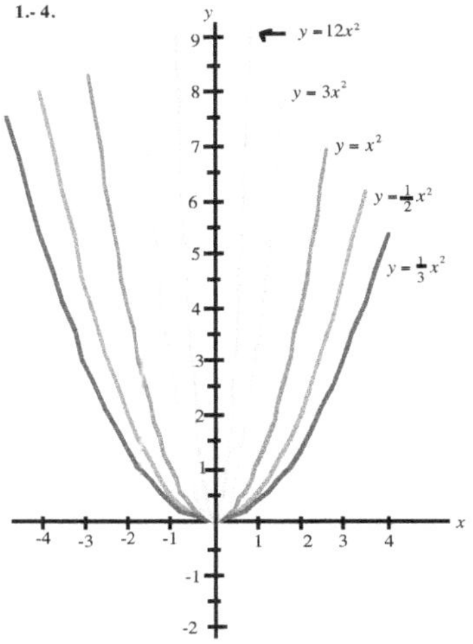

5.- 8.

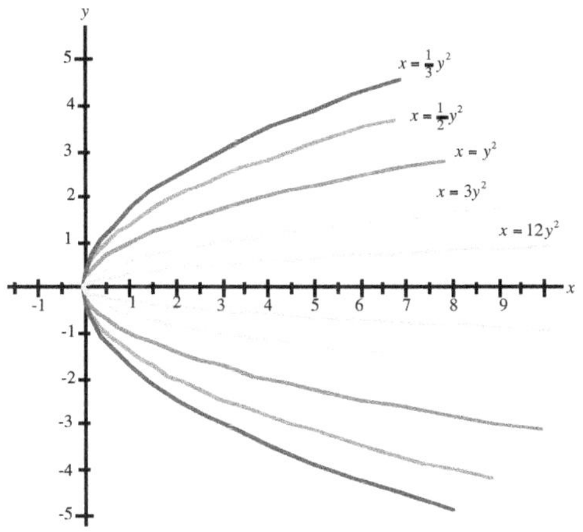

Lesson 14

Symmetry; Even and Odd Functions

*The word "symmetric" means having exactly "congruent " parts on either side of a dividing line.

Symmetry about the *y*-axis

A curve is symmetric with respect to the *y*-axis if the equation remains equivalent when the equation is reflected (i.e., replacing *x* by -*x*, keeping *y* unchanged) about the *y*-axis. More formally, a curve is symmetric with respect to the *y*-axis if $f(x) = f(-x)$). If the page were folded along the *y*-axis, the two halves of the curve would coincide. (The equation remaining **equivalent** means that after replacing *x* by -*x* then simplifying the resulting equation, we obtain the original equation}

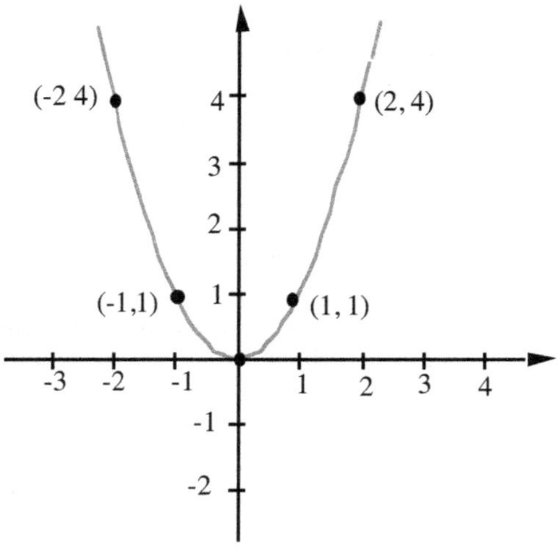

Figure 1 Graph of $y = x^2$ is symmetric about the *y*-axis

Example Is the curve $y = x^2 + 3$ symmetric about the *y*-axis?

Solution: The given curve would be symmetric about the *y*-axis if $f(x) = f(-x)$
Replacing *x* by -*x* in the given equation, we obtain

$$y = (-x)^2 + 3 \qquad \text{(A)}$$

$$y = x^2 + 3 \qquad \text{(B)}$$

On simplifying equation (A), we obtain equation (B) which is the same as the given equation, and therefore, $f(x) = f(-x)$. **Yes**, the given curve is symmetric about the *y*-axis.

Congruent figures are figures which can be made to coincide. Furthermore, the author suggests the following for symmetry about the *y*-axis: " a curve is symmetric about the *y*-axis if the *y*-axis divides the curve into two congruent parts. Perhaps, we can also say that the curve is "congruent" about the *y*-axis: usage of congruency would be consistent with the description that if the page were folded along the *y*-axis the two halves of the curve would coincide.Outside mathematics, the concept of symmetry is used heavily in structural chemistry, mineralogy and in gemology to identify precious stones such as diamond, ruby and sapphire.

Symmetry about the *x*-axis

A curve is symmetric about the *x*-axis if the equation remains equivalent when the equation is reflected about the *x*-axis . (That is replacing *y* by -*y*, keeping *x* unchanged.). More formally, a curve is symmetric about the *x*-axis if $f(y) = f(-y)$. If the page (Figure 2) were folded along the *x*-axis, the two halves of the curve would coincide.

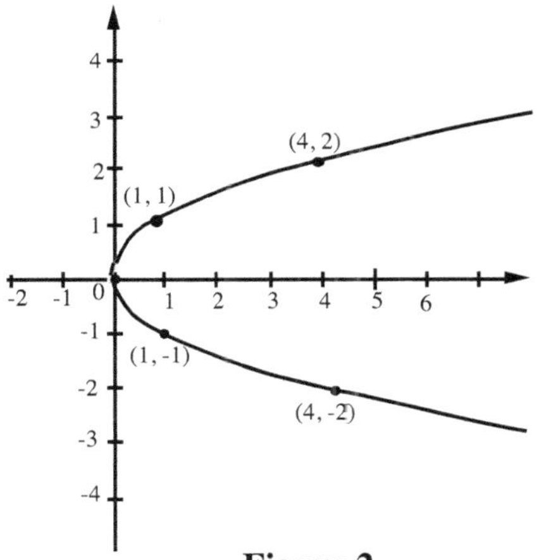

Figure 2

Example 1 Is the curve given by $x = y^2$ symmetric about the *x*-axis?

Solution The given curve would be symmetric about the *x*-axis if $f(y) = f(-y)$
Replacing *y* by -*y* in the given equation, we obtain

$$x = (-y)^2 \quad \text{(A)}$$
$$x = y^2 \quad \text{(B)}$$

On simplifying equation (A), we obtain equation (B) which is the same as the given equation. Therefore $f(y) = f(-y)$. **Yes,** the given curve is symmetric about the *x*-axis.

Example 2 Is the curve $y = x^2$ symmetric about the *x*-axis?

Solution The given curve would be symmetric about the *x*-axis if $f(y) = f(-y)$
Replacing *y* by -*y* in the given equation, and solving for *y* we obtain

$$-y = x^2 \quad \text{(A)}$$
$$y = -x^2 \quad \text{(B)}$$

On simplifying equation (A), we obtain equation (B) which is **not** equivalent to the original equation. Therefore $f(y) \neq f(-y)$. **No,** the curve is **not** symmetric about the *x*-axis.

Symmetry about the origin

A curve is symmetric about the origin if the equation of the curve remains equivalent when the equation is reflected about the origin . More formally, a curve is symmetric about the origin if on replacing x by $-x$ and y by $-y$ simultaneously, the equation remains equivalent (i.e., on simplifying, the equation remains unchanged)

If the page were folded first along the positive y-axis (lst quadrant on to 2nd quadrant), and then along the x-axis (lst and 2nd quadrants together on to the 3rd quadrant), the two halves of the curve would coincide.

In Figures 1 and 2 below, each graph is symmetric about the origin.

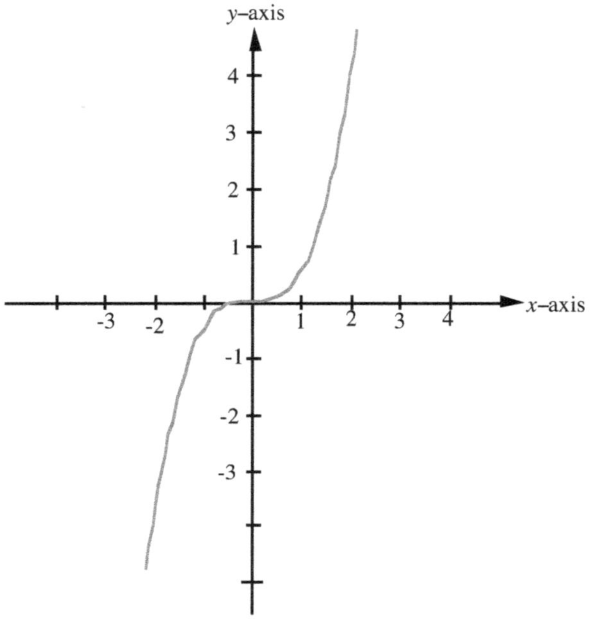

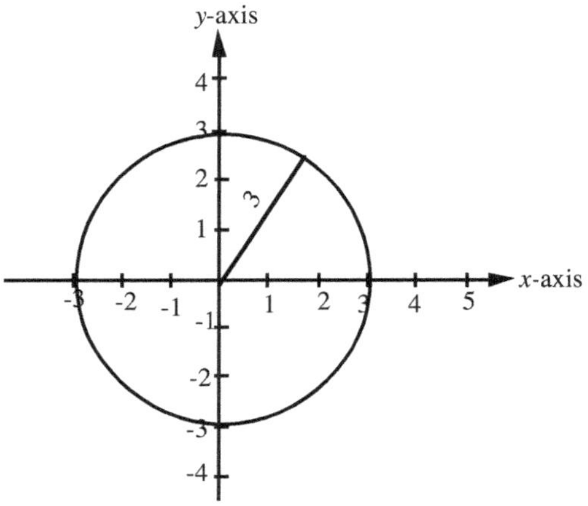

Figure 1 Graph of $y = x^3$ **Figure 2:** Graph of $x^2 + y^2 = 9$

Example Is the curve given by $x^2 + y^2 = 9$ symmetric about the origin.

Solution: Replacing x by $-x$ and y by $-y$ in the given equation,

$$(-x)^2 + (-y)^2 = 9 \qquad \text{(A)}$$

$$x^2 + y^2 = 9 \qquad \text{(B)}$$

On simplifying equation (A), we obtain equation (B) which is the same as the given equation

∴ **Yes**, the given curve is symmetric about the origin.

From the above discussions on symmetry, we make the following observations:

1. If a curve is symmetric about both the x- and y-axes, then the curve is symmetric with respect to the origin
2. However, if a curve is symmetric about the origin, it does **not** necessarily mean that it is symmetric with respect to both x-and y-axes. Example: In Figure **1** above, even though the curve is symmetric about the origin, it is **not** symmetric about the x-axis or the y-axis However Figure **2** is symmetric about both the x- and y-axes.

Even functions

The graph of an even function is symmetric about the y-axis. In sketching the graph of an even function, if we know the part of the graph on one side of the y-axis, then we can easily draw the other part of the curve.

Definition: A function $f(x)$ is **even** in x if $f(x)= f(-x)$ (A)

In words, a function $f(x)$ is even if whenever the negative value of x is substituted in the functional equation, and the resulting equation is simplified, the equation remains unchanged (i.e. the functional value for x is the same as that for $-x$).

Examples of even functions:

Polynomials: **1.** $f(x) = x^2$; **2.** $f(x) = 5x^2$; **3.** $f(x) = 5x^2 + 3$; **4.** $f(x) = -x^4$

Trigonometric functions: **1**. $f(x) = \cos x$ is an even function of x because $\cos x = \cos(-x)$.
So also is $f(x) = \sec x$, the reciprocal of the cosine function.

Exponential functions: $y = 2^{-x^2}$ (an important function in probability and statistics)

Example 1 Is the function $f(x) = x^2 + 5$ an even function?

Solution Substitute the negative of x in the above equation.

Then we obtain $f(-x) = (-x)^2 + 5$
$= (-x)(-x) + 5$
$f(-x)= x^2 + 5$ (B)

We observe that the right-hand side of $f(x)$ (the given equation) equals the right-hand side of $f(-x)$ (equation (B)), and therefore $f(x)= f(-x)$. **We can also say that each term on the right-hand side did not change sign.** Note that any constant term will not change sign.
Yes, the given function is even.

Testing some numerical values

If $x = 3$ in $f(x) = x^2 + 5$

Then, $f(3) = (3)^2 + 5$; $f(-3) = (-3)^2 + 5$
$= 9 + 5$ $= 9 + 5$
$= 14$ $= 14$

We observe above that $f(3) = f(-3)$

Odd functions

The graph of an odd function is symmetric about the origin. In sketching the graph of an odd function, we can use this symmetric property to complete the graph by reflections in the origin.

Definition: A function $f(x)$ is odd in x if
$f(-x) = -f(x)$ (C)

In words, a function $f(x)$ is odd if whenever the negative value of x is substituted in the functional equation, the new equation obtained is the negative of the original functional equation. (i.e. the functional value for x is the negative of the functional value for $-x$..). **We can also say that each term on the right-hand side changed sign.** Note that any constant term will not change sign.

Examples of odd functions:

Polynomial functions : **1**. $f(x) = x^3$; **2**. $f(x) = x^5$; **3**. $f(x) = x$; Rational functions: $f(x) = \frac{1}{x}$

Trigonometric functions: **1**. $f(x) = \sin x$; **2**. $f(x) = \tan x$; **3.** $f(x) = \cot x$ and **4**. $f(x) = \csc x$.

We may note that $\cot x$ and $\csc x$ are derived from the sine function which is odd.

Example 1 Is the function $f(x) = 2x$ odd?

Solution Substitute the negative of x in the given equation.

Then $f(-x) = 2(-x)$

$f(-x) = -2x$ (D) (The term on the right-hand side changed sign)

We observe that the right-hand side of $f(-x)$ (equation D) is the negative of the right-hand side of $f(x)$ (the given equation), and therefore $f(-x) = -f(x)$

Yes, the given function is odd.

Testing some numerical values

If $x = 4$ in $f(x) = 2x$

Then $f(4) = 2(4)$ $\quad$ $f(-4) = 2(-4)$

$= 8$ $\quad$ $= -8$

Thus $f(4) = 8$ but $f(-4) = -8$, and therefore $f(-4) = -f(4)$.

Functions which are neither even nor odd

Some functions are neither even nor odd.. **Here, there are sign changes in some of the terms.**

Example The function given by $f(x) = 2x^5 - x^2 + x$ is neither even nor odd. Why?

Lesson 14 Exercises

A 1. Explain what is meant by saying that a curve is symmetric about (a) the x-axis, (b) the y-axis.

Describe the symmetry of each of the following with respect to the x-axis, the y-axis and the origin.

2. $f(x) = x$; **3.** $f(x) = x^2 + 3$; **4.** $f(x) = 5x^2 + 1$; **5.** $f(x) = 2x^3 + 4$

6. $f(x) = x^3 - 5x - 2$; **7.** $f(x) = 4x^6 - x^2 - 5$

Answers: See text..

B Determine which of the following are even functions, odd functions or neither:

1. $f(x) = x$; **2.** $f(x) = x^2 + 3$; **3.** $f(x) = 5x^2 + 1$; **4.** (a) $f(x) = 2x^3 + 4$; (b) $f(x) = 2x^3 + 4x$

5. (a) $f(x) = x^3 - 5x - 2$; (b) $f(x) = x^3 - 5x$; **6.** $f(x) = 4x^6 - x^2 - 5$

7. Name one property of the graph of an even function.
8. Name one property of the graph of an odd function.
9. Can a function specified by a single equation be odd and even simultaneously? If yes, graph an example.
10. Can a function specified by piecewise rule be odd and even simultaneously? If yes, graph an example.
11. Are there any relationships between symmetry about the y-axis and even functions? Explain.
12. Are there any relationships between symmetry about the origin and even functions? Explain.
13. Repeat Exercise 11 for odd functions (instead of even functions).

Answers: 1. Odd; **2.** Even; **3.** Even; **4.** (a) Neither; (b) Odd; **5.**(a) Neither; (b) Odd; **6.** Even.

CHAPTER 5

Introduction to Trigonometry, Right Triangle Trigonometry

Lesson 15 **Basic Review for Geometry**

Lesson 16: **Right Triangle Trigonometry**

Lesson 17: **Angles of Elevation and Depression; Bearing; Linear Interpolation**

Introduction to Trigonometry

Trigonometry has applications in the study of physics, astronomy, engineering, medicine and so on. We use trigonometry in surveying, in the study of alternating electrical currents, periodic motion and stretched strings. The sine curve can be used in the study of alternating currents, periodic motion, vibration of stretched strings, mechanical vibrations, and motion of electrons in atoms.
The laws of sines and cosines can be used in determining the location of an object relative to a given point. The law of cosines is also used in air traffic control as well as in osteopathy (in Medicine).

To those students who might have been exposed previously to trigonometry and have developed some phobia towards trigonometry, now is the time to undo this phobia. The topics in trigonometry, are **not** more difficult to learn than the topics in algebra that we have previously covered. What we have to do is to "learn the rules of the game very well and play the game".

In this book, we will structure the coverage of trigonometry as follows:

1. Beginning with some basic geometric review, we will cover the trigonometry of the right triangle where we restrict the measures of the angles involved to between $0°$ and $90°$. (Chapter 5.)

2. We will then cover the trigonometry of any angle, where the measures of the angles involved may be positive or negative and of any measure. (Chapter 6)

3. In Chapter10, we will extend the trigonometry of any angle to the trigonometry of real numbers by letting the radian measure, x, of any angle equal to a real number x.

Lesson 15
Basic Review for Geometry

Triangles

A triangle is a closed geometric figure bounded by three straight line segments.
The point at which any two sides of the triangle meet is called a vertex. A triangle has three sides and three vertices. We shall call the three sides and the three angles the six parts of the triangle. We may name a triangle by using three capital letters in any order, where the letters represent t he vertices.

Example Triangle *ABC* is symbolized ΔABC (Fig.10)

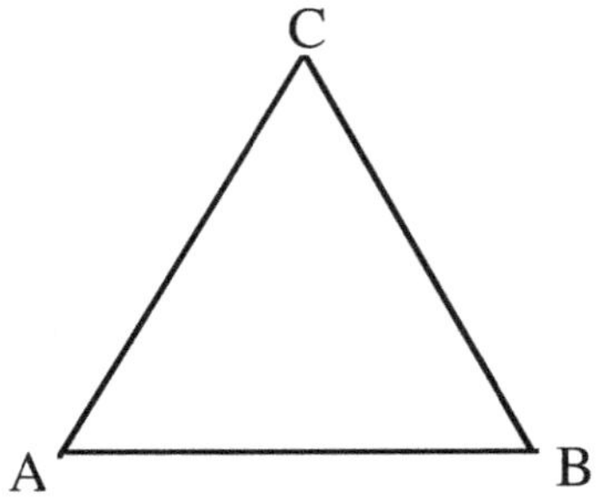

Fig.10: The points *A*, *B*, and *C* are the vertices of ΔABC.

Standard notation for labeling the sides of a triangle

Consider triangle *ABC* , below, in which the capital (or upper case) letters represent the vertices. By agreement, we shall label the sides opposite to these angles with small letters (lower case letters). These lower case letters represent the lengths of the sides of the triangle: c is the length of side opposite $\angle C$, b is the length of side opposite $\angle B$ and a is the length of the side opposite $\angle A$. We shall call this notation the **standard notation** for labeling a triangle.

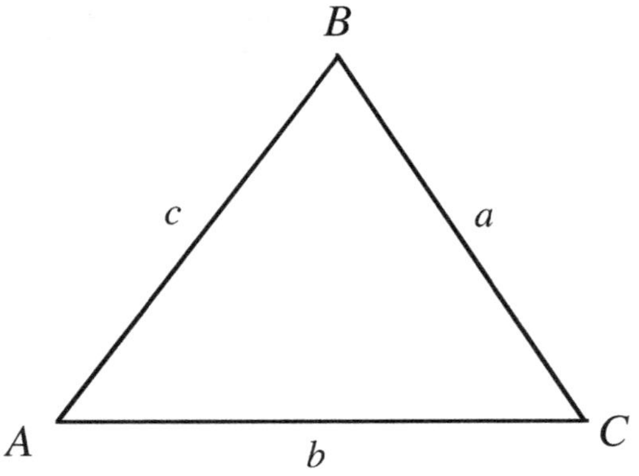

Classification of Triangles

A triangle may be classified either by its sides or by its angles.

Classification by Angles

By its angles a triangle may be classified as being acute, obtuse or right.

Acute triangle: A triangle in which all the angles are acute. The measure of each angle is between 0° and 90°.

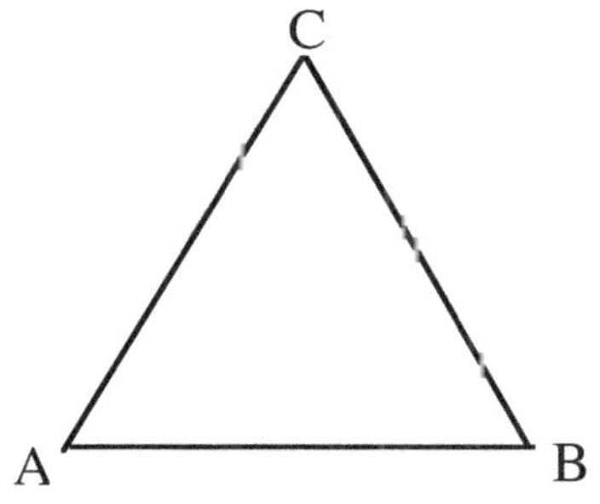

Obtuse triangle: A triangle in which there is an obtuse angle. **Note** that a triangle cannot have more than one obtuse angle. The measure of one of the angles is between 90° and 180°.

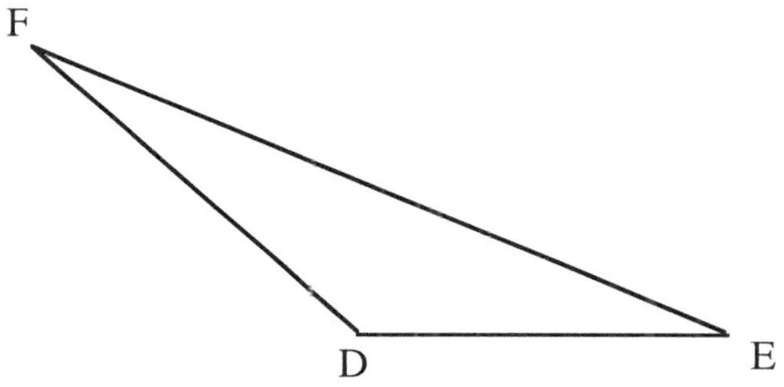

Right triangle (or right-angled triangle): A triangle which has one right angle (90°). **Note** that a triangle cannot have more than one right angle.

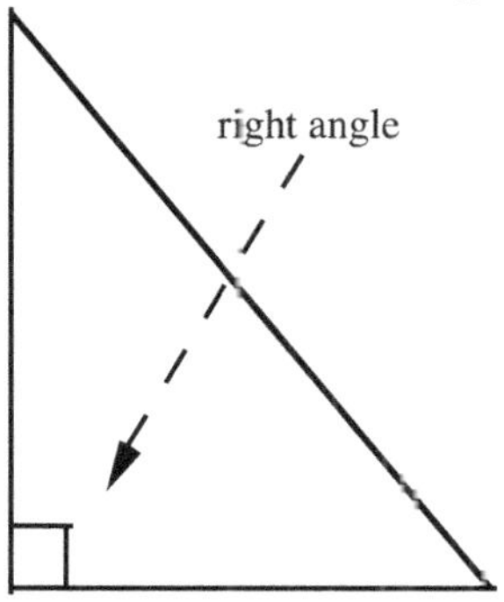

Classification by sides

We can also classify a triangle by sides. By its sides, a triangle may be classified as being scalene, isosceles or equilateral.

Equilateral triangle: A triangle in which all three sides are congruent. (The lengths of all three sides are equal)

All the interior angles of an equilateral triangle are congruent. (Each angle has a measure of 60°).

Isosceles triangle: A triangle in which two sides are congruent. In an isosceles triangle, the angles opposite to the congruent sides are congruent.

Scalene triangle

This is a triangle in which the lengths of all the three sides are different from one another (Fig.11).

How to determine if a triangle is acute, obtuse, or right, knowing the lengths of the sides

Let the longest side of a triangle be of length c units, and let the lengths of the other two sides be of lengths a units and b units.

Then, (1) if $c^2 = a^2 + b^2$ (the Pythagorean Theorem) then the triangle is a right triangle
Example: A triangle whose sides are of lengths 5, 4, and 3.

(2) if $c^2 > a^2 + b^2$, then the triangle is obtuse.
Example: A triangle whose sides are of lengths 8, 6, and 5.

(3) if $c^2 < a^2 + b^2$ then the triangle is acute.
Example: A triangle whose sides are of lengths 6, 5, and 4.

Theorem: The sum of the measures of the interior angles of a triangle is 180°.
In the figure below, $m\angle A + m\angle B + m\angle C = 180°$

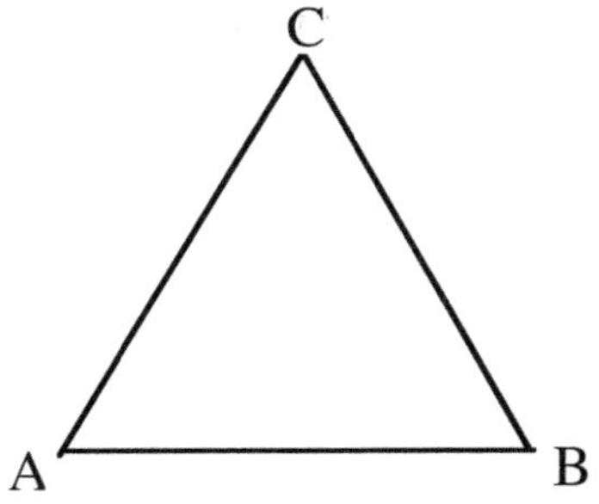

Right Triangle

A **right triangle** (Fig.18) has two acute angles and one right angle. In Fig.18, the right angle is at the vertex C. The side opposite to $\angle A$. is $\overline{BC}$; and the side adjacent to angle $\angle A$. is $\overline{AC}$. $\overline{AB}$ is the side opposite to the right angle. The side opposite the right angle is called the hypotenuse.

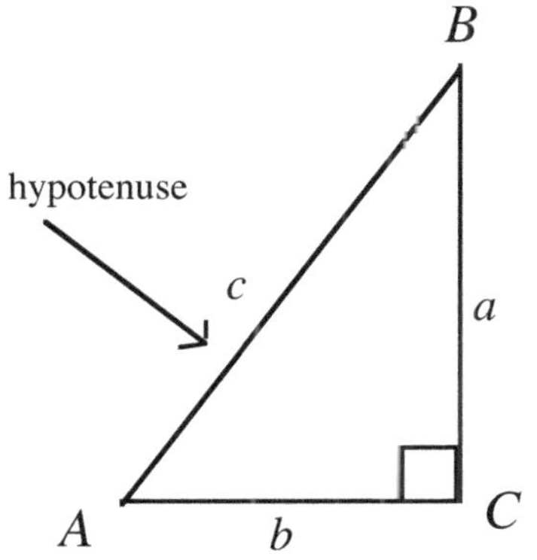

Similarly, if we refer to $\angle B$, $\overline{AC}$ is the side opposite $\angle B$ (or simply the opposite side); $\overline{BC}$ is then the adjacent side but $\overline{AB}$ is still the hypotenuse. Note that the hypotenuse remains unchanged whether we refer to angle A or angle B but the adjacent and opposite sides change depending upon which angle we are referring to.

The Pythagorean Theorem

The Pythagorean theorem states that in a right triangle, the square of the length of the hypotenuse equals the sum of the squares of the lengths of the other two sides.

Symbolically, $c^2 = a^2 + b^2$ (using standard notation)

Useful relationships concerning the relative measures of the sides and angles of a triangle

1. In any triangle, the **sum** of the **lengths** of any **two sides** is larger than the length of the **third** side.

2. In any triangle, the **longest side** is opposite to the **largest angle**, and the **shortest side** is opposite to the **smallest angle**.

3. In an **acute triangle, the** square of the length of **longest side is smaller than** the sum of the squares of the lengths of the other two sides.

4. In an **obtuse triangle**, the **square** of the length of the **longest side is larger than** the sum of the **squares** of the lengths of the other two sides.

5. In a **right triangle**, the **square** of the length of the **longest side** (the hypotenuse) **is equal** to the sum of the **squares** of the lengths of the other two sides. (**The Pythagorean theorem**)

Lesson 15 Exercises

A Define the following:
1. Triangle; 2. Acute triangle, 3. Obtuse triangle,;
4. Right triangle,; 5. Equilateral triangle,
6. Isosceles triangle, 7. scalene triangle,

B. State the Pythagorean Theorem.

Lesson 16
Right Triangle Trigonometry

Units of measuring angles in Trigonometry

As it is in the case of geometry, we may measure angles in degrees. However, in addition, we may also measure angles in another unit called the "radian". In calculus and scientific work, we will usually measure angles in radians.

Trigonometric Functions (Trigonometric ratios)

These ratios are sine (**sin**); cosine (**cos**); tangent (**tan**); cosecant (**csc**); secant (**sec**); and cotangent (**cot**).
We can derive these six ratios by comparing two similar right triangles. Each ratio is independent of the lengths of the sides involved. Each ratio is not a function of the lengths of the sides involved. However, each ratio of corresponding sides is a function of the angles involved. As the angle we refer to changes, the ratios of corresponding sides change and hence we call these relations trigonometric functions. We summarize the trigonometric functions below,.

Definitions: The following definitions are applicable to **right** triangles only.

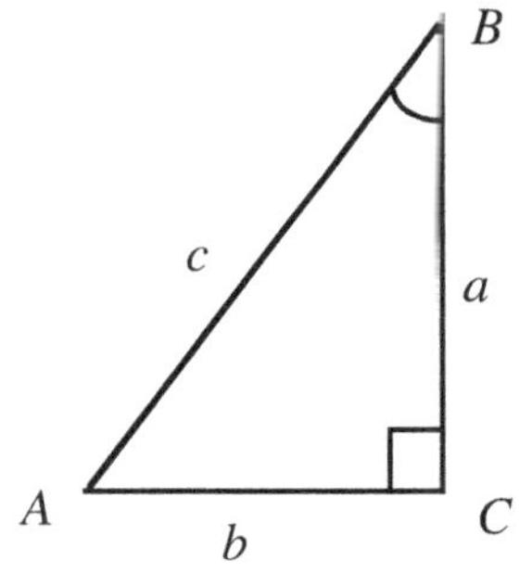

1. $\sin A = \dfrac{\text{opposite side}}{\text{hypotenuse}} = \dfrac{a}{c}$ (sin A is abbreviation for the sine of *A*)

2. $\cos A = \dfrac{\text{adjacent side}}{\text{hypotenuse}} = \dfrac{b}{c}$ (cos A is abbreviation for the cosine of A)

3. $\tan A = \dfrac{\text{opposite side}}{\text{adjacent side}} = \dfrac{a}{b}$ (tan A is abbreviation for the tangent of *A*)

4. $\csc A = \dfrac{1}{\sin A} = \dfrac{\text{hypotenuse}}{\text{opposite side}} = \dfrac{c}{a}$ (csc A is abbreviation for the cosecant of *A*)

5. $\sec A = \dfrac{1}{\cos A} = \dfrac{\text{hypotenuse}}{\text{adjacent side}} = \dfrac{c}{b}$ (sec A is abbreviation for the secant of *A*)

6. $\cot A = \dfrac{1}{\tan A} = \dfrac{\text{adjacent side}}{\text{opposite side}} = \dfrac{b}{a}$ (cot *A* is abbreviation for the cotangent of *A*)

If instead of referring to $\angle A$, we refer to $\angle B$, then we have the following

1. $\cos B = \frac{a}{c}$; $\sin A = \frac{a}{c}$
2. $\sin B = \frac{b}{c}$; $\cos A = \frac{b}{c}$ (Note also that $\cos B = \frac{a}{c} = \sin A = \frac{a}{c}$)
3. $\cot B = \frac{a}{b}$; $\tan A = \frac{a}{b}$.

Note above that for $\angle B$, the opposite side is $\overline{AC}$, and the adjacent side is $\overline{BC}$. For $\angle A$, the
opposite side is $\overline{BC}$ and the adjacent side is $\overline{AC}$
As we have already observed, certain functions of $\angle A$, are equal to certain functions of $\angle B$,
We may deduce from the above equations that:

$$\sin A = \cos B$$

$$\sin B = \cos A$$

$$\tan B = \cot A$$

(Note that $\angle A$ and $\angle B$ are complementary angles. See also page 103.

To summarize for all the six functions, we have

1. $\sin A = \frac{a}{c} \quad = \quad \cos B = \frac{a}{c}$

2. $\cos A = \frac{b}{c} \quad = \quad \sin B = \frac{b}{c}$

3. $\tan A = \frac{a}{b} \quad = \quad \cot B = \frac{a}{b}$

4. $\csc A = \frac{c}{a} \quad = \quad \sec B = \frac{c}{a}$

5. $\sec A = \frac{c}{b} \quad = \quad \csc B = \frac{c}{b}$

6. $\cot A = \frac{b}{a} \quad = \quad \tan B = \frac{b}{a}$

A memory **device** for remembering the reciprocal relationships for sine and cosine is as follows. Usually, students remember that the secant and the cosecant are reciprocals of some functions. The recall problem is: Is the secant for the sine or for the cosine, and is the cosecant for the sine or for the cosine?
You might think that because each of **s**ine and **s**ecant begins with the letter " **s** " the reciprocal of the sine function (sin) is the secant function (sec) , but this is **not** so, they are **not** reciprocals of each other. Similarly, you might also think that because each of **c**osine and **c**osecant begins with the letter " **c** " the reciprocal of the cosine function (cos) is the cosecant function (csc) , but this also is **not** so, they are **not** reciprocals of each other. Therefore, note that, the name of each given function and its reciprocal do not rhyme with each other. As such, always think the opposite: The " reciprocal of **s**ine is **c**osecant " and the " reciprocal of **c**osine is **s**ecant". It is rather unfortunate that mathematicians did not choose the names to facilitate recall.. Perhaps, for recall purposes, we can use" S-C " for sine and cosecant; and " C-S "for cosine and secant.

Important: Remember that all the above trigonometric definitions for the sine, cosine, tangent, cosecant, secant and cotangent apply only to a triangle which has a right angle. The ratios do **not** apply to a triangle which does not have a right angle.

How to use the electronic calculator as well as trigonometric tables

We use trigonometric tables or the trigonometric functions on an electronic calculator for two main tasks. One task is to find the trigonometric functional value for an angle when the measure of the angle is known (or given). The other task (the opposite task) is to find the measure of an angle when given the functional value of the angle.

Many scientific calculators have built-in trigonometric functions. To use the calculator, we push or depress the appropriate buttons (or keys).

Given the measure of an angle, how to find the value of the trigonometric function

Most calculators have buttons for the sine, the cosine and the tangent functions. Recall that we measure the angles either in degrees or in radians. As such, we may have to determine which units we are using in operating the calculator. We do this by pushing the button for degree measure (degree mode) or radian measure (radian mode) according to which unit of measurement we want to use. However, some old calculators may not have buttons (keys) for degrees and radians.

In this case, we would want to change from one unit using the relation $\boxed{180^\circ = \pi \text{ radians}}$.

Example 1 Find $\sin 60°$

Solution

Step 1: Put the calculator in degree mode by pushing the button for degrees. Some calculators may use the same button for both degrees and radians (R/D). In this case, one push of button may indicate radian and a second push of button may indicate degrees and vice-versa. In some calculators a red dot (depending on color of display panel) on the label "radian" on the display panel indicates radians measure and no dot is understood to mean degree measure, however, some calculators can display both the degrees and the radians on the panel.

Step 2: Press the buttons in the following order from left to right.

Thus, press (the button) 6, then 0, and finally press sin.
and then read 0.8660254.

$\boxed{6}\ \boxed{0}\ \boxed{\sin}$

Then $\sin 60° = 0.8660254$

Note: On some calculators, the order of pressing the buttons may be reversed. For the above example, we press "sin" followed by 6 and 0. This order is sometimes labeled DAL (Direct Algebraic Logic). Therefore, experiment with the order. Before using a calculator on an exam, practice (and review) how to use your calculator, since this will save you time on the exam. Even, if you knew how to use your calculator, in the past, it is good practice to review its usage before the exam.

Example 2 Find $\cos 60°$.

Solution

Step 1: Put calculator in degree mode.

Step 2: Press the buttons for 6, 0, and then cos and read 0.5.

$\boxed{6}\ \boxed{0}\ \boxed{\cos}$

Thus, $\cos 60° = 0.5$.

Note: For calculators with the label "**D.A.L.**" (Direct Algebraic Logic), the sequence for example, for finding $\cos 60°$ is: Press the button for **cos** first, followed by **60°**
Thus:

$\boxed{\cos}\ \boxed{6}\ \boxed{0}$

Example 3 Find sec $60°$.

Solution

sec key on calculator

If your calculator has key for the sec function, then press 6, 0, then sec to obtain sec $60° = 2$.

No sec key on calculator

If your calculator has no key (button) for the sec function, then recalling tha t sec $60° = \frac{1}{\cos 60°}$, we can find $\cos 60°$, and then find the reciprocal of $\cos 60°$.

Step 1: Put calculator in degree mode.

Step 2: Press the buttons in the following order 6, 0, and cos.
This gives you 0.5

Step 3: Press on the button for $\frac{1}{x}$ to obtain 2.
Thus sec $60° = 2$.

Similarly for csc $60°$ we can use the sin and $\frac{1}{x}$ buttons; for cot $60°$, we can use the tan and $\frac{1}{x}$ keys.

Example 4 Find sin $390°$

Solution

Press the buttons in the following order 3, 9, 0, and then sin, to obtain sin $390° = 0.5$.

Examples using radian measure

Example 5 Find sin 1.5

Solution

Since no unit is indicated, the unit is understood to be the radian unit.

Step 1: Put calculator in radian mode (Follow operating instructions of calculator.).
Step 2: Push the button in the following order: 1.5, sin and read 0.997495.
Thus, sin $1.5 = 0.997495$

Example 6 Find sec 1.5

Step 1: Put calculator in radian mode (Since 1.5 is understood to be in radians)
Step 2: Press 1.5, then cos.
Step 3: Then push the $\frac{1}{x}$ (Recall sec $1.5 = \frac{1}{\cos 1.5}$ and read 14.1368329.
Therefore sec $1.5 = 14.1368329$. If you have the sec key, then use it directly.

If we are given the radian measure and we want to operate in degree mode, we may change the radian measure to degree measure by multiplying by $\frac{180°}{\pi}$. On the other hand, given the degree measure, to change to radian measure, we multiply the degree measure by $\frac{\pi}{180°}$. However, if the calculator has an automatic conversion button then use it.

Finding the measure of an angle given the value of the trigonometric function, using a calculator

Here, we shall use the following keys if available: the buttons $\sin^{-1}$, $\cos^{-1}$, $\tan^{-1}$. If these are not available, you may have to use a combination of the keys: "inv or arc together with sin, cos and tan.

Example 7 If sin = .97, find θ in degrees.

Step 1: Since we want θ to be in degrees we put the calculator in degree mode.

Step 2: Press .97.

Step 3: Push the "2nd" button, then the $\sin^{-1}$ button and read 75.93013°

[.] [9] [7] then [$\sin^{-1}$]

Therefore, $\theta = 75.93013$. (to the nearest tenth, $\theta \approx 75.9$)

In step 3, if your the calculator does not have $\sin^{-1}$, $\cos^{-1}$, $\tan^{-1}$ keys, press inv (inverse) or arc, then sin as:

[.] [9] [7] [inv] or [arc] [sin]

How to find the values of trigonometric functions by using trigonometric tables

On occasion, we may not have access to a calculator. When this is the case, then we may resort to using trigonometric tables. Trigonometric tables are usually in the back pages of most books. In most representative trigonometric tables, the trigonometric functions are for angles from 0° to 90° for degree measures and from 0 to 1.57 for radian measures. The angles and the corresponding functional values are not listed continuously in one column from 0° to 90° (0 to 1.57 radian) but the readings for the angles between 0° and 45° (0 to .7854 radians) are in the left-hand column while the readings for the angles between 45° and 90° (.7854 and 1.57) are located in the right-hand column. Note this discontinuous arrangement as it will save you time and frustration. Some books, however, may have separate tables for the angles measured in radians and in this case, the angles may be listed.

Partial Table of values for Trigonometric Functions

↓ Angle θ									
Degrees	Radians	Sin	Cos	Tan	Csc	Sec	Cot		
....	....	....	....	....	....	....	....	....	....
29° 00'	.5061	.4848	.8746	.5543	2.063	1.143	1.804	1.0647	61° 00'
29° 10'	.5091	.4874	.8732	.5581	2.052	1.145	1.792	1.0617	60° 50'
29° 20'	.5120	.4899	.8718	.5619	2.041	1.147	1.780	1.0588	60° 40'
29° 30'	.5149	.4924	.8704	.5658	2.031	1.149	1.767	1.0559	60° 30'
29° 40'	.5178	.4950	.8689	.5696	2.020	1.151	1.756	1.0530	60° 20'
29° 50'	.5207	.4975	.8675	.5735	2.010	1.153	1.744	1.0501	60° 10'
--------	....	....	....	....	....	....	....	....	-------
		Cos	Sin	Cot	Sec	Csc	Tan	Radians	Degrees
								Angle θ ↑	

Example 8 Find the functional value for angles between **0° and 45°**.(table on p.508)

(a) $\sin 29°\ 50'$ (b) $\cos 29°\ 50'$

(c) $\tan 29°\ 50'$ (d) $\text{Sec } 29°\ 50'$

(a) Step 1: From the table p.508, we locate the angle $29°\ 50'$ in the left column, headed "Degrees". (Since it is between 0° and 45°).

Step 2: We read across this row to the right until this row meets (or intersects) the column for sin (sin being at the top of the table)

The value at the intersection of this row and the column for the sin function is the value we are looking for and we would read .4975. Therefore, $\sin 29°\ 50' = 0.4975$.

(b) For $\cos 29°\ 50'$, we read .8675 = 0.8675

(c) For $\tan 29°\ 50'$, we read .5735 = 0.5735

(d) For $\sec 29°\ 50'$ we read 1.153.

We must note that if the angle whose functional value we want to find is more than 90° , then we will find the reference angle first and then the functional value as on pages 121-123.

Example 9 Angles between 45° and 90°

Find the functional value for $\cos 61°\ 00'$.

Reading down the right-hand column.

Step 1: We locate the angle $61°\ 00'$ in the right-hand (right-most) column.

Step 2: We then read across the row to the left until we intersect the column with the heading $\cos\theta$ (or $\cos x$). The intersection of the row for $61°\ 00'$ and the column for $\cos\theta$ (at the bottom of the table) gives the desired functional value. In this case, note the headings $\cos x$, $\sin x$, $\tan x$ are at the **bottom** of the table.

Step 3: We read .4848. Therefore $\cos 61°\ 00' = 0.4848$

Similarly we shall obtain the following

$\sin 61°\ 00' = 0.8746$

$\csc 61°\ 00' = 1.143$.

Given the functional values we can also use the trigonometric tables to find the corresponding angles. In this case, we will locate the functional value first ,and then move across the table row-wise either to the left or right to read the corresponding angles. Note that we are reversing the steps in this case.

Complementary Angles and Cofunctions

Complementary angles : If the sum of the measures of two angles is 90°, then the two angles are complementary. The complement of a 40° angle is a 50° angle.

$\angle 1$ and $\angle 2$ are complementary angles.(i.e., m $\angle 1$ + m $\angle 2 = 90°$)

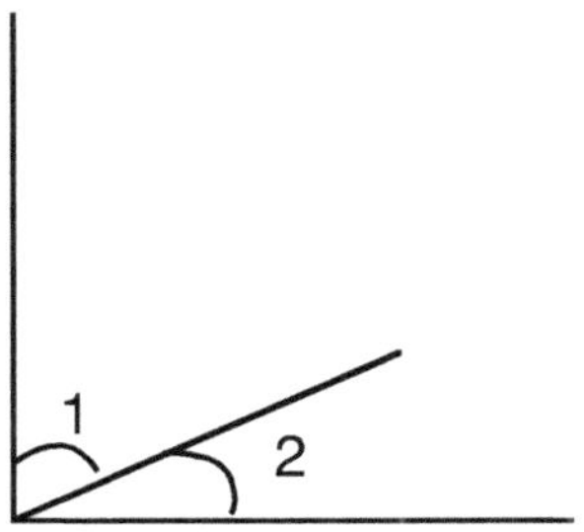

Figure 15

Note that it is **not** necessary that the angles have the same vertex in order to be complementary.

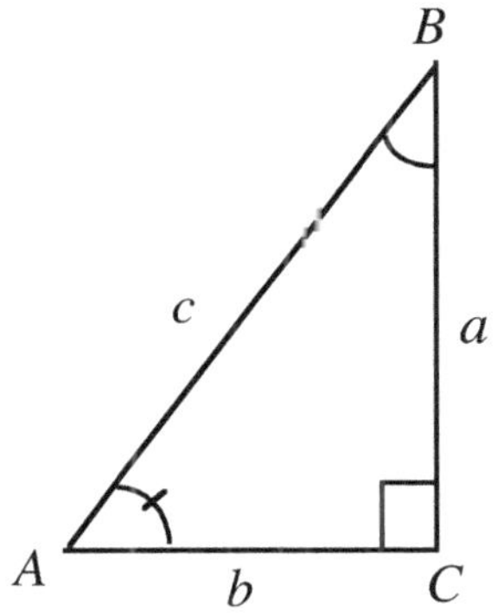

We know from geometry that the sum of the measures of $\angle$ A, $\angle$ B, and $\angle$C is 180°.
Since m$\angle$C = 90°,
m$\angle A$ + m$\angle B$ = 90° .... (7)

Equation (7) shows that angle *A* and angle *B* are complementary (i.e. their sum = 90°).

From equation (7) $m\angle A = 90^o - m\angle B$, and

$$m\angle B = 90° - m\angle A$$

We know that $\sin A = \cos B$ (since $\sin A = \frac{a}{c}$, and $\cos B = \frac{a}{c}$.)

Therefore, if we replace $m\angle A$ *by* $90° - m\angle B$

$$\sin(90° - B) = \cos B \text{........................(8)}$$

Similarly, $\sin(90° - A) = \cos A$

$$\cot(90° - A) = \tan A$$

We summarize the above relations which are functions below:

Cofunctions (Complementary functions)

$\sin A = \cos(90^\circ - A)$
$\cos A = \sin(90^\circ - A)$ } ← The sine and the **co**sine are **co**functions.

$\tan A = \cot(90^\circ - A)$
$\cot A = \tan(90^\circ - A)$ } ← The tangent and the **co**tangent are cofuntions.

$\sec A = \csc(90^\circ - A)$
$\csc A = \sec(90^\circ - A)$ } ← The secant and the **co**secant are **co**funtions.

We call the above relations cofunctions and we also say that the **cofunctions** of **complementary angles are equal in value.**

Example 10 Given ΔABC in which $m\angle C = 90^\circ$ and $m\angle A = 60^\circ$, we derive the following:

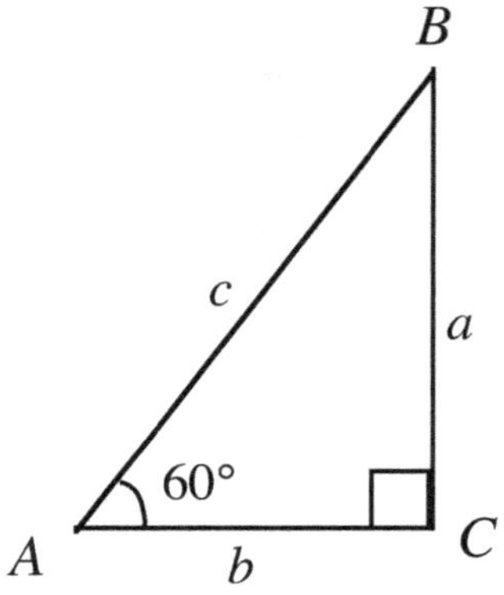

$\sin 60^\circ = \cos(90^\circ - 60^\circ)$

$\sin 60^\circ = \cos 30^\circ$ (Note : $\sin 60^\circ = .866$; $\cos 30^\circ = .866$, from calculator or tables)

$\sin 30^\circ = \cos(90^\circ - 30^\circ)$

$\sin 30^\circ = \cos 60^\circ = .5$

$\tan 30^\circ = \cot(90^\circ - 30^\circ)$

$\tan 30^\circ = \cot 60^\circ = .58$

$\sin 10^\circ = \cos 80^\circ$.

If θ is in radians:

$60^\circ = \frac{\pi}{3}$; $90^\circ = \frac{\pi}{2}$, and

$\sin\frac{\pi}{3} = \cos\left(\frac{\pi}{2} - \frac{\pi}{3}\right)$ Note : $\frac{\pi}{2} - \frac{\pi}{3} = \frac{\pi}{6}$

$\sin\frac{\pi}{3} = \cos\frac{\pi}{6}$ (equivalent to $\sin 60^\circ = \cos 30^\circ$)

Similarly, $\sin\frac{\pi}{6} = \cos\frac{\pi}{3}$ (equivalent to $\sin 30^\circ = \cos 60^\circ$)

Special triangles: The 30°-60°-90° triangle and the 45°-45°-90° triangle

The trigonometric functions for 30°, 45°, 60°, and 90° occur very frequently in mathematics that students are usually required to memorize them.

The 30°-60°-90° triangle

For the 30°-60°-90° triangle (Figure 2), the dimensions are obtained by considering an equilateral triangle whose sides are each 2 units long. From geometry, the length of the hypotenuse is twice the length of the shortest side. The third side is calculated by the Pythagorean theorem.
Thus if AB = 2, then AC = 1 and $2^2 = 1^2 + BC^2$ and from which BC = $\sqrt{3}$. Usually, after having gone through the derivation, using geometric considerations and the Pythagorean theorem, students become familiar with the dimensions 1, 2, $\sqrt{3}$. The problem of recall that remains then is how to place these dimensions on the 30°-60°-90° triangle, if one does not draw the equilateral triangle, but draws only Δ ABC. The following mnemonic device will be helpful:

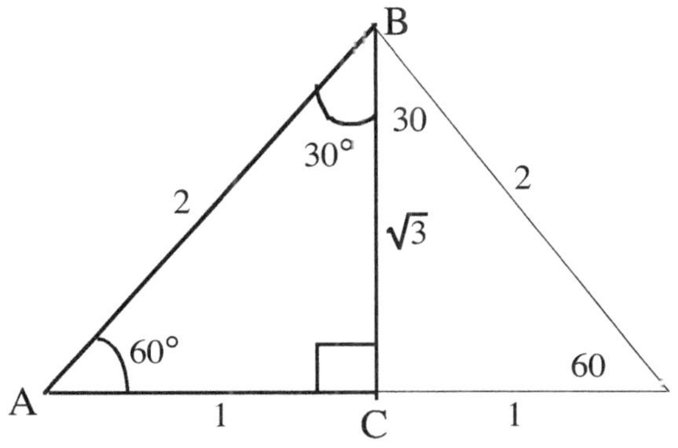

Figure 2

Mnemonic device: If you can remember the numbers 1, 2, $\sqrt{3}$ as being the lengths of the sides of this triangle, and also remember that in any triangle, the longest side is opposite the largest angle, and that the shortest side is opposite the smallest angle, you should be able to place these dimensions accordingly on the triangle.Thus, 1 is opposite the 30° angle, 2 is opposite the 90° angle (the right angle) , and $\sqrt{3}$ is opposite the 60° angle. ($\sqrt{3} \approx 1.73$)

By applying the mnemonic devices, SOH , CAH, TOA we can write down some functional values.

Examples

$$\sin 30° = \frac{\text{opposite side}}{\text{hypotenuse}} = \frac{1}{2}$$

$$\cos 30° = \frac{\text{adjacent side}}{\text{hypotenuse}} = \frac{\sqrt{3}}{2}$$

$$\sin 60° = \frac{\text{opposite side}}{\text{hypotenuse}} = \frac{\sqrt{3}}{2}$$

$$\tan 60° = \frac{\text{opposite}}{\text{adjacent side}} = \frac{\sqrt{3}}{1} = \sqrt{3}\,.$$

The 45°-45°-90° triangle

By sketching an isosceles right triangle with measures of 45°, 45°, and 90° (Figure 1) we will be able, using the basic definitions, write down the trigonometric functional value for 45°.

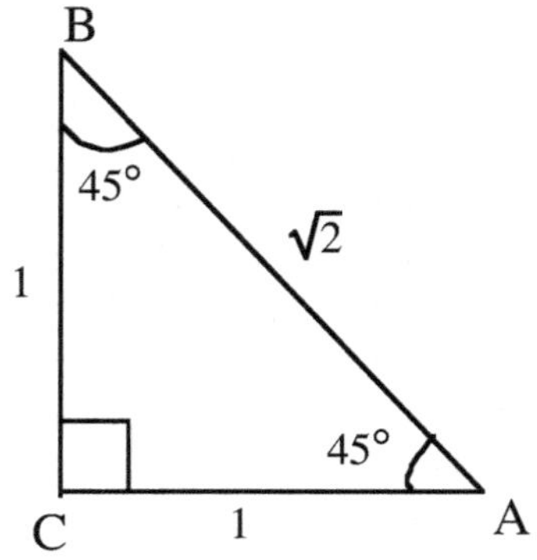

Figure 1

In an isosceles triangle, the sides opposite the congruent angles have the same length. If we let the length of the sides opposite the congruent angles be 1 unit each, we can calculate the length of the hypotenuse using the Pythagorean theorem.

$$AB^2 = AC^2 + BC^2$$
$$AB^2 = 1^2 + 1^2$$
$$AB^2 = 2$$
$$AB = \sqrt{2}.$$

Example 11 $\sin 45° = \frac{\text{opposite side}}{\text{hypotenuse}} = \frac{\mathbf{1}}{\mathbf{\sqrt{2}}} = \frac{1}{\sqrt{2}} \cdot \frac{\sqrt{2}}{\sqrt{2}} = \frac{\mathbf{\sqrt{2}}}{\mathbf{2}}$ (Rationalizing the denominator)

$$\cos 45° = \frac{\text{adjacent side}}{\text{hypotenuse}} = \frac{\mathbf{1}}{\mathbf{\sqrt{2}}} = \frac{1}{\sqrt{2}} \cdot \frac{\sqrt{2}}{\sqrt{2}} = \frac{\mathbf{\sqrt{2}}}{\mathbf{2}}$$

$$\tan 45° = \frac{\text{opposite side}}{\text{adjacent side}}\ \frac{1}{1} = 1.$$

Solutions of Right Triangles

In any triangle, there are six parts, namely, three sides and three angles. When we want to "solve a triangle" what we want to do is to find the measures of all the unknown parts. These parts may be sides or angles or a combination of angles and sides. Usually, the values of three parts are given or known and we are to find the other three parts.

How to determine which function to use in solving right triangles

The determination of which trigonometric relationship to use depends on what we are given and what we want to find. A good practice is to scribble down the following three relationships:

$\sin\theta = \dfrac{\text{opposite side}}{\text{hypotenuse}}$ $\quad\left(\dfrac{\textbf{O}\text{pposite}}{\textbf{H}\text{ypotenuse}}\right.$: "**SOH**" <-------- mnemonic device)

$\cos\theta = \dfrac{\text{adjacent. side}}{\text{hypotenuse.}}$ $\quad\left(\dfrac{\textbf{A}\text{djacent}}{\textbf{H}\text{ypotenuse}}\right.$: "**CAH**" <-------- mnemonic device)

$\tan\theta = \dfrac{\text{opposite side}}{\text{adjacent side}}$ $\quad\left(\dfrac{\textbf{O}\text{pposite}}{\textbf{A}\text{djacent}}\right.$: "**TOA**" <-------- mnemonic device)

and use the relationship in which we know all but one quantity. Sometimes, we may use two relationships simultaneously to find two unknowns.

Example 1 In right triangle *ABC*, below, find the length of *BC* by using the appropriate trigonometric function.

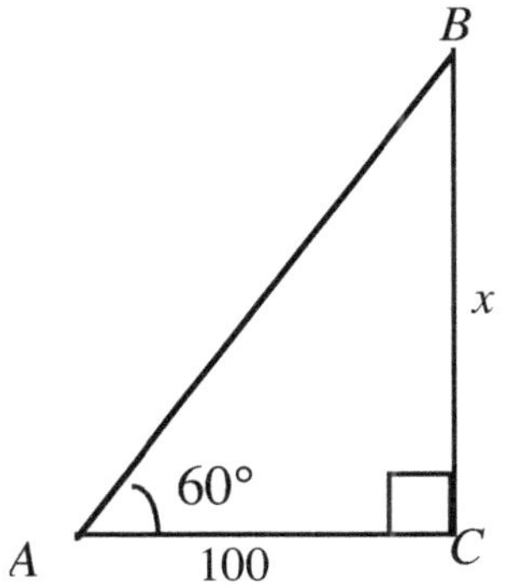

Step 1: Find out which parts are given. We are given m∠A = 60°, length of $AC = 100$ units.

Step 2: Note the part that we want to find. In this problem it is the side $\overline{BC}$ which we represent by x

Step 3: Determine how the side we want to find is related to the given (known) angle, and also how the given side, $\overline{AC}$, is related to the given angle (m∠A = 60°), by using any of the mnemonic devices: **S**OH, **C**AH, **T**OA. $\overline{BC}$ is opposite to ∠A and AC is adjacent to ∠A. Thus, opposite side (say ,"O") and adjacent side (say, "A") are involved and we have " OA". Now, "**OA**" is contained only in the device **T**OA which translates to: Tangent = $\dfrac{\textbf{Opposite}}{\textbf{Adjacent}}$

Hence, the trigonometric function to use in finding *BC* is that of the tangent. Therefore,

$$\tan A = \frac{x}{100} \quad \left(\frac{\text{opposite side}}{\text{adjacent side}}\right)$$

$$\tan 60^\circ = \frac{x}{100}$$

Solving for x, we obtain $x = 100\tan 60^\circ$

From tables, calculator, or from memory tan 60° = 1.73 approximately.

$x = 100(1.73)$

$= 173$ units.

Therefore the length of BC is 173 units..

In the last example, if what we were **given** and what we were to **find** involved the Opposite side and the Hypotenuse,with reference to the given angle, we would then have "OH" and we would have used the relationship

$\sin A = \frac{BC}{AB}$ (**O***pposite side* and **H***ypotenuse* → **OH** → *S***OH** → sin $= \frac{O}{H}$)

Similarly, if the Adjacent side and the Hypotenuse were involved, then we would have "AH" and we would have used the cosine relationship

$\cos A = \frac{AC}{AB}$ (**A**djacent *side* and **H***ypotenuse* → **AH** → *C***AH** → cos $= \frac{A}{H}$)

Example 2 Solve triangle *ABC*.

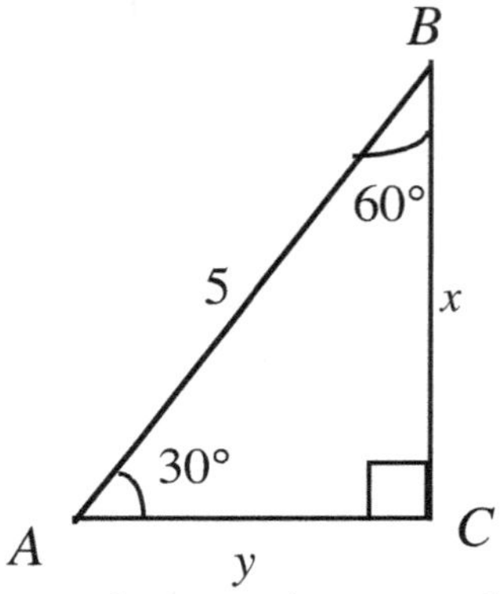

Solution There are only two parts that are unknown, and therefore we have to find only two parts namely, the sides $\overline{AC}$ and $\overline{BC}$.

Step 1: Finding x

The approach here is to find a ratio which involves the known (or given) side and the unknown side. With reference to ∠A, we will use the opposite side, $\overline{BC}$ and the hypotenuse, $\overline{AB}$ (Note: **O***pposite side* and **H***ypotenuse* → **OH** → *S***OH** → sin $= \frac{O}{H}$

$\sin 30 = \frac{x}{5}$ $\left(\frac{\text{Opposite}}{\textit{Hypotenuse}}\right)$

$x = 5\sin 30$ (solving for x)

$= 5\left(\frac{1}{2}\right)$ ($\sin 30 = \frac{1}{2}$, from tables or from memory)

$= 2.5$

Step 2: Finding y

With reference to ∠A, the sides involved here are the adjacent side (unknown) and the hypotenuse (known). Note: **A**djacent *side* and **H***ypotenuse* → **AH** → *C***AH** → cos $= \frac{A}{H}$.

$\cos 30° = \frac{y}{5}$ $\left(\frac{\text{Adjacent}}{\text{Hypotenuse}}\right)$

$y = 5\cos 30°$

$y = 5(.866)$ ($\cos 30° = .866$, from tables or calculator)

$= 4.33$

We could have found y by referring to $\angle B$; and in which case, the opposite side and hypotenuse would be involved. i.e.

(**O***pposite side* and **H***ypotenuse* → **OH** → *S***OH** → $\sin = \frac{O}{H}$

$\sin 60° = \frac{y}{5}$

$y = 5\sin 60°$

$= 5(.866)$ ($\sin 60° = .866$)

$= 4.33$

Again, we obtain the same value as when we referred to $\angle A$.

Example 3 Solve the right triangle ABC , given that $b = 4$, m $\angle$ B = 30°

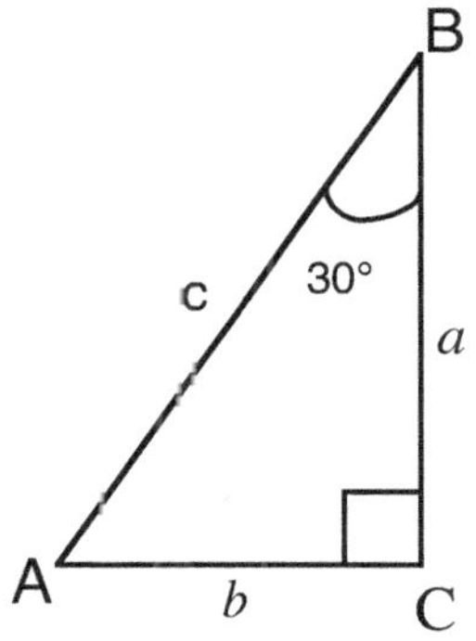

Solution

(a) m $\angle$ A = 180° - (30° + 90°) (m$\angle$ C = 90°; sum of the measures of the $\angle$'s of a Δ is 180°)

= 180° - 120°

m $\angle$ A = 60°

(b) To find c:

c is the hypotenuse, $b = 4$ is opposite $\angle$ B.

$\sin 30 = \frac{4}{c}$ $\left(\frac{\text{opposite}}{\text{hypotenuse}}\right.$, also m$\angle$ B = 30°, b = 4)

$c \sin 30° = 4$

$c = \frac{4}{\sin 30°} = \frac{4}{\frac{1}{2}} = 8$ ($\sin 30° = \frac{1}{2}$)

(c) To find a :

Now $c = 8$, $b = 4$

Apply the Pythagorean theorem (We may apply this theorem since Δ ABC is a right triangle)

$8^2 = 4 + a^2$

$64 = 16 + a^2$

$48 = a^2$

$a = \sqrt{48} = \sqrt{16}\sqrt{3} = 4\sqrt{3}$

Note: In (b), we could have found a first using the tangent relationship

$\tan A = \dfrac{a}{4}$ $\left(\dfrac{\textbf{O}\text{pposite side}}{\textbf{A}\text{djacent side}}\right.$ "**TOA**" <-------- mnemonic device$\Big)$

Solving, $a = 4 \tan A$

$a = 4 \tan 60°$

$a = 4\sqrt{3}$

Also, $\cos 30° = \dfrac{a}{c}$

$\dfrac{\sqrt{3}}{2} = \dfrac{a}{8}$ and $a = \dfrac{8(\sqrt{3})}{2} = 4\sqrt{3}$

We conclude that, sometimes, we have the option of choosing from two or more of the trigonometric functions.

Finding other trigonometric ratios given one functional value

In a right triangle, if we are given one trigonometric ratio for an acute angle we can determine the other trigonometric ratios.

Example: If $\sin\theta = \frac{3}{7}$, (a) find $\cos\theta$; (*b*) find $\tan\theta$.

Step 1: Sketch the a right triangle *ABC*. Let angle $A = \theta$

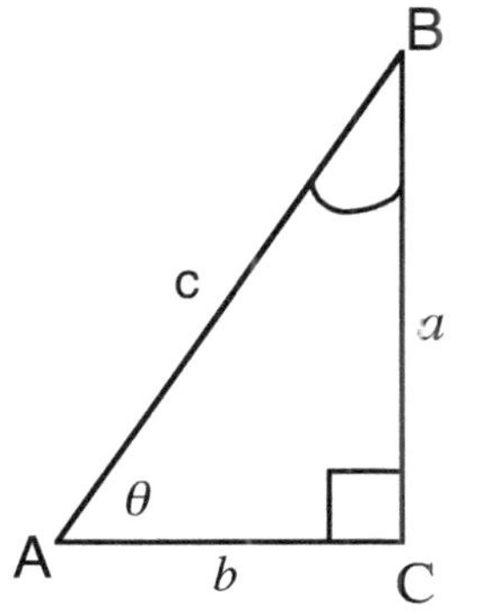

Step 2: Since $\sin\theta = \frac{3}{7} = \frac{\text{Opposite side}}{\text{Hypotenuse}} = \frac{a}{c} = \frac{3}{7}$ and therefore, a =3, c = 7.

We first calculate b., using the Pythagorean theorem

$$\text{From } c^2 = a^2 + b^2$$

$$7^2 = 3^2 + b^2$$

$$49 = 9 + b^2$$

Calculating b.

$$40 = b^2$$

$$\sqrt{40} = b$$

$$2\sqrt{10} = b$$

$$b = 2\sqrt{10}$$

Step 3: Now, $a = 3$, $b = 2\sqrt{10}$, and, $c = 7$.

$$\cos\theta = \frac{b}{c}$$

$$= \frac{2\sqrt{10}}{7}$$

$$\tan\theta = \frac{a}{b}$$

$$= \frac{3}{2\sqrt{10}}$$

$$= \frac{3\sqrt{10}}{20} \quad \text{(rationalizing the denominator)}$$

Lesson 16 Exercises

A Without using tables or calculator answer the following questions

1. If $\sin 49° = 0.7547$, then $\cos 41° = ...?$
2. If $\tan 50° = 1.1918$, then $\cot 40° = ...?$
3. If $\cot 60° = 0.5774$, then $\tan 30° = ...?$
4. If $\cot 43° = 1.0724$, then $\tan 47° = ...?$
5. If $\cot 40° = t$, then $\tan 50° = ...?$
6. If $\sec 60° = k$, then $\csc 30° = ...?$

Answers: 1. 0.7547; **2.** 1.1918; **3.** 0.5774; **4.** 1.0724; **5.** t; **6.** k.

B Solve the following right triangles:

1. $m\angle A = 38°$, $b = 4$; **2.** $m\angle A = 30°$, $b = 10$; **3.** $a = 8$, $c = 10$; **4.** $a = 40$, $b = 32$.
5. $b = 6$, and $m\angle A = 60°$

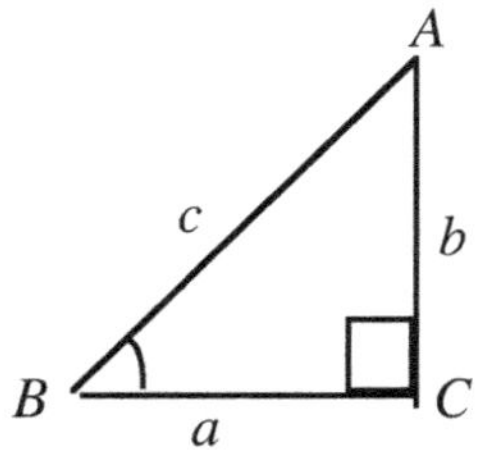

Answers: 1. $a = 3.13$, $c = 5.08$, $m\angle B = 52°$; **2** $a = 5.77$, $c = 11.54$, $m\angle A = 60°$
3. $b = 6$, $m\angle A = 53.13°$; $m\angle B = 36.87°$; **4.** $c = 51.22$; $m\angle A = 51.34°$; $m\angle B = 38.66°$.
5. $m\angle B = 30°$; $a = 6\sqrt{3}$; $c = 12$.

C If $\sin\theta = \frac{3}{5}$, find (a) $\cos\theta$; (b) $\tan\theta$; (c) $\sec\theta$.

Answers: (a) $\frac{4}{5}$(if θ in quad I) or $-\frac{4}{5}$(if θ in quad II); **(b)** $\frac{3}{4}$(if θ in quad I) or $-\frac{3}{4}$(if θ in quad II)
(c) $\frac{5}{4}$(if θ in quad I) or $-\frac{5}{4}$(if θ in quad II).

D **1.**. If $\sin\theta = -\frac{3}{5}$, find $\cos\theta$; (b) $\csc\theta$; $\tan\theta$ and $\cot\theta$.

2.. If $\cos\theta = \frac{2}{3}$, find (a) $\sin\theta$; (b) $\tan\theta$, and $\cot\theta$.

Answers: **1.** If θ is in the **third** quadrant: $\cos\theta = -\frac{4}{5}$; $\csc\theta = -\frac{5}{3}$; $\tan\theta = \frac{3}{4}$; $\cot\theta = \frac{4}{3}$.

If θ is in the **fourth** quadrant: $\cos\theta = \frac{4}{5}$; $\csc\theta = -\frac{5}{3}$; $\tan\theta = -\frac{3}{4}$; $\cot\theta = -\frac{4}{3}$.

2. If θ is in the **first** quadrant: $\sin\theta = \frac{\sqrt{5}}{3}$; $\tan\theta = \frac{\sqrt{5}}{2}$; $\cot\theta = \frac{2\sqrt{5}}{5}$.

If θ is in **fourth** the quadrant: $\sin\theta = -\frac{\sqrt{5}}{3}$; $\tan\theta = -\frac{\sqrt{5}}{2}$; $\cot\theta = -\frac{2\sqrt{5}}{5}$.

Lesson 17

Application of Trigonometric Ratios; Angles of Elevation and Depression; Bearing; Linear Interpolation

Angle of elevation

Definition: For an observer sighting (looking) at something above the observer, **the angle of elevation** is the angle between the horizontal line (x-axis) from the observer's eye and the line of sight to the object, the angle being measured from the horizontal to the line of sight.

Example Find the height reached by a kite, if the length of the kite is 400 ft, and the angle of elevation of the kite is 60°.

Step 1: Sketch a diagram for the system under consideration.
Assume that the hand holding on to the kite and the eye coincide at one point.

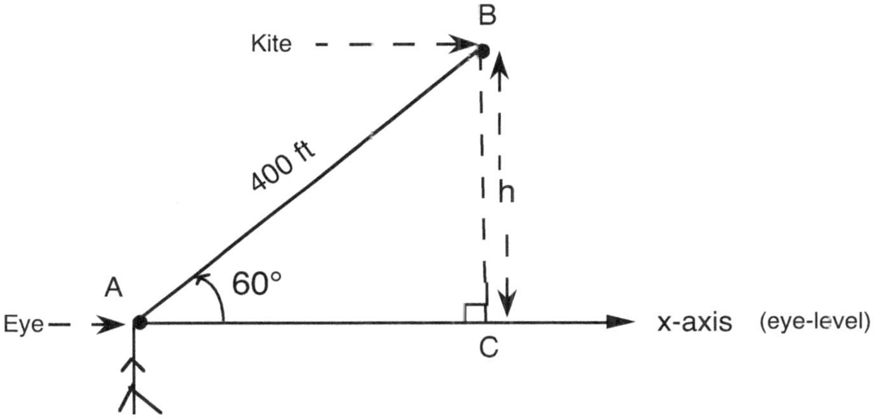

AB = Length of kite = 400 ft.

BC = h = height of kite (vertical distance from eye-level to the kite)

Step 2: Δ ABC is a right triangle, and therefore we may apply the trigonometric relationships.

$$\sin 60° = \frac{h}{400} \qquad \left(\text{SOH} : \frac{\textbf{O}\text{pposite}}{\textbf{H}\text{ypotenuse}}\right)$$

$$h = 400 \sin 60°$$

$$= 400\left(\frac{\sqrt{3}}{2}\right) \text{ ft.}$$

$$= 200\sqrt{3} \text{ ft.} \qquad (\sqrt{3} \approx 1.73)$$

$$\approx 346.41 \text{ ft.}$$

The height reached is $200\sqrt{3}$ ft. ($\approx$346.41 ft.)

Angle of depression

Definition**:** For an observer sighting (looking) at something below the observer, the **angle of depression** is the angle between the horizontal line (x-axis) from the observer's eye and the line of sight to the object, the angle being measured from the horizontal to the line of sight. In the figure below: $\angle\alpha$ is the angle of elevation of B from A.

$\angle\beta$ is the angle of depression of A from B.

The angle of elevation is congruent to the angle of depression.

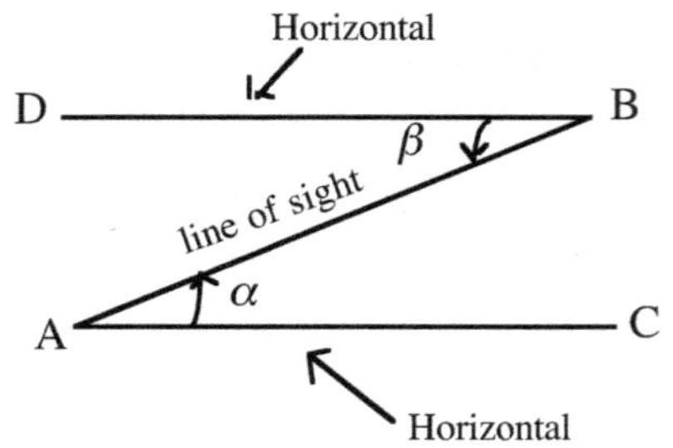

Bearing

The **bearing** of a given line in a horizontal plane is the acute angle between the given line and the North-Side (N-S) line. The angle is measured from a point on the N-S line to a a point on the given line. In finding bearings, the reference line is the North-South line which is a vertical line.

To indicate a bearing, we first specify whether North (N) or South (S), specify the measure of the angle and then specify whether East (E) or West (W). Note that the angle is always between the N-S line and the given line. Note also that the first letter is either N or S and the second letter after the angle is either E or W. (Contrast this angle with the reference angle of an angle in standard position in an x-y coordinate system of axes) Furthermore, note that if the angle is above the E-W line, the f irst letter is N; and below the E-W line, the first letter is S.

Examples:

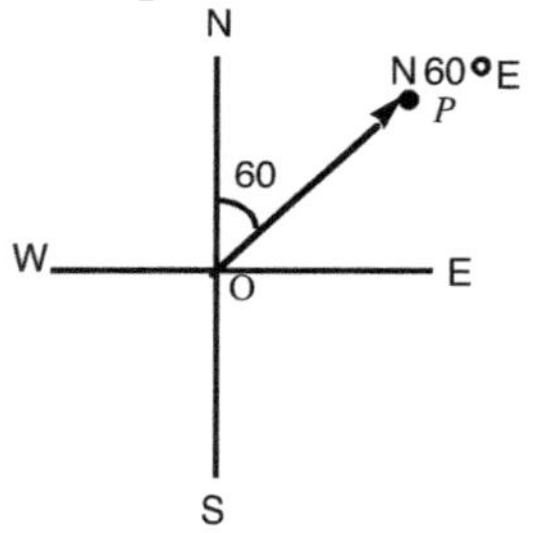

Bearing of P from O is $N60^{\circ}E$

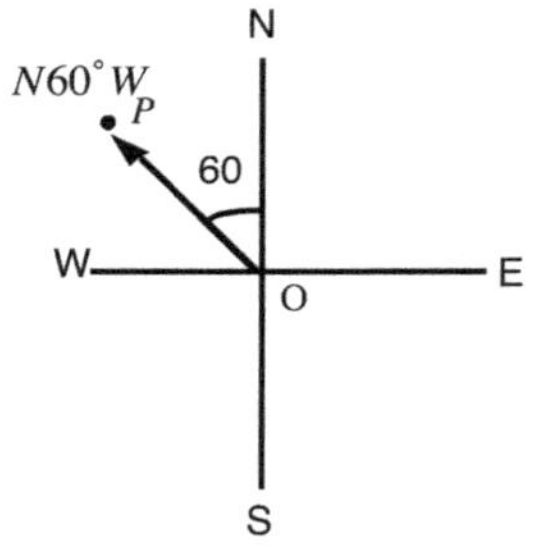

Bearing of P from O is $N60^{\circ}W$

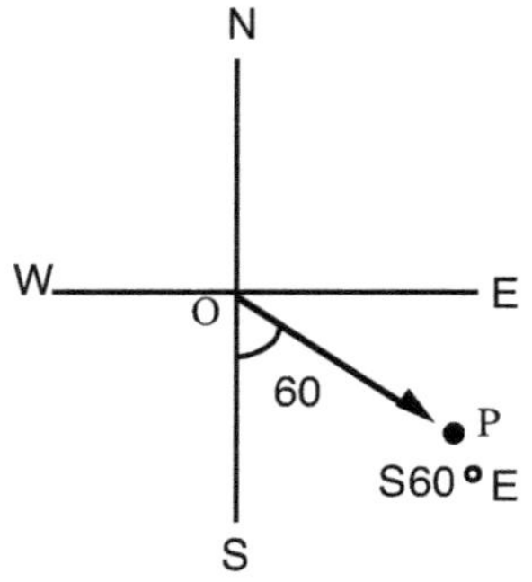

Bearing of P from O is $S60^{\circ}E$

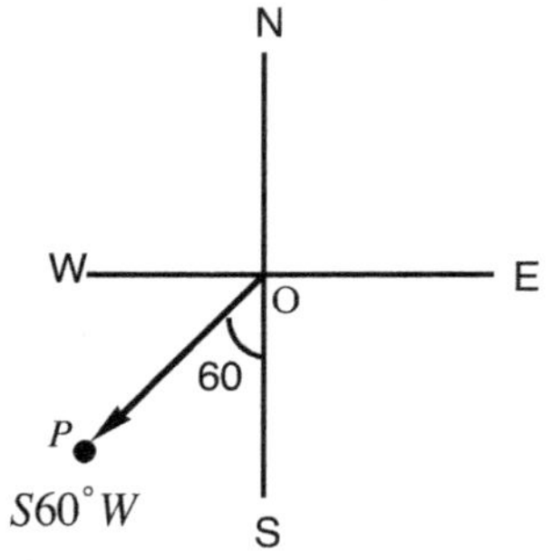

Bearing of P from O is $S60^{\circ}W$

Linear Interpolation

Sometimes, we may know two data points or corresponding values (x_1, y_1) and (x_2, y_2) and we would like to know either the y-value or the x-value of a point between the two known data points. This situation arises in applications such as below:

(1) In using logarithmic, trigonometric tables (and or similar tables) to approximate an unknown value between two known values.

(2) To approximate an unknown root between two other known roots in finding the real roots of polynomial equations for which there are no exact formulas for calculating the real roots.

The method of linear interpolation assumes that small changes in. say, x results in proportional changes in y , and therefore, the graph between the two points (x_1, y_1) and (x_2, y_2) is a straight line. Geometrically, there are two popular methods for setting up the proportion which will yield the required value:

(1) By similar triangles.
(2) By using the two-point-slope form of the equation of a straight line, which is given by

$$(y - y_1) = \frac{y_2 - y_1}{x_2 - x_1}(x - x_1) \text{ or } \frac{y - y_1}{x - x_1} = \frac{y_2 - y_1}{x_2 - x_1}$$

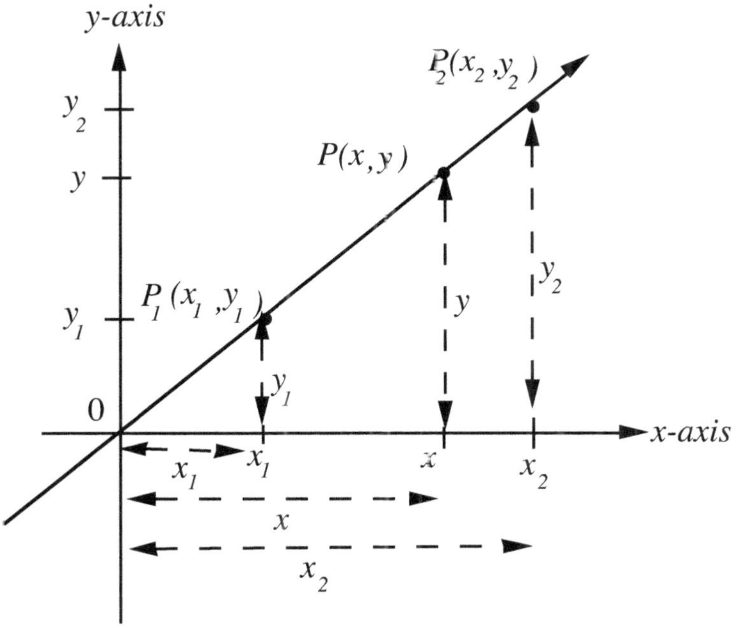

We will use the two-point-slope form, since this can be applied by mere substitution, and moreover, may be easier to remember (from elementary mathematics).
In using the above equation, we would know the values of five quantities in the proportion and we will calculate the sixth quantity.

Example Consider the incomplete table, below, and find y.

x	y
$x_1 = .42$	$y_1 = .4078$
$x = .43$	$y = ?$
$x_2 = .44$	$y_2 = .4259$

We will calculate the y-coordinate of the point $P(.43, y)$ between $P_1(x_1, y_1)$ and $P_2(x_2, y_2)$ by substituting the known values x_1, y_1, x_2, y_2, and x in $\frac{y_2 - y_1}{x_2 - x_1} = \frac{y - y_1}{x - x_1}$, we obtain

$$\frac{.4259 - .4078}{.44 - .42} = \frac{y - .4078}{.43 - .42}$$

Solving, $y = \frac{(.43-.42)(.4259 - .4078)}{.44-.42} + .4078$

$y = .4169$

Lesson 17 Exercises

A 1. A 20-meter ladder leans against the wall of a room. If the angle of elevation of the top of the ladder is $30°$, how far above the floor does the top of the ladder reach the wall?

2. From a point on the ground 30 feet from the base of a tree, the angle of elevation of the top of the tree is $60°$. Find the height of the tree.

3. The angle of depression of a boat from a cliff 100 meters above the shore line is $40°$. Find the distance from the boat to a point directly below the cliff.

4. Find the height reached by a kite, if the length of the kite is 600 ft, and the angle of elevation of the kite is 30°.

Answers: **1.** 10 meters; **2.** height = $30\sqrt{3}$ ft (≈ 52 ft); **3.** ≈ 119 meters; **4.** 300ft.

B 1. City A is 400 miles due east of City B. A third city, City C, due north of City B, is $N60°E$. How far is City C from City B?

2. Two ships A and B depart a port at 1pm. Ship A travels at 20 mph in direction $N30°W$. Ship B travels in direction $S60°W$ at 15 mph. Find the bearing of ship A from ship B at 3 pm?

3. Give the bearing for A, B, and C.

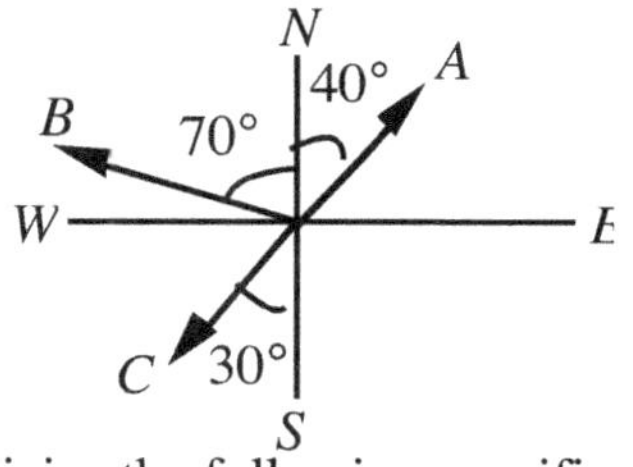

4. Criticize the following specifications for bearings:
(a) $W15°N$; (b) $E15°N$.; (c) $S40°N$

5. Draw the bearing for each of the following:
(a) $N15°W$; (b) $N15°E$.; (c) $S40°E$; (d) $S75°E$; (e) $S75°W$

Answers:

1.

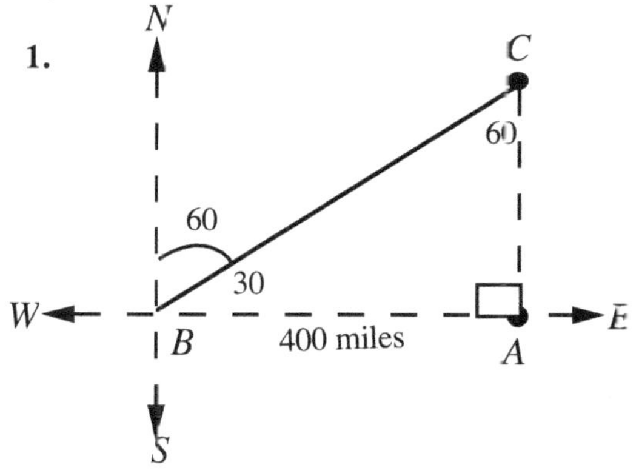

City C is 462 miles from City B.

2.

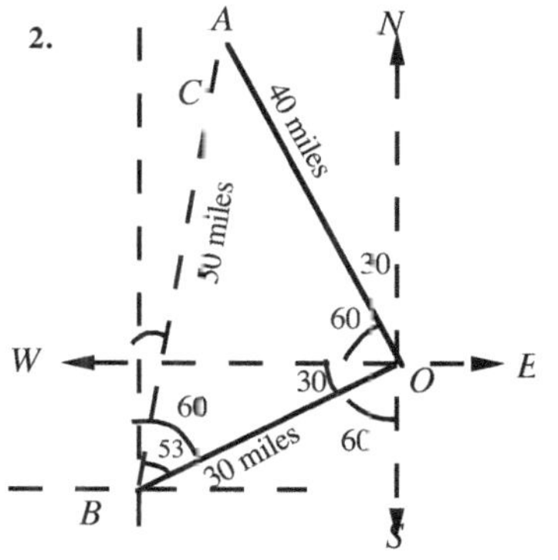

The bearing of A from B at 3 p.m.is $N7°E$.

3. A: $N40°E$; B: $N70°W$; C: $S30°W$;

4. (a) First letter must be N or S and second letter must be E or W.
(b) First letter must be N or S and second letter must be E or W.
(c) Second letter must be E or W.

Answers continued.

5.

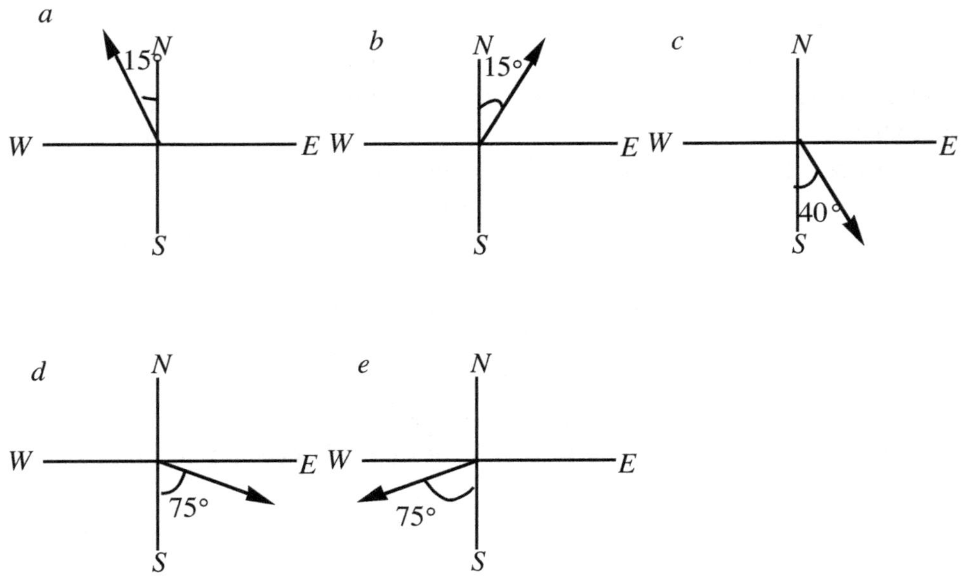

C **1.** Using linear interpolation, find the y -value corresponding to x = .6500 in the following table:

x	y
.6421	.4460
.6500	?
.6612	.4613

2. In the following, interpolate if necessary:
(a) Find $\sin 15° \ 18'$; (b) Find $\cos 60° \ 15'$;

3. Find the measure of θ given that
(a) $\cos\theta = .4000$; (b) $\sin\theta = .9623$

Answers: 1. 0.4523; **2.** (a) 0.2639; (b) 0.4962; 66.42° (or 1.16 rad); (b) 74.22° (or 1.30 rad).

CHAPTER 6

Trigonometrical Functional Value of any Angle

Lesson 18: **Definitions; Functional Value Given the Measure of the Angle**

Lesson 19: **Given a Point on the Terminal Side of the Angle**

Lesson 20: **Trigonometric Functional Values of Quadrantal Angles**

So far, we have defined the trigonometric ratios (functions) in terms of the sides of a right triangle and it did not matter where the vertex and the initial side of the angle were located, provided we obtained the required angular rotation. In this chapter, we will define two additional terms so that we can apply the trigonometric functions much more widely and we will not be restricted to angles between 0° and 90°.

Lesson 18

Basic Definitions; Functional Values given the Measure of the Angle

Basic Definitions

Angle

In trigonometry, we define an **angle** as being formed by rotating a line segment (a ray) about one of its end points, from an initial position to a final or terminal position. We call the end point, the **vertex**. We call the initial position, the initial side and the terminal position, the terminal side. The angle formed is measured by the amount of rotation. We shall refer to the angles formed by these rotations as **directed angles**. By a directed angle we mean if the rotation (used in forming the angle) is in a **counterclockwise direction**, then we shall say that the measure of t he angle formed is **positive** (Figure. 7). However, if the rotation is in a **clockwise direction**, then we say that the measure of the angle formed is **negative** (Figure 8). A curved arrow is used to denote the direction and amount of rotation. If no direction is indicated, then it is to be understood that the angle involved is the smallest positive angle.

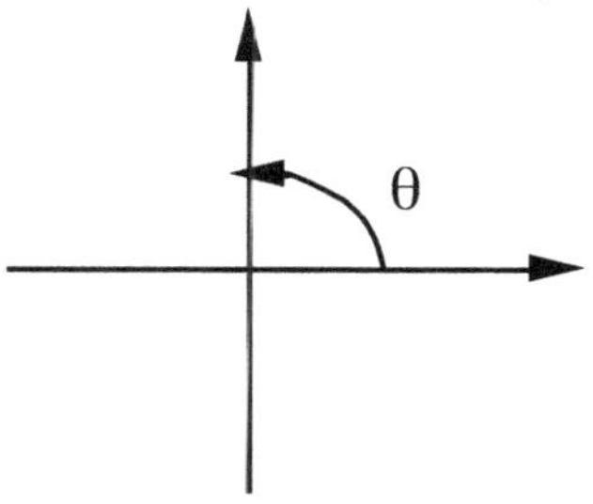

Fig. 7: θ is a positive angle

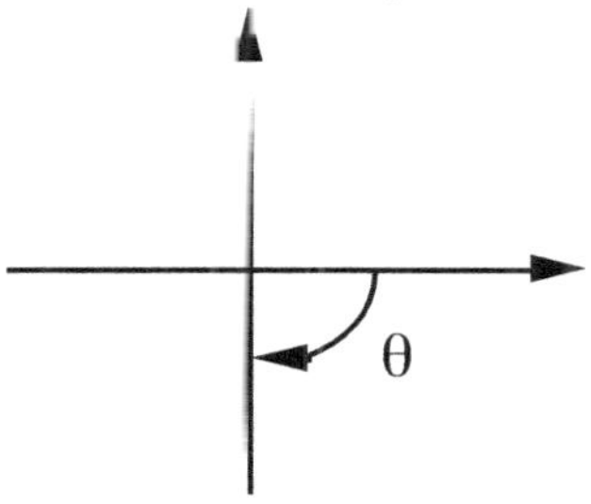

Fig. 8: θ is a negative angle

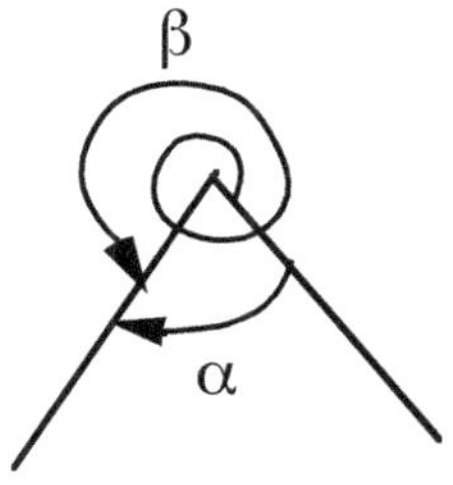

Fig. 9: β is a positive angle
α is a negative angle

A ray or a side can be rotated clockwise or counter-clockwise indefinitely. Consequently, the measure of an angle has an unlimited number of values. However, with some restrictions we can have unique values for the angles.

Angle in standard position

Definition: An angle is in standard position in an x-y rectangular coordinate system if it's vertex is at the origin and its initial side is along positive x -axis.

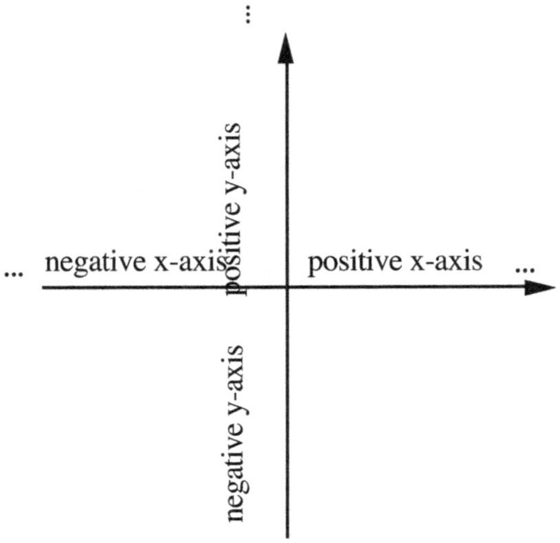

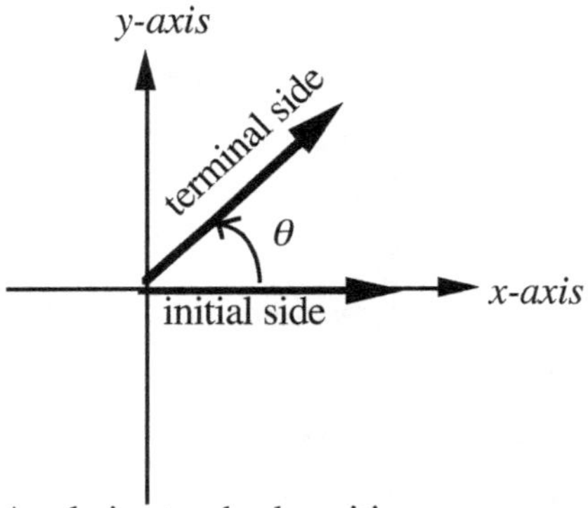

Angle in standard position

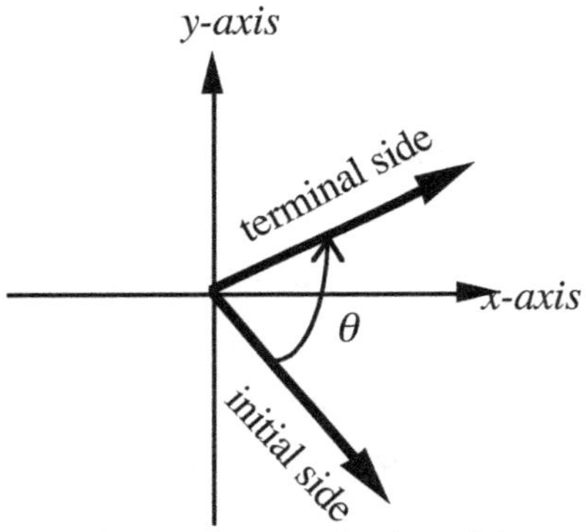

Angle **not** in standard position
(Initial side not along the positive x-axis)

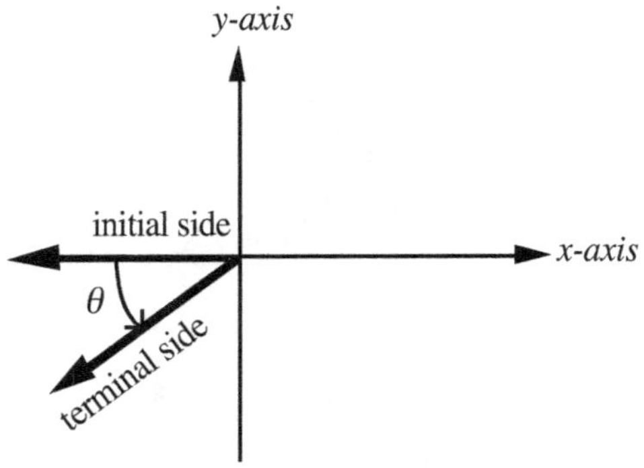

Angle **not** in standard position
(Initial side not along the positive x-axis)

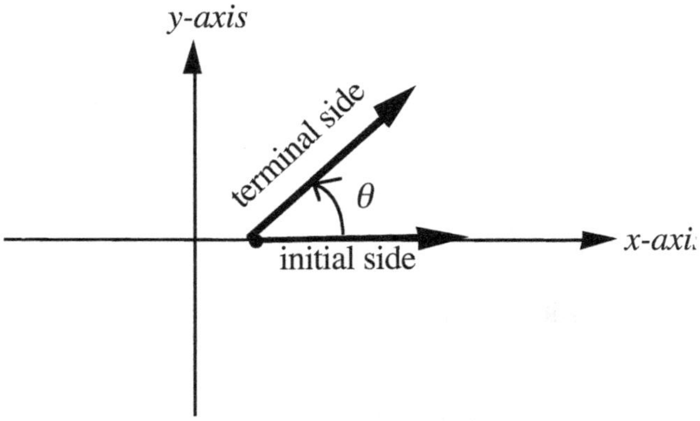

Angle **not** in standard position
(Vertex not at the origin)

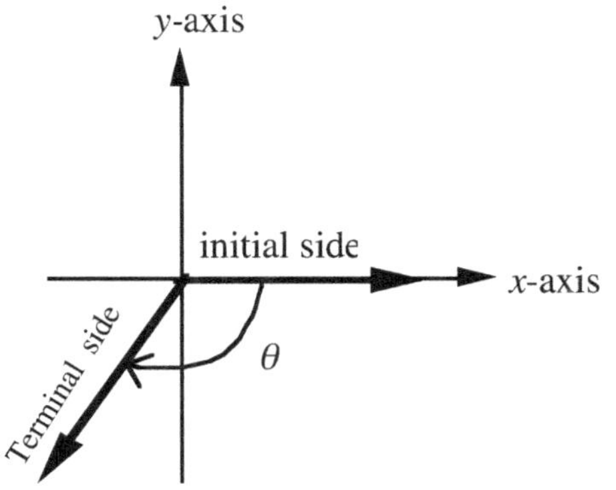

Angle in standard position

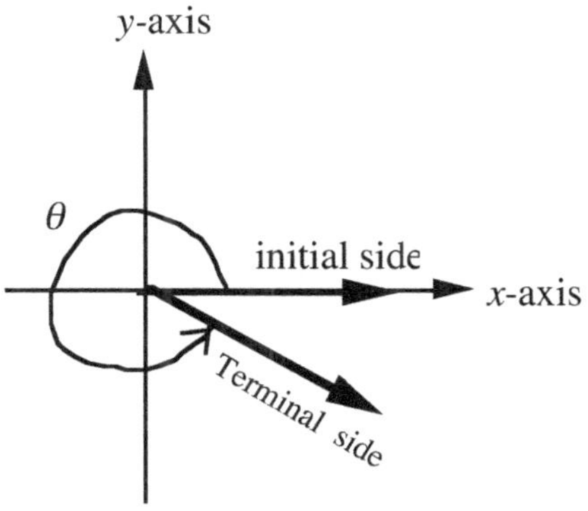

Angle in standard position

Reference Angle

Definition: If an angle, θ, is in standard position, then the reference angle is the positive acute angle between the terminal side of θ and the *x*- axis (the horizontal axis).

Example 1 Find the reference angle for 230°.

Solution

Step 1: Draw (or sketch) $\theta = 230^\circ$ in standard position.
To draw the arc, begin from the positive *x*-axis and draw the arc counterclockwise.

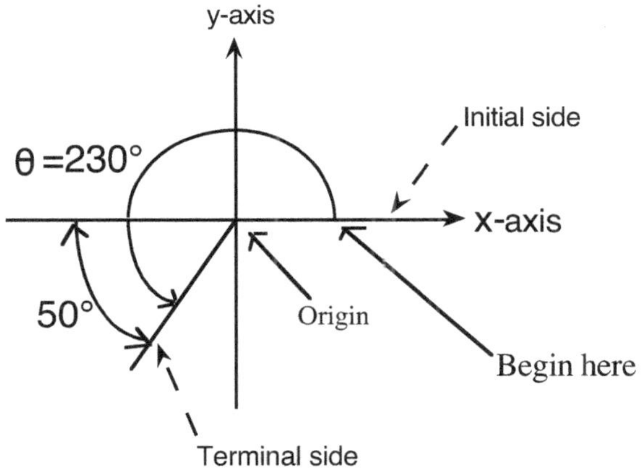

Step 2 :Determine the reference angle by difference. The reference angle for 230° is 230 - 180 i.e. 50°
The reference angle for 230° is 50° .

Note: The reference angle is always between the terminal side and the *x*-axis.

Example 2 Find the reference angle for 460°

Solution

Step 1: Draw (sketch) $\theta = 460°$ in standard position.

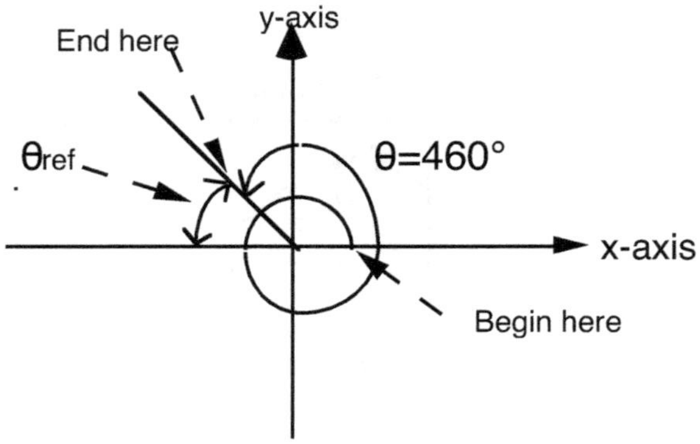

Step 2: The reference angle is the difference between 180° and 100°.

i.e., $180° - 100° = 80°$

Note that 360° will bring you back to the initial side, leaving 460 - 360 = 100 to consider.

Example 3 Find the reference angle for $\theta = -135°$

Solution

Step 1: Draw (sketch) $\theta = -135°$ (Draw this clockwise, since θ is negative.)

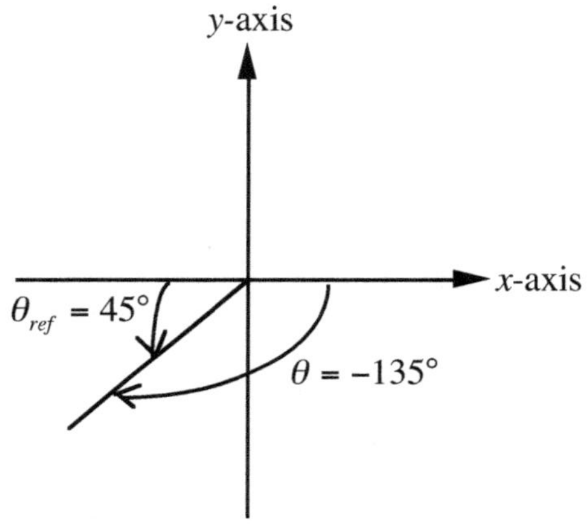

Step 2: The reference angle is $180° - 135° = 45°$, but note that the reference angle is always between the terminal side and the x-axis (never between the y-axis).

Relationship between θ (standard-position angle) and θ_{ref}

1. If θ is in the **first quadrant**, then $\theta_{ref} = \theta$

2. If θ is in the **second quadrant**, then $\theta_{ref} = 180° - \theta$ or equivalently $\theta = 180° - \theta_{ref}$

3. If θ is in the **third quadrant**, then $\theta_{ref} = \theta - 180°$ or equivalently $\theta = \theta_{ref} + 180°$

4. If θ is in the **fourth quadrant**, then $\theta_{ref} = 360° - \theta = \theta$ or equivalently $\theta = 360° - \theta_{ref}$

Note that θ and θ_{ref} are always in the **same** quadrant.

Trigonometric Functional Value of any Angle (Given the measure of the angle)

Example (a) Find the value of $\sin 210^\circ$ without using a calculator: use trigonometric tables.
(b) Verify your answer using a calculator.

Solution (a)

Step 1: Draw (or sketch) the given angle in standard position.

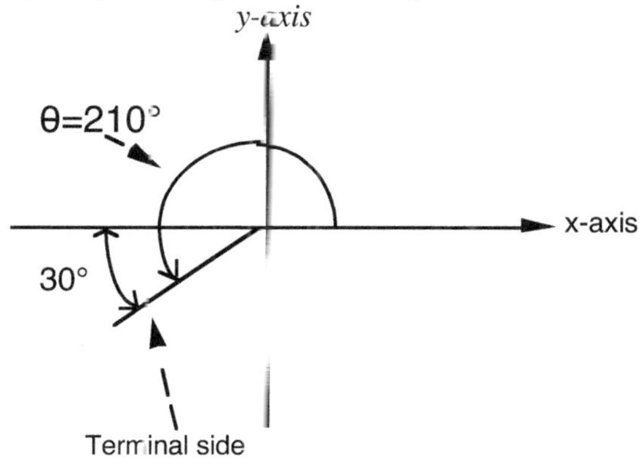

Figure 1

Step 2: Find the reference angle.
The reference angle is $210^\circ - 180^\circ = 30^\circ$

Step 3: From tables or from memory, find $\sin 30^\circ$. This value will be positive, and we must then determine the sign of $\sin 210^\circ$ according to the quadrant in which the terminal side of this angle lies. (See Figure 2 below) The terminal side of 210° is in the 3rd quadrant where the sine function is negative.

$$\therefore \ \sin 210 = -\sin 30 = -\frac{1}{2}$$

Signs of Trigonometric Functions According to Quadrants

Mnemonic device " **CAST**" for remembering in which quadrant a trigonometric function is positive or negative

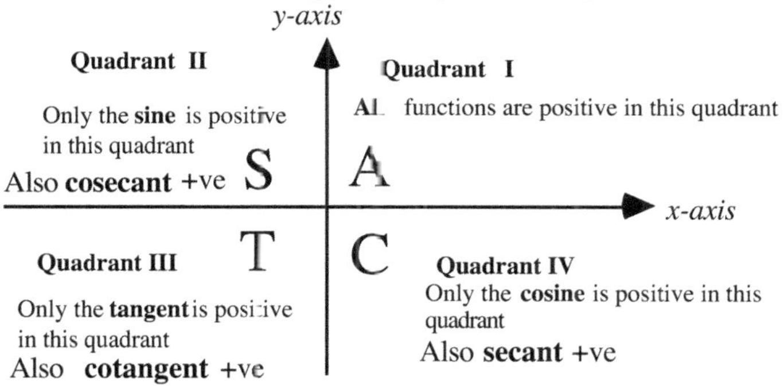

Figure 2

In the above device: The **A** in "**CAST**" means **A**ll functions are positive in this quadrant; **S** for only sine is positive in this quadrant; and similarly **T** for tangent; and **C** for cosine.

Note: The order of the above "CAST" is counterclockwise. There are other mnemonic devices such as "**A**ll **S**tudents **T**ake **C**alculus".

(b) With the calculator in degree mode, press 210; then press "the sin" key and read -0.5

Note above: The sign of a trigonometric function and its reciprocal are the same.

Lesson 18 Exercises

A Find the reference angles for the following: **1.** 250° ; **2.** 295°; **3.** 560°; **4.** 75°; **5.** 360°

6 −35°; **7.** −125°; **8.** −265°

Answers. **1.** 70° ; **2.** 65° ; **3.** 20°; **4.** 75° ; **5.** 0°; **6** 35°; **7.** 55°; **8.** 85°.

B In the following problems, first use tables; and then verify your answers using a calculator.

Find the values of the following: **1.** sin 250° ; **2.** cos 250°; **3.** sin 460°; **4.** sin 75°

Answers: **1.** - 0.94 ; **2.** -0.34; **3.** 0.98; **4.** 0.96

C Find the following:

(a) $\sin(-60°)$; (b) $\cos(-60°)$; (c) $\sin(-220°)$; (d) $\cos(-380°)$; (e) $\sin\frac{5\pi}{4}$; (f) $\cos\left(-\frac{\pi}{3}\right)$.

(g) $\sin(-150°)$; (h) $\cos(-250°)$; (i) $\sin(-460°)$; (j) $\sin(-75°)$

Answers: (a) -.86; (b) .5; (c) .64; (d) .94; (e) -.71; (f) .5; (g) -.5; (h) -.34; (i) -.98; (j) -.97

Lesson 19

Trigonometric Functional Value of any Angle

(Given the coordinates of a point on the terminal side of an angle)

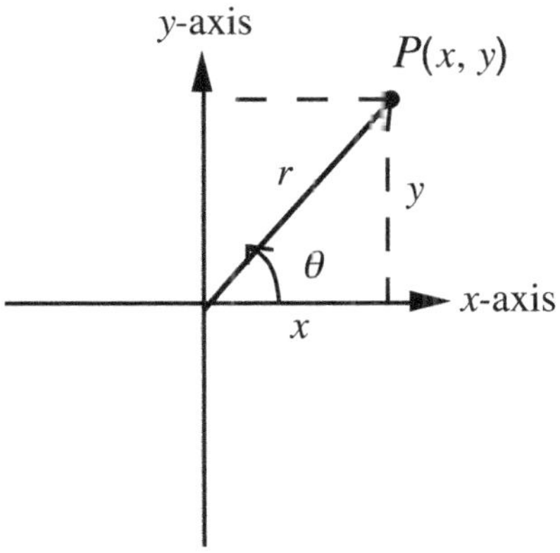

Let θ be an angle in standard position in a rectangular coordinate plane. Let $P(x,y)$ be a point on the terminal side of the angle θ and at a positive distance r from the origin (**Figure**).

Applying Pythagorean theorem to the above figure,

$$r^2 = x^2 + y^2$$

$$r = \sqrt{x^2 + y^2}$$

With reference to θ, r is the hypotenuse, x the adjacent side, y is the opposite side.
We define the six trigonometric functions (see also page 97.) for θ in terms of r, x and y.
(Recall: SOH , CAH, TOA)

$\sin\theta = \dfrac{y}{r}$; $\csc\theta = \dfrac{r}{y}$.

$\cos\theta = \dfrac{x}{r}$; $\sec\theta = \dfrac{r}{x}$.

$\tan\theta = \dfrac{y}{x}$; $\cot\theta = \dfrac{x}{y}$.

Note above that r is always positive while x and y may be positive or negative.

Example Find the six trigonometric values of θ , (if θ is an angle in standard position) given that the terminal side of θ passes through the point $(-3, 2\sqrt{3})$.

Step 1: Draw (or sketch) θ with terminal side passing through the point $(-3, 2\sqrt{3})$
(x-coordinate = -3; y-coordinate = $2\sqrt{3}$)

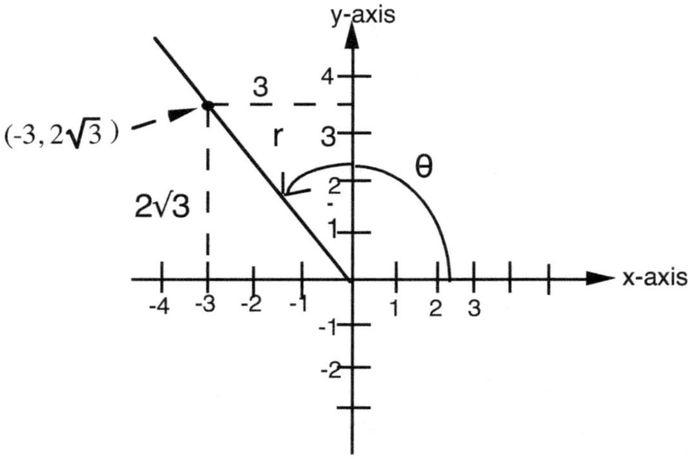

Step 2: Find r using the Pythagorean Theorem.

$$r^2 = (-3)^2 + (2\sqrt{3})^2 \quad (r^2 = x^2 + y^2)$$
$$r^2 = 9 + 4\sqrt{9}$$
$$r^2 = 9 + 12 \qquad \text{(Note: } 4\sqrt{9} = 4(3) = 12))$$
$$r^2 = 21$$
$$r = \sqrt{21}$$

Step 3: Now, $r = \sqrt{21}$, $x = -3$, $y = 2\sqrt{3}$

(a) $\cos\theta = \dfrac{x}{r}$

$$= \frac{-3}{\sqrt{21}}$$
$$= -\frac{3}{\sqrt{21}} \bullet \frac{\sqrt{21}}{\sqrt{21}} \quad \text{(Rationalizing the denominator)}$$

$$\therefore \cos\theta = -\frac{3\sqrt{21}}{21}$$

(b) $\sin\theta = \dfrac{y}{r}$ $\quad (y = 2\sqrt{3},\ r = \sqrt{21})$

$$= \frac{2\sqrt{3}}{\sqrt{21}}$$

$= \frac{2\sqrt{3}}{\sqrt{21}} \cdot \frac{\sqrt{21}}{\sqrt{21}}$ (Rationalizing the denominator)

$= \frac{2\sqrt{63}}{21}$

$= \frac{2\sqrt{9}\sqrt{7}}{21}$

$= \frac{2(3)\sqrt{7}}{21_7}$ $(\sqrt{9} = 3)$

$\sin\theta = \frac{2\sqrt{7}}{7}$

(c) $\tan\theta = \frac{y}{x}$ $(y = 2\sqrt{3}, x = -3)$

$= \frac{2\sqrt{3}}{-3}$

$\therefore \tan\theta = -\frac{2\sqrt{3}}{3}$

(d) $\sec\theta = \frac{1}{\cos\theta}$

$= \frac{1}{\frac{x}{r}} = \frac{r}{x} = \frac{\sqrt{21}}{-3}$

$\therefore \sec\theta = -\frac{\sqrt{21}}{3}$

(e) $\csc\theta = \frac{1}{\sin\theta}$

$= \frac{r}{y}$

$= \frac{\sqrt{21}}{2\sqrt{3}}$

$= \frac{\sqrt{21}}{2\sqrt{3}} \cdot \frac{\sqrt{3}}{\sqrt{3}}$ (Rationalizing the denominator)

$= \frac{\sqrt{63}}{2\sqrt{9}}$

$= \frac{\sqrt{63}}{2(3)}$ $(\sqrt{9} = 3)$

$= \frac{\sqrt{9}\sqrt{7}}{6}$

$= \frac{3\sqrt{7}}{6}$

$\therefore \csc\theta = \frac{\sqrt{7}}{2}$

(f) $\cot\theta = \frac{1}{\tan\theta}$

$$= \frac{1}{\frac{y}{x}}$$

$$= \frac{x}{y}$$

$$= \frac{-3}{2\sqrt{3}} \qquad (x = -3, y = 2\sqrt{3})$$

$$= -\frac{3}{2\sqrt{3}} \cdot \frac{\sqrt{3}}{\sqrt{3}} \qquad \text{(Rationalizing the denominator).}$$

$$= -\frac{3\sqrt{3}}{2\sqrt{9}}$$

$$= -\frac{\cancel{3}\sqrt{3}}{2(\cancel{3})} \qquad (\sqrt{9} = 3)$$

$$\therefore \cot\theta = -\frac{\sqrt{3}}{2}$$

Comparatively, note in the last two examples that:
A. On page123 we had to determine the sign of the function according to the quadrant (in which the terminal side of the angle lies) if we obtain the values of the acute angles from tables or from memory.

B.On page125 , we did not have to determine the sign of the function. The signs were obtained solely from the signs of the *x*- and *y*-coordinates; but note also that *r* is always positive.
Note also however that we may use the method in **A** to do the problems referred to in **B**.

Lesson 19 Exercises

A Find the six trigonometric values of θ , if θ is an angle in standard position, given that the terminal side of θ passes through the point (5, 2).

Solutions: $\sin\theta = \frac{2\sqrt{29}}{29}$; $\cos\theta = \frac{5\sqrt{29}}{29}$; $\tan\theta = \frac{2}{5}$; $\csc\theta = \frac{\sqrt{29}}{2}$; $\sec\theta = \frac{\sqrt{29}}{5}$; $\cot\theta = \frac{5}{2}$

B Redo the Example on page 126 by applying the mnemonic devices SOH, CAH, TOA together with "CAST" referred to on page 123.

Lesson 20

Trigonometric Functional Values of Quadrantal Angles

Quadrantal angles are angles whose terminal sides lie on either the x-axis or the y-axis. These angles are multiples of 90°. Examples of quadrantal angles are 0° , 90°, 180, -180°, 270°, and 360°.

Trigonometric functional values for the quadrantal angle 0°

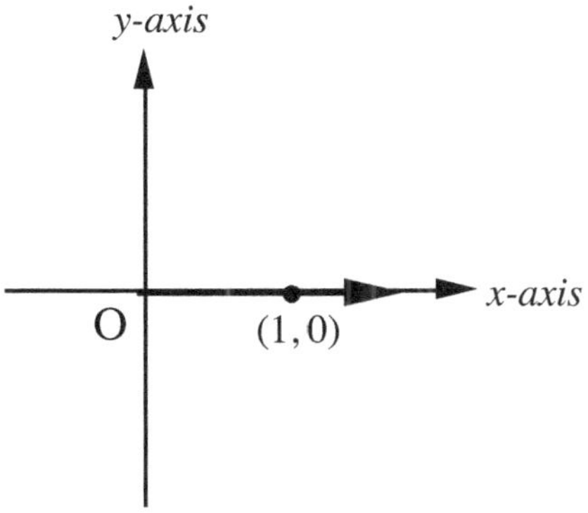

Consider a point $P(1, 0)$ on the terminal side of an angle $\theta = 0°$ in standard position and its distance from the origin to this point is 1. From the figure, $x = 1$, $y = 0$, and $r = 1$.

(a) $\sin\theta = \frac{y}{r}$

$\sin 0 = \frac{0}{1}$ $\quad (y = 0,\ r = 1)$

$\sin 0° = 0$

(b) $\cos\theta = \frac{x}{r}$

$\cos 0° = \frac{1}{1}$ $\quad (x = 1,\ r = 1)$

$\cos 0° = 1$

(c) $\tan\theta = \frac{y}{x}$

$\tan 0° = \frac{0}{1}$ $\quad (y = 0,\ x = 1)$

$\tan 0° = 0$

The reciprocal relationships are obtained by inverting (a), (b), b and (c) above.

(e) $\csc 0° = \frac{r}{y} = \frac{1}{0}$ **is undefined**.

(f) $\sec 0° = \frac{r}{x} = \frac{1}{1}$

$\sec 0° = 1$

(g) $\cot 0° = \frac{1}{0}$ **is undefined**

Trigonometric functional values for the quadrantal angle 90°

Consider a point $P(0, 1)$ on the terminal side of an angle $\theta = 90°$ in standard position. Then the distance from origin to P is 1. Thus, $r = 1$, $x = 0$ and $y = 1$.

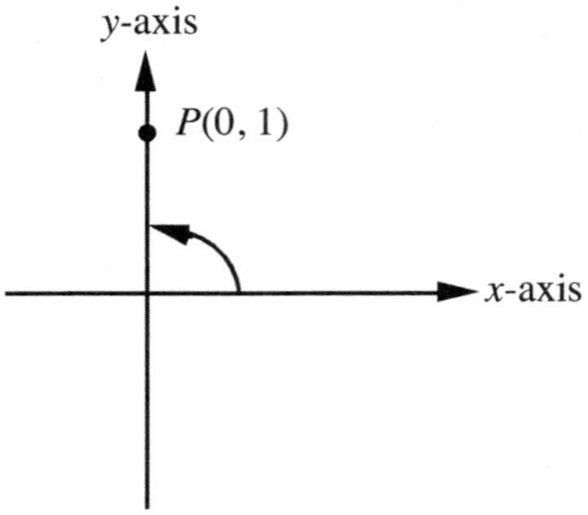

(a) $\sin\theta = \dfrac{y}{r}$

$\sin 90° = \dfrac{1}{1}$ $\quad (y = 1,\ r = 1)$

$\mathbf{\sin 90° = 1}$

(b) $\cos\theta = \dfrac{x}{r}$

$\cos 90° = \dfrac{0}{1}$ $\quad (x = 0,\ r = 1)$

$\mathbf{\cos 90° = 0}$

(c) $\tan\theta = \dfrac{y}{x}$

$\tan 90° = \dfrac{1}{0}$ $\quad (y = 1,\ x = 0)$

tan 90° is undefined

To find the functional values for the cosecant, the secant, and the cotangent , invert the right-hand sides of the sine, cosine, and tangent functions, respectively, and deduce the results.

Trigonometric functional values for the quadrantal angle 180°

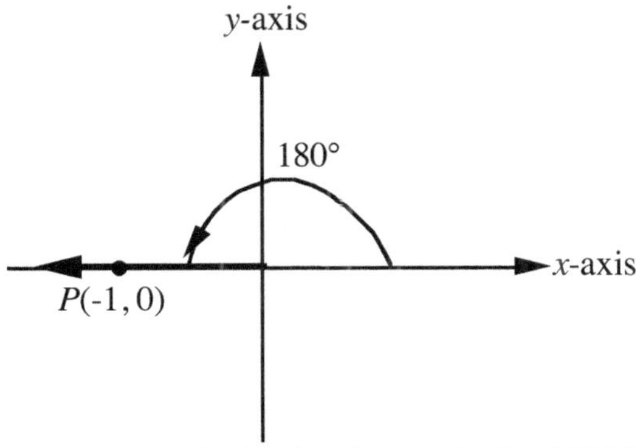

Consider a point $P(-1, 0)$ on the terminal side of an angle $\theta = 180°$ in standard position. Then the distance from the origin to P is 1. Thus, $r = 1$, $x = -1$ and $y = 0$.

(a) $\sin 180° = \dfrac{y}{r}$

$\sin 180° = \dfrac{0}{1}$ $(y = 0,\ r = 1)$

sin 180° = 0

(b) $\cos 180° = \dfrac{x}{r}$

$= \dfrac{-1}{1}$ $(x = -1,\ r = 1)$

cos 180° = −1

(c) $\tan 180° = \dfrac{y}{x}$

$= \dfrac{0}{-1}$ $(y = 0,\ x = -1)$

tan 180° = 0

To find the functional values for the cosecant, the secant, and the cotangent, invert the right-hand sides of the sine, cosine, and tangent functions, respectively, and deduce the results.

$\csc 180° =$ undefined
$\sec 180° = -1$
$\cot 180° =$ undefined

Trigonometric functional values for the quadrantal angle 270°

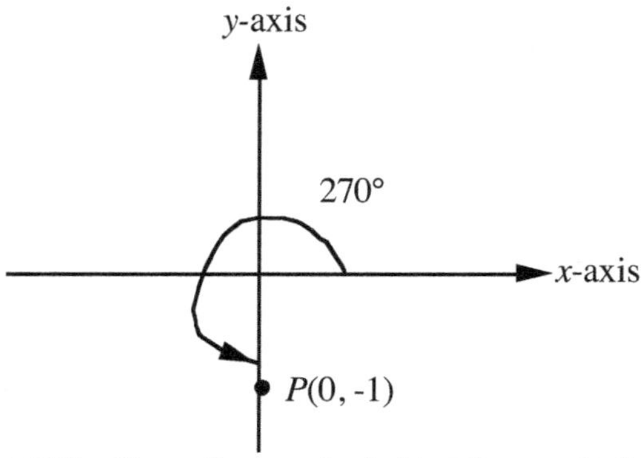

Consider a point $P(0, -1)$ on the terminal side of an angle $\theta = 270°$ in standard position. Then the distance from origin to P is 1. Thus, $r = 1, x = 0$ and $y = -1$.

(a) $\sin 270° = \frac{y}{r}$

$\sin 270° = \frac{-1}{1}$ $\quad (y = -1,\ r = 1)$

$\sin 270° = -1$

(b) $\cos 270° = \frac{x}{r}$

$= \frac{0}{1}$ $\quad (x = 0,\ r = 1)$

$\cos 270°$ $= 0$

(c) $\tan 270° = \frac{y}{x}$

$= \frac{-1}{0}$ $\quad (y = -1,\ x = 0)$

$\tan 270°$ is undefined

To find the functional values for the cosecant, the secant, and the cotangent, invert the right-hand sides of the sine, cosine, and tangent functions, respectively, and deduce the results.

Note In the above derivations, the value of r does not matter.
For example, for $\theta = 0°$, if $r = a, x = a,$ and $y = 0$ we obtain the following:

$$\sin 0° = \frac{y}{r} = \frac{0}{a} = 0$$

$$\cos 0° = \frac{x}{r} = \frac{a}{a} = 1$$

$$\tan 0° = \frac{y}{x} = \frac{0}{a} = 0$$

The trigonometric functional values for the quadrantal angle 360° are the same as those for 0°.

You may also observe that the values for sine, cosine, tangent, cotangent, secant, and cosecant are either -1, 0, 1, or undefined. For sine and cosine, the values are either -1, 0, 1. A mnemonic device for remembering these values is to locate the appropriate axis and choose a convenient "r" (say, $r = 1$) and then apply the fundamental definitions.

Functional Values for Coterminal Angles

Coterminal angles are angles which have the same terminal sides when both angles are placed in standard position on the same coordinate system of axes. Since the location the terminal side of an angle completely determines the trigonometric functional value, **coterminal angles have the same functional values**.

Examples

(a) $270°$ and $-90°$ are coterminal, and therefore have the same functional value.

(b) $-270°$ and $90°$ are coterminal, and therefore have the same functional value.

(c) $180°$ and $-180°$ are coterminal, and therefore have the same functional value.

(d) $30°$and $390°$ are coterminal, and therefore have the same functional value.

(e) $40°$and $760°$ are coterminal, and therefore have the same functional value.

Trigonometric functional values for negative quadrantal angles

Since coterminal angles have the same functional values, the functional values for $270°$and $-90°$ are the same, since these two angles are coterminal.

Therefore for $-90°$ see values for $270°$. Similarly, for -180° see values for 180° and for $-270°$ see values for $90°$.

Functional Values for sin (180° – A), cos (180° – A), and tan (180° – A)

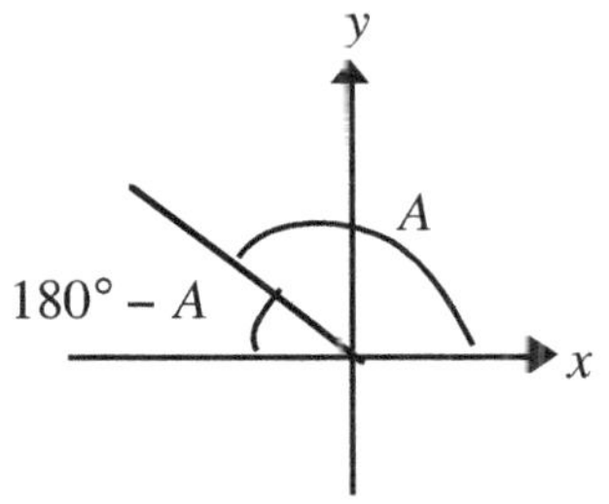

$\sin A = +\sin(180° - A)$ (the sine is positive in the second quadrant)

$\therefore \sin(180° - A) = \sin A$

$\cos A = -\cos(180° - A)$ (the cosine is negative in the second quadrant)

$\therefore \cos(180° - A) = -\cos A$

$\tan A = -\tan(180° - A)$ (the tangent is negative in the second quadrant)

$\therefore \tan(180° - A) = -\tan A$.

Lesson 20 Exercises

Derive the trig functional values for the following:
$0°$, $-180°$, $-270°$

Answers: $\sin 0° = 0$, $\cos 0° = 1$; $\tan 0° = 0$; $\csc 0°$ is undefined; $\sec 0° = 1$; $\cot 0°$ is undefined
$\sin(-180°) = 0$; $\cos(-180°) = -1$; $\tan(-180°) = 0$; $\sec(-180°) = -1$; $\csc(-180°)$ is undefined.; $\cot(-180°)$ is undefined.
$\sin(-270°) = 1$; $\cos(-270°) = 0$; $\tan(-270°)$ is undefined; $\csc(-270°) = 1$; $\sec(-270°)$ is undefined; $\cot(-270°) = 0$.

CHAPTER 7

Finding Other Trigonometric Functional Values

Lesson 21: **Given a Functional Value and the Quadrant of the Angle**

Lesson 22: **Given a Functional Value and no Specification of the Quadrant of the Angle**

Lesson 21

Given a Functional Value and the Quadrant of the Angle

Example If $\sin\theta = -\frac{4}{5}$ and the terminal side of θ is in the fourth quadrant (θ being in standard position)

(a) Find $\cos\theta$, (b) $\tan\theta$, (c) $\csc\theta$

We will determine r, x and y and apply the definitions of the trigonometric functions

Step 1: Since we know that the terminal side is in the 4th quadrant, we will draw θ accordingly.

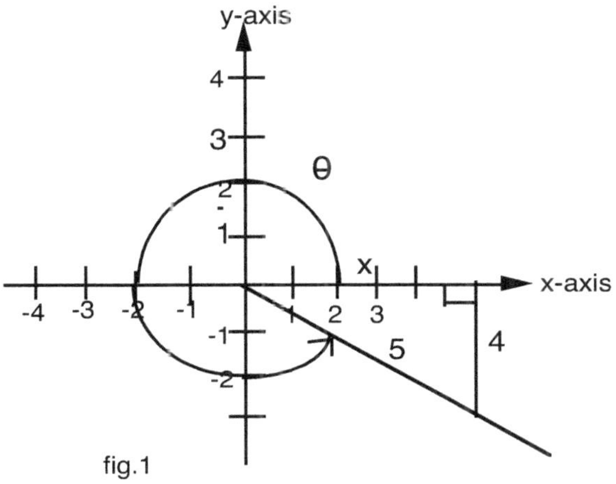

fig.1

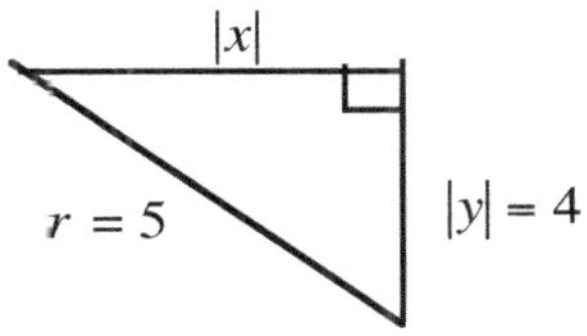

fig.2

Step 2: In this quadrant , the y-coordinate is negative and the x-coordinate is positive.

$$\sin\theta = \frac{y}{r} = -\frac{4}{5}$$

$y = -4$ since r is always positive (We give the minus sign to the "4") .

$r = 5$

Step 3: To find x, we use the Pythagorean theorem.

$$5^2 = (-4)^2 + x^2$$
$$25 = 16 + x^2$$
$$9 = x^2$$
$$3 = x$$
$$\therefore \quad x = 3,\ y = -4,\ r = 5$$

Step 4:

(a) $\cos\theta = \frac{x}{r}$

$\cos\theta = \frac{3}{5}$ $(x = 3, r = 5)$

(b) $\tan\theta = \frac{y}{x}$

$= \frac{-4}{3}$ $(y = -4, x = 3)$

$= -\frac{4}{3}$

(c) $\csc\theta = \frac{1}{\sin\theta}$

$= \frac{r}{y}$

$= \frac{5}{-4}$

$= -\frac{5}{4}$

Note: We could have obtained $\csc\theta$ from $\sin\theta = -\frac{4}{5}$ by inversion.

We may observe from the last two examples that the value of a trigonometric function does not depend upon the direction of rotation used in generating the angle. Thus if we know the coordinates of a point on the terminal side of the angle (and hence the quadrant in which the terminal side is) we can find the values of the trigonometric functions.

Lesson 21 Exercises

1. If $\sin\theta = -\frac{3}{7}$ and the terminal side of θ is in the third quadrant (θ being in standard position)
(a) Find $\cos\theta$, (b) $\tan\theta$, (c) $\csc\theta$

2. If $\sin\theta = -\frac{3}{5}$, and the terminal side of θ passes through the point $(-4,-3)$, find the other five trigonometric functional values.

Solutions: **1.** $\cos\theta = \frac{-2\sqrt{10}}{7}$; (b) $\tan\theta = \frac{3\sqrt{10}}{20}$; (c) $\csc\theta = -\frac{7}{3}$

2. $\cos\theta = -\frac{4}{5}$; . $\tan\theta = \frac{3}{4}$; $\csc\theta = -\frac{5}{3}$; $\sec\theta = -\frac{5}{4}$; $\cot\theta = \frac{4}{3}$.

Lesson 22

Finding Other Trigonometric Functional Values

(Given a functional value and no specification of the quadrant of the angle.)

Example: If $\sin\theta = -\frac{3}{7}$, (a) find $\cos\theta$; (*b*) find $\tan\theta$.

Step 1: The sine function is negative in the 3rd and 4th quadrants. Since the exact quadrant is not specified in this problem, we shall consider two cases here.

Case 1 Terminal side of θ is in 3rd quadrant.

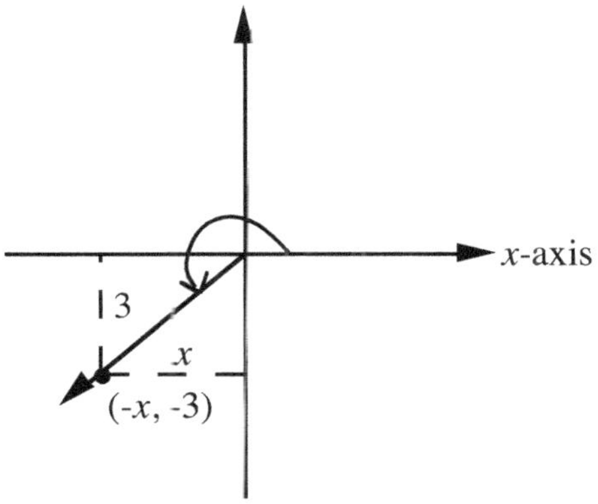

In the third quadrant, the *x* and *y*-coordinates are negative.

From the definition $\sin\theta = \frac{y}{r} = \frac{-3}{7}$. $y = -3$, $r = 7$, since *r* is always positive,

Plot (-*x*, -3), where *x* is the distance from the point (on the terminal side of the angle) to the *y*-axis.

Step 1: Find *x*.

$$\text{From } r^2 = x^2 + y^2$$
$$7^2 = (-x)^2 + (-3)^2$$
$$49 = x^2 + 9$$
$$40 = x^2$$
$$\sqrt{40} = x$$
$$2\sqrt{10} = x$$

Step 2: Now, the *x*-coordinate is $-2\sqrt{10}$, $y = -3$, $r = 7$.

$$\cos\theta = \frac{x}{r}$$
$$= -\frac{2\sqrt{10}}{7} \qquad (\text{the } x\text{-coordinate is } -2\sqrt{10})$$

$$\tan\theta = \frac{y}{x}$$

$$= \frac{-3}{-2\sqrt{10}}$$

$$= \frac{3\sqrt{10}}{20}$$

Case 2 Terminal side of θ is in 4th quadrant.

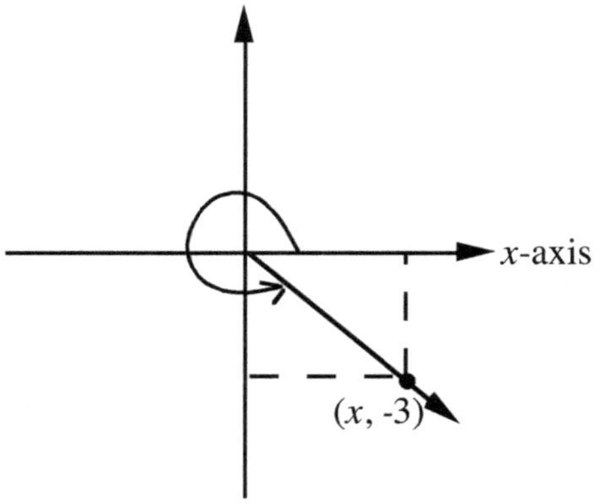

In this quadrant, the x-coordinate is positive and it is $2\sqrt{10}$, $y = -3$, $r = 7$.

$$\cos\theta = \frac{x}{r}$$

$$= \frac{2\sqrt{10}}{7}$$

$$\tan\theta = \frac{y}{x}$$

$$= \frac{-3}{2\sqrt{10}}$$

$$= -\frac{3\sqrt{10}}{20}$$

You can check the signs of the functional values by using the mnemonic device "CAST" see p.123

Lesson 22 Exercises

Redo Step 2 of the last Example by applying the mnemonic devices SOH, CAH, TOA together with "CAST" referred to on page 123 . .

CHAPTER 8

Trigonometry of Oblique Triangles

Lesson 23: **The Law of Cosine; The Law of Sines**

Lesson 24 **Applications of the Laws of Cosine and Sines**

An oblique triangle is a triangle in which no angle is a right angle. We may consider the right triangle as a special case of the oblique triangle and that the laws which are applicable to oblique triangles are also applicable to the right triangle.

In solving oblique triangles, we are usually given three parts of the oblique triangle and we are required to find the remaining three parts. We will usually seek a unique solution (that is only one solution). There may be a unique solution, two distinct solutions or no solutions at all depending on the relative lengths of the sides and the measures of the angles

Lesson 23

The Law of Cosine; The Law of Sines

The Law of Cosines.

Direct use of the Law of Cosines applies to the following cases:

(1) Given two sides and an included angle (abbreviated SAS), we are therefore to find the remaining side and two angles.

(2) Given all three sides (abbreviated SSS) we are therefore to find the remaining three angles. The above cases yield unique solutions. A mnemonic device for the applicability of this law of cosines is **Co-SAS -SSS** - pronounced Kosaszz.

Statement of the Law of Cosines

Using standard notation for corresponding angles and sides, the law of cosines states that :

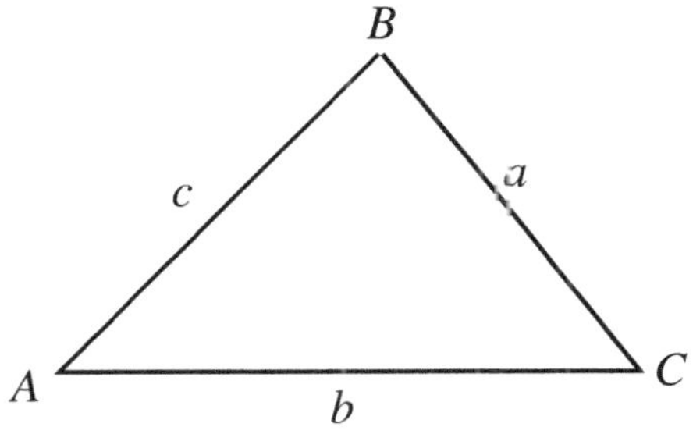

$$a^2 = b^2 + c^2 - 2bc\cos A$$

or

$$b^2 = a^2 + c^2 - 2ac\cos B$$

or

$$c^2 = a^2 + b^2 - 2ab\cos C$$

In words, the square of the length of one side is equal to the sum of the squares (of the lengths) of the other two sides minus twice the product (of the lengths) of the other two sides multiplied by the cosine of the angle between these two sides.

Derivation of the law of cosines

We will consider (1) an acute triangle; (2) an obtuse triangle; (3) a right triangle as a special case.

Case 1 For an acute triangle

Given Acute triangle *ABC* with sides a, b, c

Required: To derive $a^2 = b^2 + c^2 - 2bc\cos A$

Derivation

Consider the acute triangle *ABC* shown labeled with the standard notation for corresponding sides and angles.

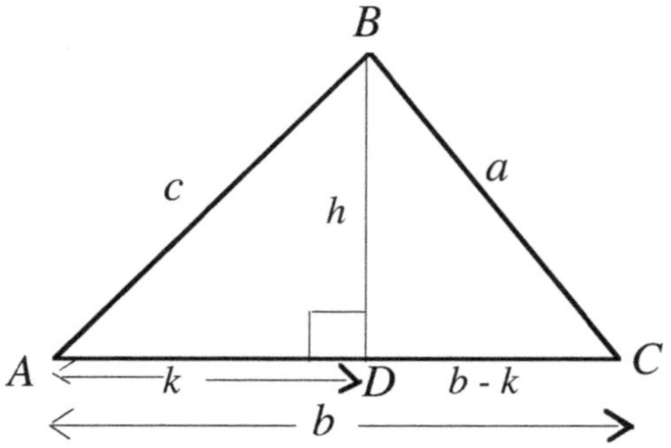

Draw the height h of triangle *ABC* by drawing $\overline{BD}$ perpendicular to $\overline{AC}$

Let $AD = k$. Then $DC = b - k$ We will consider subtriangles *ABD* and *CBD*; apply the Pythagorean theorem to both triangles and then eliminate h and k in terms of the given sides.

In rt. $\Delta\, ABD$, $c^2 = h^2 + k^2$ ....(1)

In rt. $\Delta\, BCD$, $a^2 = h^2 + (b - k)^2$ ....(2)

$a^2 = h^2 + b^2 - 2bk + k^2$ ....(3)

Equation (3) - Equation (1): $a^2 - c^2 = h^2 + b^2 - 2bk + k^2 - h^2 - k^2$

$a^2 = c^2 + b^2 - 2bk$

$a^2 = b^2 + c^2 - 2bc\cos A$ $\quad (k = c\cos A, \text{from } \cos A = \frac{k}{c})$

Therefore, the derivation is complete.

Case 2 For an for obtuse triangle

Given Obtuse triangle *ABC* with sides a, b, c

Required: To derive $a^2 = b^2 + c^2 - 2bc\cos A$

Derivation

Consider the obtuse triangle *ABC* shown labeled with the standard notation for corresponding sides and angles.

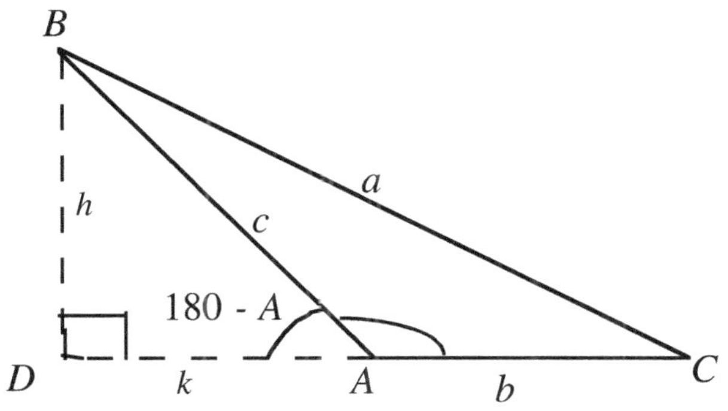

Draw the height h of triangle *ABC* by drawing $\overline{BD}$ perpendicular to $\overline{CA}$ extended (produced), and note that the height h of the triangle falls outside the given triangle .
Let $AD = k$. Then $DC = (b + k.)$ We will consider right triangles *BAD* and *BCD* ; apply the Pythagorean theorem to both triangles and then eliminate h and k in terms of the given sides.

In rt. Δ *BAD*, $c^2 = h^2 + k^2$ ....(1)

In rt. Δ *BCD*, $a^2 = h^2 + (b+k)^2$ ....(2)

$a^2 = h^2 + b^2 + 2bk + k^2$ ....(3)

Eqn. (3) - Eqn (1): $a^2 - c^2 = h^2 + b^2 + 2bk + k^2 - h^2 - k^2$

$a^2 = c^2 + b^2 + 2bk$

$a^2 = b^2 + c^2 + 2b(-c\cos A)$ $\quad (k = -c\cos A$, from $\cos(180 - A) = \frac{k}{c})$

$a^2 = b^2 + c^2 - 2bc\cos A$

Expressing k in terms of C and A. (Case 2): Since the given angle $\angle A$ is not in right triangle *BAD* , we have to express $\angle A$ indirectly in terms of its supplement (180 - A). $\cos(180 - A) = -\cos A$.

In triangle *BAD*, $\cos A = -\cos(180 - A)$ or $\cos(180 - A) = \mathbf{-\cos A = \frac{k}{c}}$, and from which

$k = -c\cos A$

Again, we obtain the same result as for the acute triangle.

One of the objectives in the above derivation (and subsequent derivations) is to introduce the student to the techniques of deducing statements from other statements as well as in proving statements in mathematics, generally.

Law of Cosine for the Right Triangle (a special case)

For a right triangle , let $m\angle A = 90°$. Then the law of cosines, $a^2 = b^2 + c^2 - 2bc\cos A$ becomes

$$a^2 = b^2 + c^2 - 2bc\cos 90°$$

$$a^2 = b^2 + c^2 - 2bc\,(0) \qquad (\cos 90° = 0)$$

$$a^2 = b^2 + c^2 \leftarrow \text{- - -The Pythagorean theorem}$$

We may conclude that the law of cosines is a generalization of the Pythagorean theorem.

The Law of Sines

Use of the Law of Sines.

The direct use of the law of sines will apply in the following cases:

(1) Given **two angles** and an **included side** (ASA) and therefore, we are to find the remaining sides and angle. This case yields a unique solution.

(2) Given two sides and an angle (ASS or SSA) there may be only one solution (one triangle), two solutions (two triangles) or there may be no solution at all (no triangle) depending upon the relative lengths of the sides and the measures of the angles. This case is commonly known as the ambiguous case of the law of sines.

A mnemonic device for its application is:
(1) sin ASA - pronounced "sinasa" **unique case**
(2) ASS - t**he ambiguous case**

Statement of the Law of Sines

The law of sines states that in any triangle, the sides are proportional to the sines of the angles opposite (facing) these sides. That is, the ratio of the length of a side to the sine of the angle opposite this side is equal to the ratio of the length of any other side to the sine of the angle opposite its side

Using standard notation for angles and sides,

a **is to sin A** as ***b*** **is to sin B** as ***c*** **is to sin C**

Symbolically, $\frac{a}{\sin A} = \frac{b}{\sin B} = \frac{c}{\sin C}$ (equivalent to $\frac{\sin A}{a} = \frac{\sin B}{b} = \frac{\sin C}{c}$)

Derivation of the Law of Sines

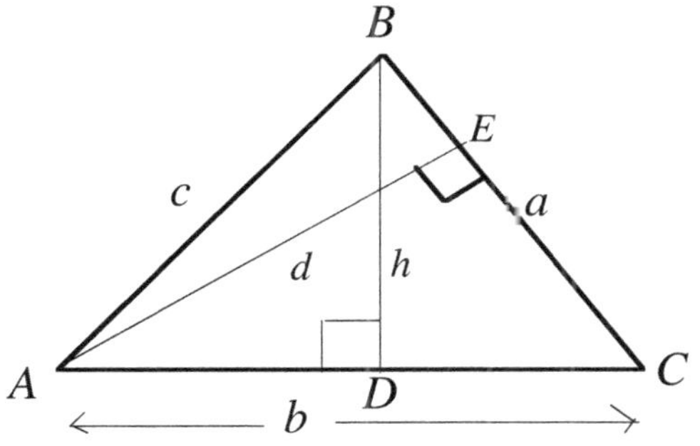

Given: ΔABC

Derivation From B, draw $\overline{BD}$ perpendicular to $\overline{AC}$.

In right ΔABD, $\sin A = \frac{h}{c}$

$$h = c \sin A \text{.................}(1)$$

Similarly, in right ΔCBD, $\sin C = \frac{h}{a}$

$$h = a \sin C \text{..................}(2)$$

Set right - hand sides of (1) and (2) equal to each other. Then

$$c \sin A = a \sin C$$

$$\frac{c \sin A}{\sin C \sin A} = \frac{a \sin C}{\sin A \sin C}$$ (Divide both sides of the equation by $\sin A$ and by $\sin C$;

$$\frac{c}{\sin C} = \frac{a}{\sin A} \quad (3)$$ (or solving for $\frac{c}{\sin C}$)

Similarly, we draw $\overline{AE}$ perpendicular to $\overline{BC}$.

In right ΔAEC, $\sin C = \frac{d}{b}$

$$d = b \sin C \text{...................}(4)$$

and in right ΔAEB, $\sin B = \frac{d}{c}$

$$d = c \sin B \text{..................}(5)$$

Set right - hand sides of (4) and (5) equal to each other. Then

$$b \sin C = c \sin B$$

$$\frac{b \sin C}{\sin B \sin C} = \frac{c \sin B}{\sin C \sin B}$$ (Divide both sides of the equation by $\sin C$ and by $\sin B$)

$$\frac{b}{\sin B} = \frac{c}{\sin C} \quad (6)$$ (or solving for $\frac{b}{\sin B}$)

From (3), $\frac{a}{\sin A} = \frac{c}{\sin C}$

Therefore, $\frac{a}{\sin A} = \frac{b}{\sin B} = \frac{c}{\sin C}$ <-------The law of sines.

Lesson 23 Exercises

1. State the Law of Cosines;	**2.** State the Law of Sines
3. Derive the Law of Cosines;	**4.** Derive the Law of sines.

Lesson 24

Applications of the Laws of Cosine and Sines

Example 1: Solve Δ ABC given that $m\angle C = 150°$, $b = 3, c = 8$

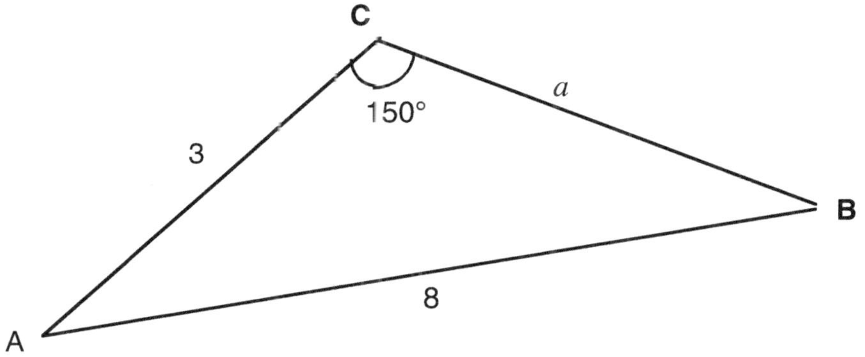

To **solve the triangle**, we will find the remaining three parts, namely the length a, m $\angle$ A, and m $\angle$ B.

Solution: We **cannot** apply the Pythagorean theorem, since Δ ABC is **not** a right triangle.

We will try the law of sines, or the law of cosines.

We apply the law of sines: $\dfrac{a}{\sin A} = \dfrac{b}{\sin B} = \dfrac{c}{\sin C}$

Step 1: We apply $\dfrac{c}{\sin C} = \dfrac{b}{\sin B}$

$$\frac{8}{\sin 150°} = \frac{3}{\sin B}$$

$$\frac{8}{\sin 30°} = \frac{3}{\sin B} \qquad (\sin 150° = \sin 30° = \tfrac{1}{2})$$

$$\frac{8}{\frac{1}{2}} = \frac{3}{\sin B} \qquad (b = 3;\ c = 8;\ \sin 150° = \sin 30° = \tfrac{1}{2})$$

$$8 \sin B = \frac{3}{2} \qquad \text{(by cross-multiplication or otherwise)}$$

$$\sin B = \frac{3}{2} \times \frac{1}{8} = \frac{3}{16}$$

$$= .1875$$

Step 2: Find the angle whose sine is .1875 (From tables or calculator)

then, m $\angle$ B $= 10.81°$

m $\angle$ A $= 180° - (150° + 10.8°) = 19.19°$.

Finding a

Method 1 Given: $c = 8$, m $\angle$ C= 150º , $b = 3$

Using the law of cosines:

$$c^2 = a^2 + b^2 - 2ab \cos C \qquad (\angle\ C \text{ is an included angle})$$

$$8^2 = a^2 + (3)^2 - 2a(3) \cos 150^\circ$$

$$64 = a^2 + 9 - 2a(3)\left(-\frac{\sqrt{3}}{2}\right) \qquad \left(\cos 150^\circ = -\cos 30^\circ = -\frac{\sqrt{3}}{2}\right)$$

$$64 = a^2 + 9 + 2a(3)\frac{\sqrt{3}}{2}$$

$$64 = a^2 + 9 + 3\sqrt{3}\,a$$

$0 = a^2 + 3\sqrt{3}\,a - 55$ <--------------quadratic equation

Solve by the quadratic formula.

$a^2 + 3\sqrt{3}a - 55 = 0$

$$a = \frac{-3\sqrt{3} \pm \sqrt{(3\sqrt{3})^2 - 4(1)(-55)}}{2} \qquad (a = 1, b = 3\sqrt{3}, c = -55)$$

$$= \frac{-3\sqrt{3} \pm \sqrt{9(3) + 220}}{2}$$

$$= \frac{-3\sqrt{3} \pm \sqrt{247}}{2}$$

$$= \frac{-3\sqrt{3} \pm 15.71}{2} = \frac{-3\sqrt{3} + 15.71}{2} \text{ or } \frac{-3\sqrt{3} - 15.71}{2}$$

$a = 5.26$ or -10.45

we reject the negative root since the length of a side of a triangle is always positive.

$\therefore a = 5.26$.

Finding a

Method 2 We could also have found a, using the law of sines after having found.

$m\angle A = 19.19°$

$$\frac{a}{\sin A} = \frac{c}{\sin C}$$

$$\frac{a}{\sin 19.19°} = \frac{8}{\sin 150°}$$ (sin 19.19 = .3287)

(sin 150° =sin 30° = .5)

$$a = \frac{8 \sin 19.19}{\sin 150}$$

$$= \frac{8(.32870)}{.5}$$

$a = 5.26$

Again, we obtain the same value for a. Method 2 is less involved than Method 1. However, the disadvantage of Method 2, in this particular problem, is that we used a value, $m\angle A = 19.19°$, which we calculated; and if there were error in $m\angle A$, there would be error in the value of a.
The advantage of Method 1 is that we used only given values and no calculated values ($c = 8$, $m\angle C = 150$, $b = 3$); but then, we had to solve a quadratic equation.

Example 2 Solve Δ ABC given that $m\angle C = 130°$, $a = 6$, $b = 9$.

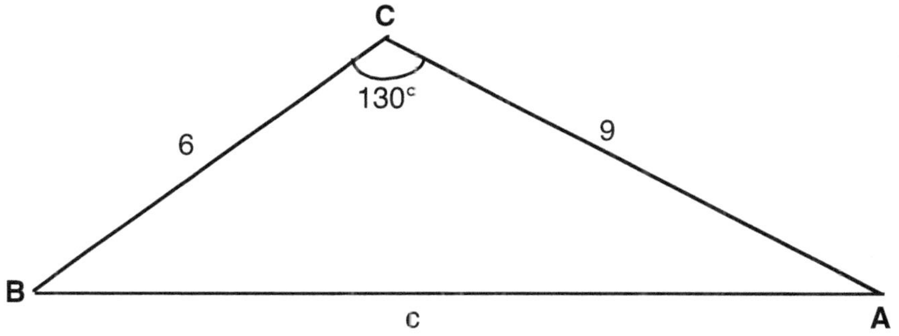

We will find $m\angle B$, $M\angle A$ and length c.

Finding c

$c^2 = a^2 + b^2 - 2ab \cos C$ <-------Law of cosines

$c^2 = 6^2 + 9^2 - 2(6)(9) \cos 130°$

$= 36 + 81 - 108(-.642)$ (cos 130° = - .642)

$= 36 + 81 + 69.42$

$c^2 = 186.42$

$c = \sqrt{186.42}$

$c = 13.65$

(Note above that if we had applied the law of sines to find c, we would have had one equation with two unknowns; and we would not be able to obtain a numerical value for c without an additional equation with the same unknowns.)

Finding M $\angle$ A:

$$\frac{c}{\sin 130^\circ} = \frac{a}{\sin A} \qquad \text{(law of sines).}$$

$$\frac{13.65}{\sin 130} = \frac{6}{\sin A} \quad (c = 13{,}65, a = 6)$$

We solve for sine A.

$$\sin A = \frac{6 \sin 130}{13.65}$$

$$= \frac{6(.766)}{13.65}$$

$$\sin A = .3367$$

Find inverse sine of .3367 (from tables or a calculator)

(That is, find the angle whose sine is .3367)

m $\angle$ A = 19.68

m $\angle$ B = 180 - (130 + 19.68)

m $\angle$ B = 30.32

(We could also use the law of sines to find m $\angle$ B, and then check that the sum of the measures of the angles equals 180°)

$\therefore$ $c = 13.65$; m $\angle$ A = 19.68°; m $\angle$ B = 30.32°

When do we apply the Pythagorean theorem, the law of cosines, or the law of sines?

We have learned how to solve triangles using the Pythagorean theorem, the law of sines and the law of cosines. Given a triangle to solve, how do we know which theorem or law to apply? Which theorem or law to apply depends on what we are given and what we want to find.

1. We may apply the Pythagorean theorem only if the triangle is a right triangle. However, if by construction, we are able to obtain right triangles, then we can apply the theorem to any such triangles obtained.

2. For the law of cosines:

(a) Given **two sides and an included angle** (SAS), use of the law of cosines yields a unique solution (i.e. one triangle). In this case, we will find the remaining side and the measures of two angles.

(b) Given all **three sides** (SSS), the application of the law of cosines yields a unique solution. In this case, we will find the measures of all the angles.

3. **For the law of sines**:

(a) Given **two angles** and **any side** (ASA, AAS, SAA), the use of the law of sines will yield a unique solution (one triangle).

Note that given any two angles, we can easily find the third angle, since the sum of the measures of the angles of a triangle equals 180°.

(b) **Ambiguous Case:** Given **two sides** and **an angle opposite one of the sides** (ASS or SSA), there may be only one solution (one triangle), two solutions (two Δ's), or there may be no solution at all (no triangle) depending on relative lengths of the sides and the measures of the angles

Note also that if the measure of any angle given is 180° or more there is no solution, since the sum of the measures of the three angles of a triangle is 180°.

Also, if the sum of the measures of any two angles given is 180° or more there is no solution.

Note also that if the triangle is a **right triangle**, we can consider applying any of the six trigonometric functions (ratios): sine, cosine, tangent, cosecant, secant, and cotangent.

Question: How do we remember the above cases? Answer: By practice, repetition and finding mnemonic devices.

Lesson 24 Exercises

A Solve the triangle:

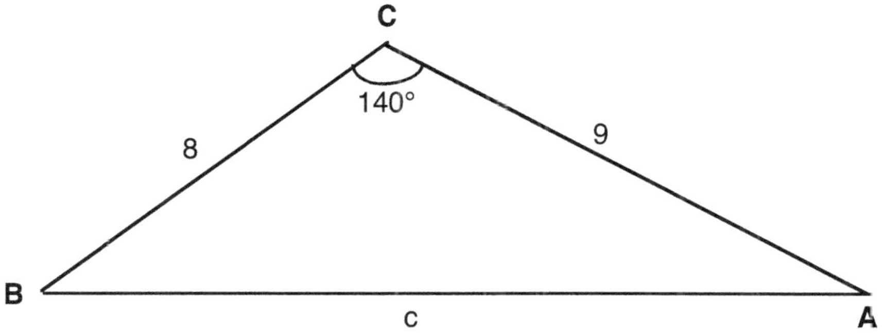

Solutions: $c = 15.98$; $m\angle B = 21.2°$; $m\angle A = 18.8°$
≈ 16; $\approx 21°$; $\approx 19°$

B Solve the following triangles:

1.

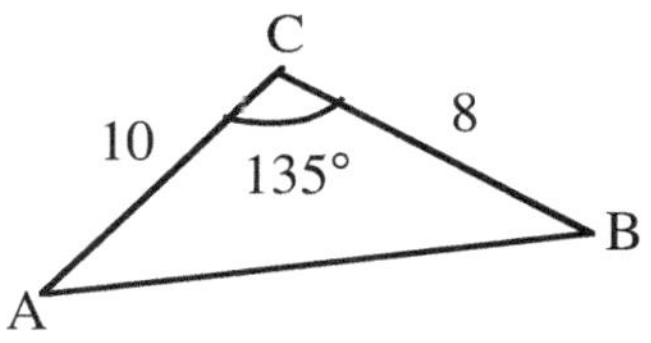

2.

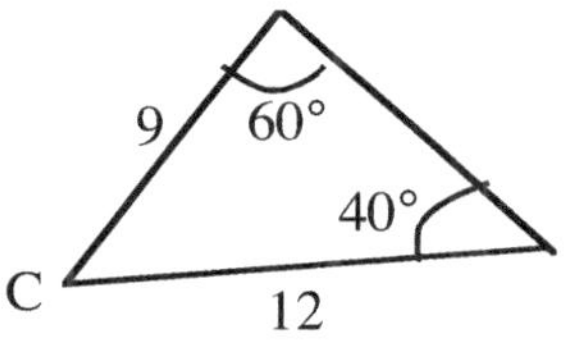

Solve the following triangles using the given dimensions:

3. $a = 15$; $m\angle A = 50°$; $m\angle B = 110°$; **4.** $b = 8$; $m\angle B = 75°$; $m\angle C = 55°$

Answers: **1.** $c = 17$, $m\angle A = 20°$; $m\angle B = 25°$; **2.** $c = 13.7$, $m\angle C = 80°$
3. $m\angle C = 20°$; $b = 18$; $c = 6.7$; **4.** $m\angle A = 50°$; $a = 6.3$; $c = 6.8$

C Determine the measure of the largest angle to the nearest degree.

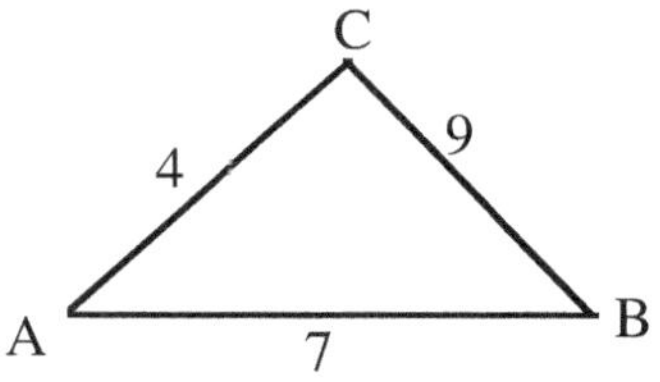

Answer: $m\angle A = 107°$ (Hint: Largest angle is opposite longest side)

CHAPTER 9

Application of Trigonometry to Vectors

Lesson 25: **Basic Definitions; Representation of Vectors**

Lesson 26 **Addition (Sum, Resultant, or Composition) of Vectors**

For anyone who will study physics, mechanics or engineering, this chapter is a very important one The mastery of the principles and techniques for resolving vectors into components as well as finding the resultant of a number of vectors will relieve you of the mathematical aspects in your first year physics and mechanics courses.

Lesson 25

Basic Definitions; Representation of Vectors

Measurable quantities can be classified either as scalars or as vectors.

Scalar Quantity

Definition A **scalar quantity** is a measurable quantity which has only magnitude and it is specified by a number and units.

Examples (1) **Speed**

A car travels at $\underbrace{\text{20 miles per hour}}_{\text{magnitude}}$.

(2) **Mass**

The mass of a block of wood is $\underbrace{\text{6 kilograms}}_{\text{magnitude}}$.

(3) **Distance**

The distance between the points A and B is $\underbrace{\text{8 meters}}_{\text{magnitude}}$.

Addition of Scalar Quantities

We add scalar quantities by ordinary arithmetic addition.

Example 3 kilograms + 5 kilograms = 8 kilograms

Vector Quantity

Definition: A **vector quantity** is a quantity that has both magnitude and direction. We specify a vector quantity by giving its magnitude and its direction.

Examples (1) **Displacement**.

A car travels $\underbrace{\text{20 miles}}_{\text{magnitude}}$ $\underbrace{\text{northwards}}_{\text{direction}}$

(2) **Force** of $\underbrace{\text{40 lbs}}_{\text{magnitude}}$ acts on a body $\underbrace{\text{vertically downwards.}}_{\text{direction}}$

(3) **Velocity**

A train travels at $\underbrace{\text{50 mph}}_{\text{magnitude}}$ $\underbrace{60^\circ\text{east of north.}}_{\text{direction}}$

Graphical Representation of Vector Quantities

We represent a vector quantity by a line segment drawn to scale and with an arrow at one end. The length of the arrow represents the magnitude of the vector and the direction of the arrow head represents the direction of the vector. The above drawing, shown below, can be done by using a ruler and a protractor.
We must specify the scalar unit used.

Examples (a) Represent graphically a force of 10 lbs in a direction 60° with the positive x-axis

Solution See Figure below

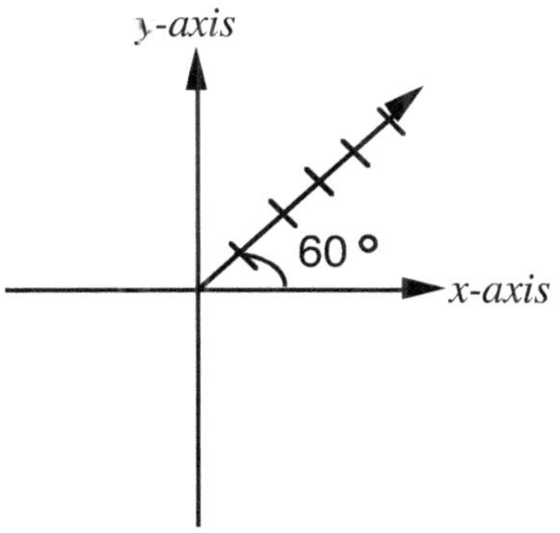

Scale: 1 unit = 2 lb

(b) Represent graphically a displacement of 40 meters in direction -45° below the horizontal.

Solution See Figure

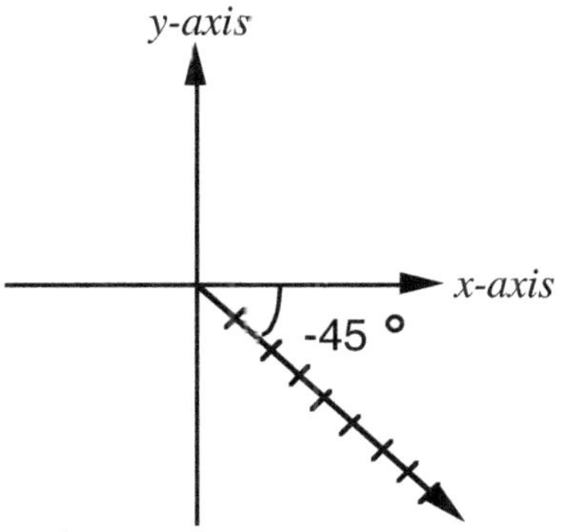

Scale: 1 unit = 5 meters

Symbolic Representation of Vectors

We must note the following differences carefully when reading and writing vectors symbolically:

In writing as done on chalkboard or in notebook by hand

We indicate that a quantity is a vector by placing an arrow over the symbol(s) (letter (s)).

Examples: (1) Using the letter F to represent a vector, we would write $\vec{F}$.

(2) Using the letter d to represent a vector we would write $\vec{d}$.

(3) Similarly in using the letter x we write $\vec{x}$.

(4)- Using two upper-case letters O and P, we write $\overrightarrow{OP}$ (Fig. ...)

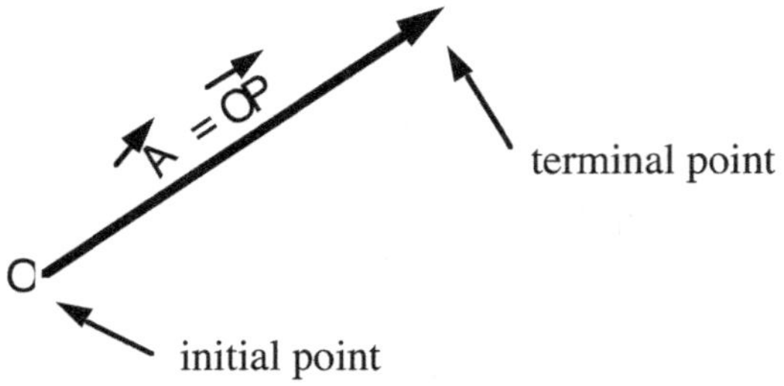

The **magnitude** of the vector, $\vec{F}$, symbolized by $|\vec{F}| = F$
Written without the arrow, the letter represents the magnitude of the vector.

In printing as it is done in textbooks

A bold face letter is used to indicate that a quantity is a vector.

Examples (1) The vector $\vec{F}$ would be printed **F** (bold face), with magnitude F.

(2) The vector $\vec{V}$ would be printed **V** (bold face)m with magnitude V.

Lesson 25 Exercises

1. Distinguish between a scalar quantity and a vector quantity.

2. How do we add scalar quantities, and how do we add vector quantities?

Lesson 26

Addition (Sum, Resultant, or Composition) of Vectors

We add vectors by geometric methods. However, if the vectors have the same line of action then the magnitude of the resultant can be found algebraically. The geometric methods are either graphical or analytic.

Graphically, we can use:

1. The vector polygon method (for any number of vectors) ;

2. The parallelogram law method (for two vectors at a time).

Analytically, we can use:

1. The law of cosines and the law of sines (for two vectors at a time) and

2. The component method (for any number of vectors). The component method is the one that we shall use most since it can be applied to any number of vectors at a time.

Definition **Resultant**

The resultant or the sum of a number of vectors is that single vector which when acting alone (so far as motion is concerned) has the same effect as the original vectors acting together.

How to Resolve ("break-up") a Vector into Rectangular components

Definition A **component** of a vector in a specified direction is the effective value of the vector in the specified direction. By **rectangular components**, we mean the components must be at right angles (perpendicular) to each other.

Procedure: We will draw a rectangle on the vector being resolved.

Step 1: Draw the rectangle so that the vector being resolved forms the diagonal of the rectangle and that the vector components will form the sides (adjacent sides) of the rectangle.

Step 2: Use the properties of a right triangle to find the components as illustrated in the following examples.

Example 1 Given a vector **F** acting at an angle θ_x with the positive x-axis, resolve **F** into rectangular components along the x- and y-axes.

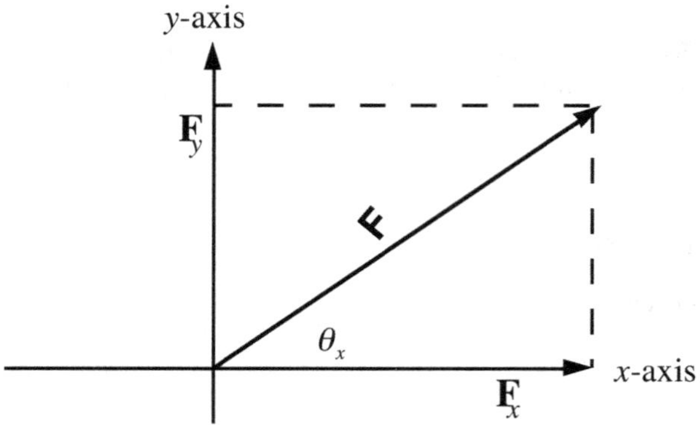

Fig 1

Solution We will call the component in the x-direction (along the x-axis), the x-component and the component in the y-direction (along the y-axis), the y-component (see Fig. 1)

Let the x-component in the positive x-direction (to the right) be positive. Then the x-component in the negative x-direction (to the left) would be negative. Similarly, let the y-component in the positive y-direction (up) be positive. Then the y-component in the negative y-direction (down) is negative. Note that the choice of positive and negative directions is arbitrary and that when convenient we may reverse the positivity or negativity of any of the directions. However we must be consistent with the choice of directions in any particular problem.

Continuing, and applying right-triangle trigonometry:
The horizontal component (x-component) $\mathbf{F}_x = +F\cos\theta_x$ (the plus sign indicates the positive direction)
The vertical component (y-component) $\mathbf{F}_x = +F\sin\theta_x$ (the plus sign indicates the positive direction)
(**Note** that if θ, say θ_y, is between **F** and the y-axis, then $\mathbf{F}_x = +F\sin\theta_y$ and $\mathbf{F}_y = +F\cos\theta_y$)
Continuing,

$$F^2 = F_x^2 + F_y^2 \quad \text{(Applying the Pythagorean Theorem)}$$

$$\tan\theta_x = \frac{F_y}{F_x} \text{ or } \theta_x = \tan^{-1}\left(\frac{F_y}{F_x}\right) \quad \text{(We will use these relationships in the next section)}$$

Example 2 Given that $F = 10$ lbs, $\theta_x = 30°$, resolve **F** into horizontal and vertical components.

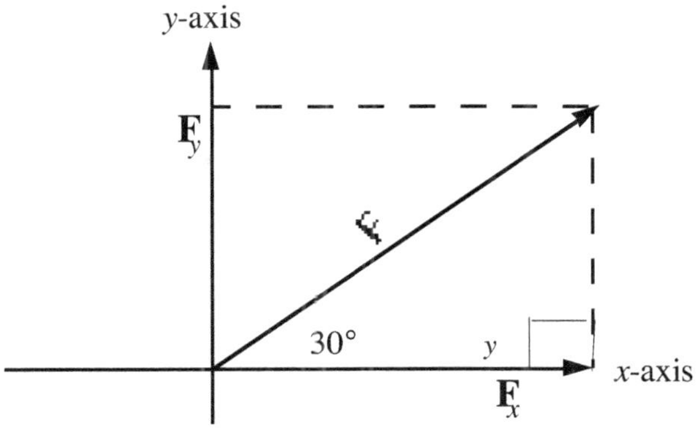

Solution

$\mathbf{F}_x = +(10\ lb)\cos 30°$

$= +10(.866)\ lb \qquad (\cos 30° = .866)$

$\mathbf{F}_x = +8.66\ lb$ (the plus sign indicates that the component is in the positive x - direction)

$\mathbf{F}_y = +(10\ lb)\sin 30°$

$= +10(.5)\ lb \qquad (\sin 30° = .5)$

$\mathbf{F}_x = +5\ lb$ (the plus sign indicates that the component is in the positive y - direction)

Therefore, the x-component = 8.66 lbs to the right; and the y-component is 5 lbs up.

Example 2 Resolve into horizontal and vertical components: F = 40 units, inclined at $\theta_x = 60°$ below the horizontal, as shown in **Figure** below.

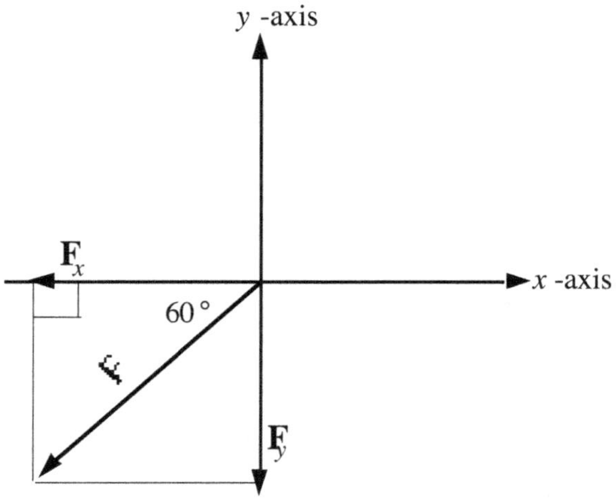

Solution Assume the horizontal components to the right are positive, and vertical components up are positive; then the horizontal components to the left are negative and the vertical components down are negative.

Horizontal component, $\mathbf{F}_x = -F\cos 60°$units

$= -(40)\cos 60°$ units

$= -(40)(.5)$ units $\qquad (\cos 60° = .5)$

$\mathbf{F}_x = -20$ units

(the minus sign shows that this component is in the negative x - direction)

Vertical component, $\mathbf{F}_y = -F\sin 60°$units

$= -(40)\sin 60°$ units

$= -(40)(.866)$ units $\qquad (\sin 60° = .866)$

$= -34.64$ units

$\mathbf{F}_y = -34.64$ units

(the minus sign shows that this component is in the negative y - direction)

The horizontal component is 20 units to the left; and the vertical component is 34.64 units down.

The main applications of the resolution of vectors into rectangular components are (a) to find the resultant (sum) of a number of vectors (displacements, forces, velocities, accelerations, etc.) and (b) to solve problems in Physics.

Component Method of Vector Addition

Procedure

Step 1: Resolve each vector into rectangular components (say, the horizontal component and the vertical component) along any convenient pair of axes.
(say x-and y-axes). We usually call the horizontal component the x-component and the vertical component the y-component.
Note that the signs of the non-zero components are either positive or negative.

Step 2: Add the x-components, noting their signs. Let the sum of the x-components be R_x.

Step 3: Add the y-components, noting their signs. Let the sum of the y-components be R_y.

Step 4: Draw R_x and R_y and let their resultant (sum) be R. The magnitude of R is given by:

$$R = \sqrt{R_x^2 + R_y^2} \qquad \text{(by the Pythagorean theorem)}$$

Step 5: Find θ, the standard-position angle giving the direction of the resultant.
Let θ_{ref} be the reference angle (the angle between the x-axis and the resultant).

$$\tan\theta_{ref} = \frac{R_y}{R_x} \quad \text{or}$$

$$\theta_{ref} = \tan^{-1}\left(\frac{R_y}{R_x}\right)$$

(θ_{ref} can be found from calculator or tables.)

Note that $\tan\theta$ is positive either in the first quadrant or in the third quadrant, but negative in either the second or fourth quadrant. We consider four cases of the location of θ_{ref} or θ.

Case 1: Both R_x and R_y are positive. In this case, θ_{ref} is in the first quadrant, and $\theta = \theta_{\text{ref}}$.

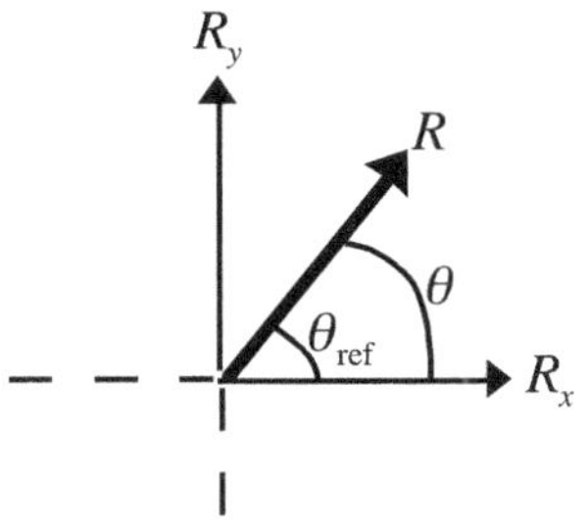

Case 2: Both R_x and R_y are negative. In this case, θ_{ref} is in the third quadrant, and $\theta = \theta_{\text{ref}} + 180°$.

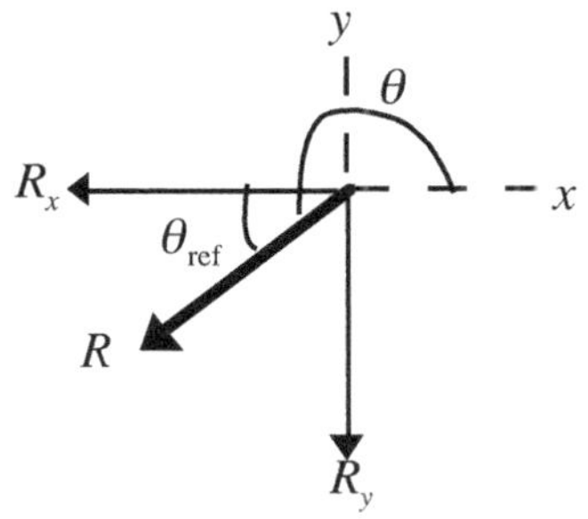

Note that saying that θ is in a specified quadrant means that the terminal side of θ lies in that quadrant.

Case 3: R_x is negative and R_y is positive. In this case, θ_{ref} is in the second quadrant and

$\theta = 180° - \theta_{\text{ref}}$

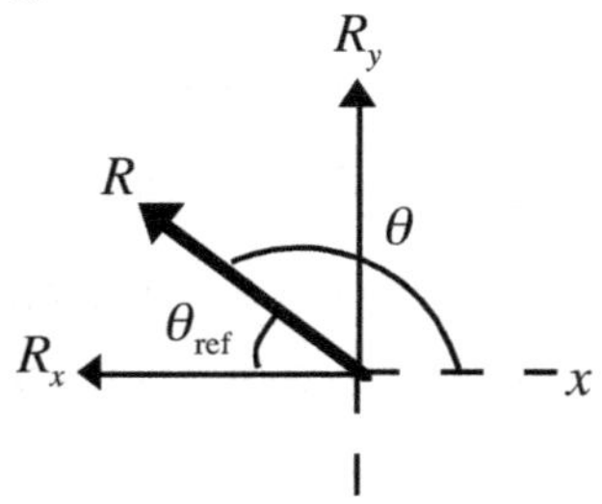

Case 4: R_x is positive and R_y is negative . In this case, θ_{ref} is in the fourth quadrant and

$\theta = 360° - \theta_{\text{ref}}$

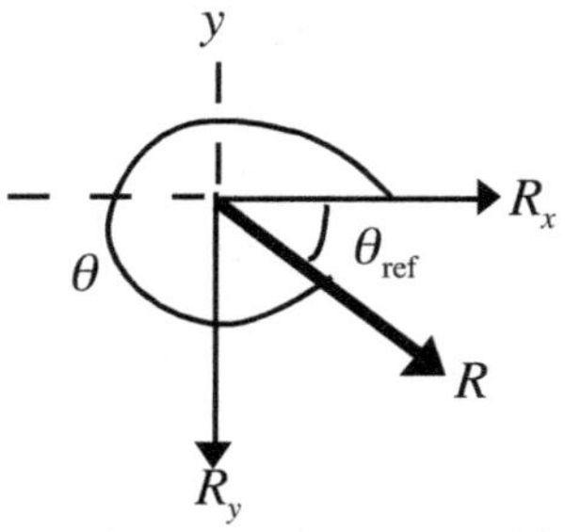

Note: Stating the direction of the resultant

Sometimes, after determining θ_{ref} (instead of stating the standard-position angle), the direction of the resultant may be specified with reference to one of the vectors (involved in the vector addition), or with respect to either the *x*- or *y*-axis.

Example 3 Find the resultant of the following forces acting on a particle at the origin. $\mathbf{F}_1 = 200$ units at 60° to the horizontal; $\mathbf{F}_2 = 120$ units along the positive x - axis; $\mathbf{F}_3 = 150$ units at 45° to the horizontal.

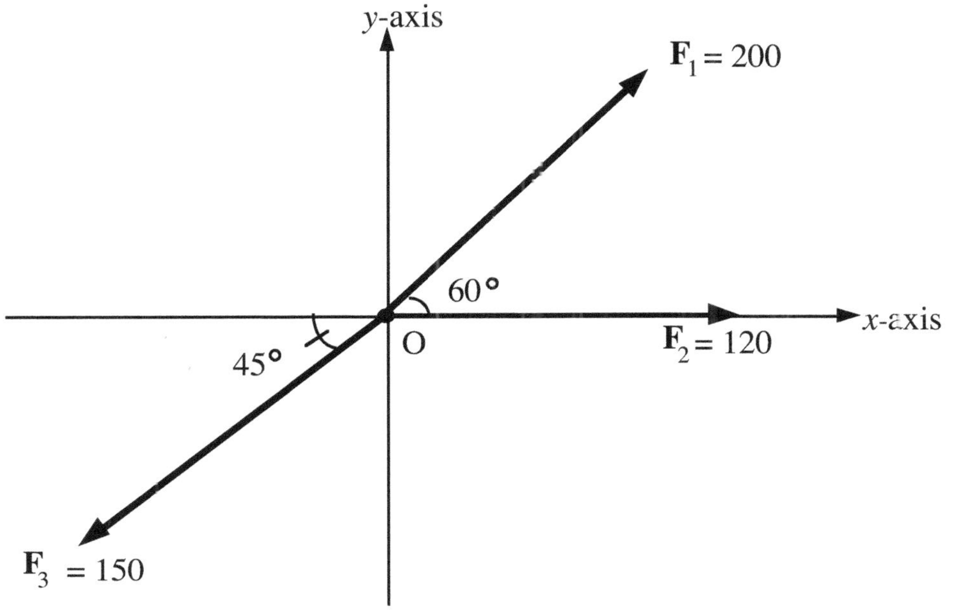

Solution Let the forces in the $+x$-direction be positive and let those in the $+y$-direction be positive.

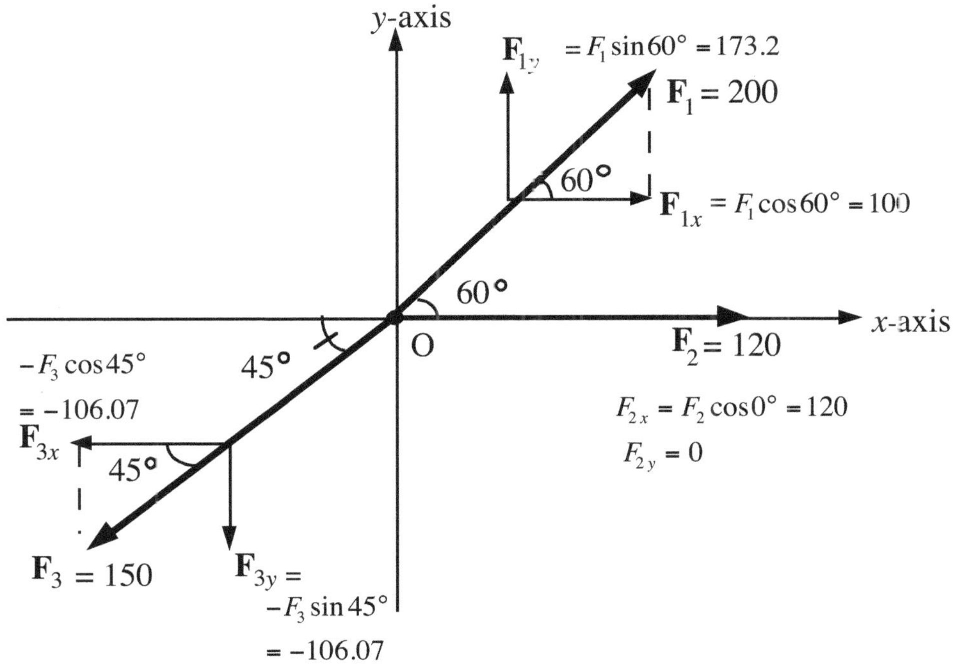

Step 1: Calculations on $\mathbf{F}_1$

$$\mathbf{F}_{1x} = +F_1 \cos 60° = +200(.5)$$
$$\mathbf{F}_{1x} = +100 \text{ units}$$

$$\mathbf{F}_{1y} = +F_1 \sin 60° = +200(.866)$$
$$\mathbf{F}_{1y} = +173.2 \text{ units}$$

Step 2: Calculations on $\mathbf{F}_2$

$$\mathbf{F}_{2x} = +F_2 \cos 0° = +120(1)$$
$$\mathbf{F}_{2x} = +120 \text{ units}$$

$$\mathbf{F}_{2y} = +F_2 \sin 0° = +120(0)$$
$$\mathbf{F}_{2y} = 0 \text{ units}$$

Note above that if the given force ($\mathbf{F}_{2y}$, in Step 2) is along the x-axis, then the y-component is zero, and we do not have to do any calculations. Similarly, if the given force were along the y-axis, the x-component would be zero.

Step 3: Calculations on $\mathbf{F}_3$

$$\mathbf{F}_{3x} = -F_3 \cos 45° = -(150)(.7071)$$
$$\mathbf{F}_{3x} = -106.07 \text{ units}$$

$$\mathbf{F}_{3y} = -F_3 \sin 45° = -(150)(.7071)$$
$$\mathbf{F}_{3y} = -106.07 \text{ units}$$

Step 4: Add the x-components

$$R_x = F_{1x} + F_{2x} + F_{3x}$$
$$= (+100) + (+120) + (-.106.07)$$
$$= 100 + 120 - 106.07$$
$$R_x = +113.93 \text{ units}$$

Step 5: Add the y-components

$$R_y = F_{1y} + F_{2y} + F_{3y}$$
$$= +(173.2) + 0 + (-106.07)$$
$$= +173.2 - 106.07$$
$$R_y = +67.13 \text{ units}$$

Step 6: Find the resultant R.

$$R = \sqrt{R_x^2 + R_y^2}$$
$$R = \sqrt{(113.93)^2 + (67.13)^2}$$
$$R = 132.24 \text{ units}$$

Step 7: Find the direction, θ., where θ is in standard position.

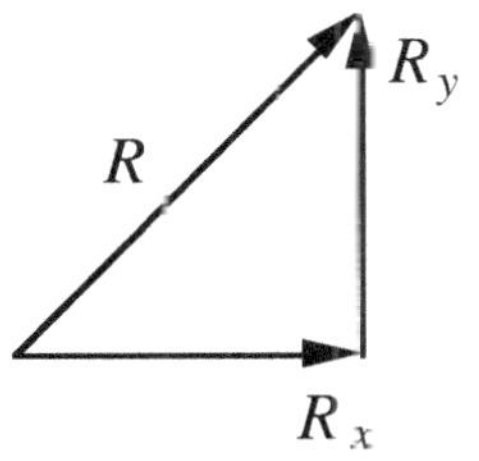

$$\tan\theta_{ref} = \frac{R_y}{R_x}$$
$$= \frac{67.13}{113.93}$$
$$= .589$$
$$\theta_{ref} = 30.5°$$

Since both R_x and R_y. are positive, $\theta = \theta_{ref}$. Therefore, $\theta = 30.5°$

The magnitude of the resultant force is 132.24 units, and is directed at standard-position angle 30.5°.

Finding the direction of the resultant

Let θ be a standard-position angle and let θ_{ref} be the reference angle.

If $\frac{R_y}{R_x}$ is positive, then θ is either in the first or third quadrant, since $\tan\theta$ is positive in these quadrants.

If $\frac{R_y}{R_x}$ is negative, θ is either in the second or fourth quadrant, since $\tan\theta$ is negative in these quadrants.

Example on finding θ, having found R_x and R_y.

Case 1: $R_x = 140$ and $R_y = 50$: For solution imitate Step 7 of Example 3 to obtain $\theta = 19.65°$.

Case 2: $R_x = -140$ and $R_y = -50$

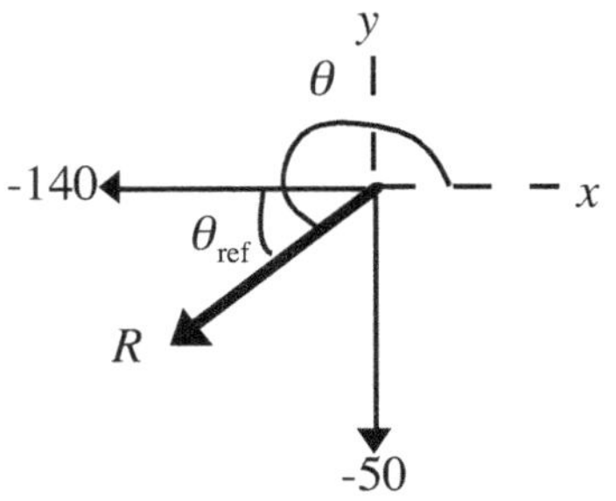

$$\tan\theta_{ref} = \frac{R_y}{R_x}$$
$$= \frac{-50}{-140}$$
$$= .36$$
$$\theta_{ref} = 19.65$$
$$\theta = 180 + 19.65 = \mathbf{199.65°}$$

Case 3 $R_x = 140$, and $R_y = -50$

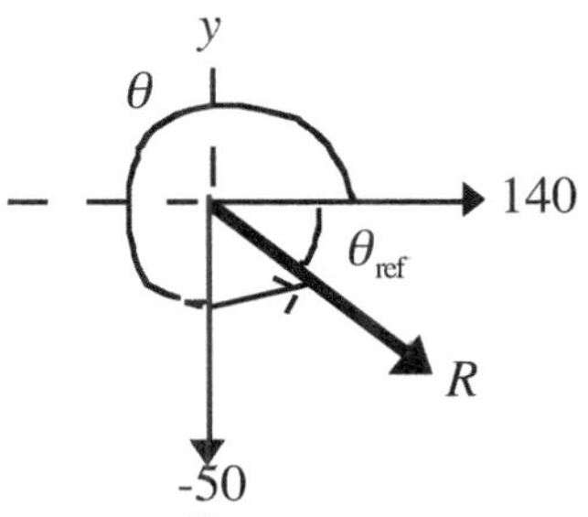

$$\tan\theta_{ref} = \frac{R_y}{R_x}$$
$$= \frac{-50}{140}$$
$$= -.36$$
$$\theta_{ref} = 19.65$$
$$\theta = 360 - 19.65 = \mathbf{340.35°}$$

Case 4: $R_x = -140$ units and $R_y = 50$ units

$$\tan\theta_{ref} = \frac{R_y}{R_x} = \frac{50}{-140} = -.36$$

$$\theta_{ref} = 19.65$$

$$\theta = 180 - 19.65 = \mathbf{160.35°}$$

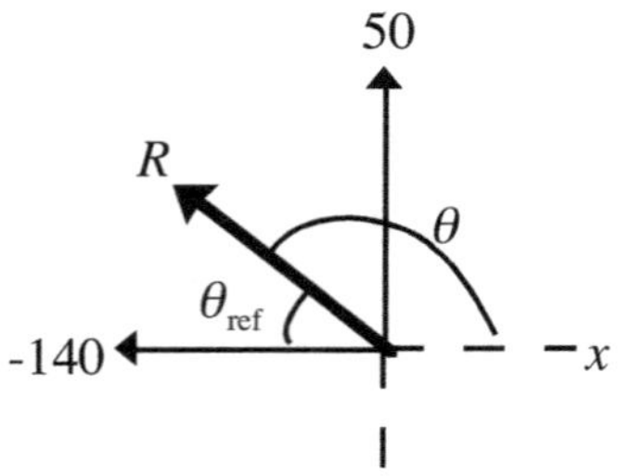

We repeat a previous note below:

Stating the direction of the resultant

Note that sometimes, after determining θ_{ref} (instead of stating the standard-position angle), the direction of the resultant may be specified with reference to one of the vectors (involved in the vector addition), or with respect to either the x- or y-axis.

Lesson 26 Exercises

A Resolve into horizontal and vertical components:
Suppose $F = 15$ lbs, $\theta_x = 60°$ resolve **F** into horizontal and vertical components.

Answer: $F_x = +7.5\ \ell b$ (to the right); $F_y = +13\ \ell b$ (up).

B Find the resultant of the following forces acting on an object at the origin.

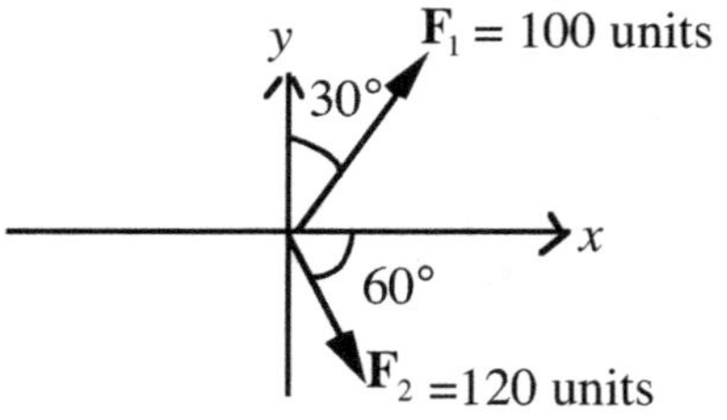

Answer: Resultant = 111 units, at $\theta_x = -9.0°$

CHAPTER 10
Trigonometry of Real Numbers

Lesson 27: **Definitions; Radian Measure; Arc Length; Reference Number**
Lesson 28: **Trig Functional Values of Angles and of Real Numbers**

Lesson 27
Definitions; Radian Measure; Arc Length; Reference Number

Definition of the "radian"

As shown in the figure below, one radian is the measure of an angle with its vertex at the center of a circle and with its sides intersecting an arc of a circle whose length is equal tc the radius of the circle.

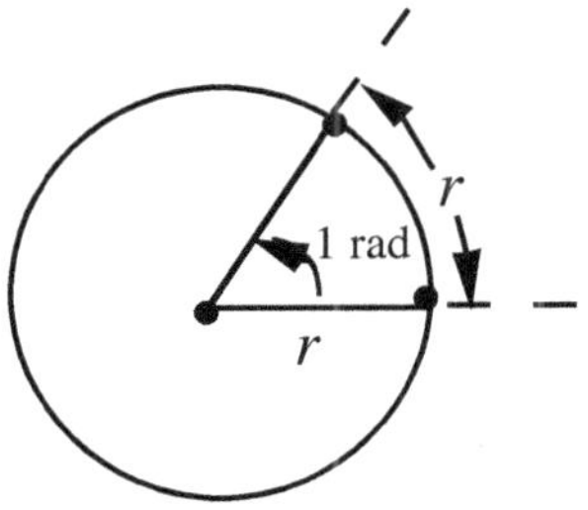

We can therefore say that radian measure = length of arc intersected by the sides of an angle divided by radius of the circle.

i.e., $\dfrac{\text{length of the arc intercepted by the sides of an angle}}{\text{the radius of the circle}}$

The circumference, C, of a circle of radius r is given by $C = 2\pi r$. Also, the circumference generates an angle of $360°$ at a center of the circle.

$$m(\theta) = \frac{\text{arc length}}{\text{radius}}, \text{ where } m(\theta) \text{ is the measure of } \theta.$$

In radians, $m(\theta) = \dfrac{2\pi r}{r}$
$= 2\pi$

Since the circumference subtends an angle of $360°$ at the center (for one revolution)

$2\pi = 360°$ (note that $\pi = 3.14$, approximately)
$\pi = 180°$
π radians $= 180°$
and 1 radian $= \dfrac{180°}{\pi}$ degrees
1 radian $= 57.3$ degrees approximately

Similarly, since 180 degrees $= \pi$ radians
1 degree $= \dfrac{\pi}{180°}$ radians
1 degree $= .017$ radians approximately.

Relationship between radian measure and arc length of a circle

Consider an arc (of a circle) of length S which subtends an angle θ at the center of a circle, where θ is in radians. By definition, θ in radians $= \dfrac{\text{arc length}}{\text{radius}}$

$$\theta = \frac{s}{r}$$

$$s = r\theta$$

(θ must be in radians. If θ is in degrees, it must be converted to radians before substituting in the formula.)

Note also that sometimes we call the angle involved in the above formula, the central angle.

Example 1

Find the arc length intercepted by the central angle of 25 radians in a circle of radius 15 cm.

Solution

$\theta = 25$ radians

$r = 15$ cm

$s = r\theta$

$s = (15)(25)$

$s = 375$ cm

Example 2

Find the arc length intercepted by the central angle of $25°$ in a circle of radius 15 in.

Step 1: Convert the angle in degrees to radians.

$$25° = \frac{25°}{180°}\pi \text{ radians}$$

Step 2: $s = r\theta$

$= (15)\left(\frac{25°}{180°}\right)\pi$ ins.

$= (15)\left(\frac{25°}{180°}\right)(3.14)$ ins. $\quad (\pi \approx 3.14)$

$= 6.54$ in.

Trigonometry of Real Numbers: Circular Functions

Some of the uses of trigonometry involve **real numbers** rather than angles. It is therefore important that we obtain definitions in trigonometry in terms of real numbers. Such definitions will reconcile the topics in calculus courses with those in trigonometry. We have already learned trigonometric definitions in terms of angles. We used these definitions in solving triangles.
However, there are problems in science to which we may apply trigonometric definitions even though triangles are not involved in these problems. It has been suggested that, perhaps, if there were no trigonometry of triangles, there would have to be trigonometry of real numbers. By first studying trigonometric definitions in terms of angles, we have become prejudiced with **angles** before we study trigonometry of **real numbers** We shall now make the transition from trigonometry of triangles to trigonometry of real numbers using the **unit circle**.

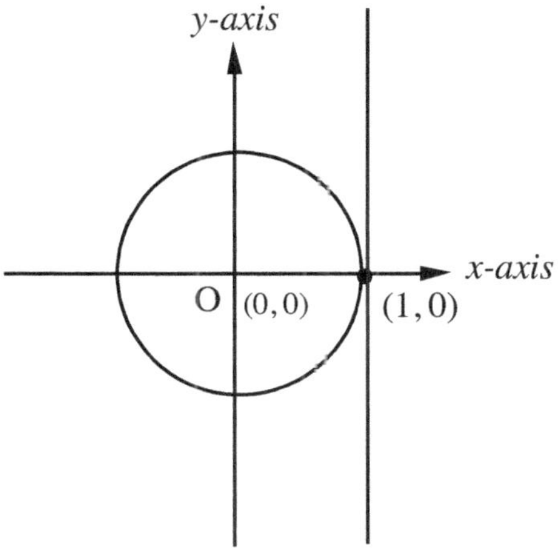

Consider a circle of radius 1 unit drawn in the x-y plane. We will use the same unit for measuring the arc length and the x - and y-axes. Consider also a real number line S with its origin at the point $(1,0)$ on the circle. Assuming that the real number line is flexible we wind the positive half of the line anticlockwise about the circle. Similarly, we wind the negative half of the line in a clockwise direction about the circle. Having completed a circle we withdraw the excess portions of the real number line. Then we obtain the figure below

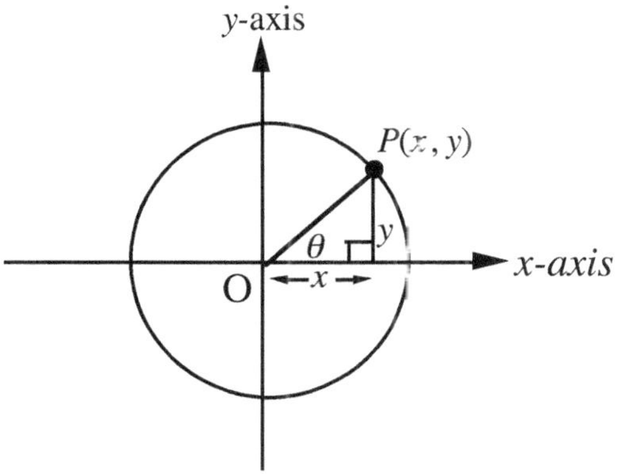

We observe that to every real number on the just wrapped real number line, there corresponds only one point on the circle.

Perhaps we may say that by wrapping the real number line around the unit circle, we obtain" the real number circle". We may thus define "the real number circle as a circle with radius 1 unit and a circumference of 6.28 units.

It is the above relationship which allows us to apply the trigonometry of angles to trigonometry of real numbers directly, when the angles are in radians.

Relationship between the angle θ in radians and the real number S.

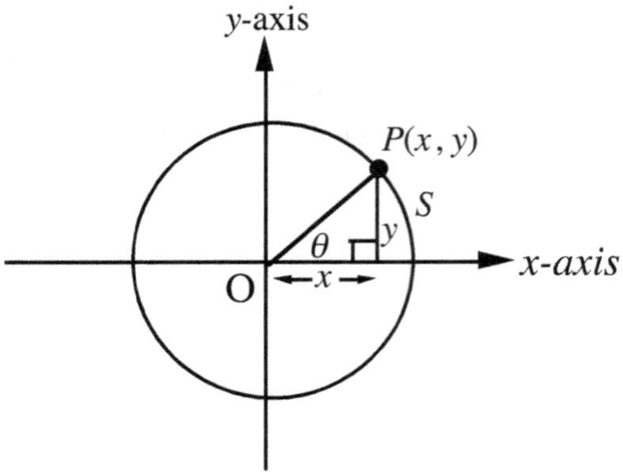

Consider a point $P(x, y)$ on the unit circle of radius r. Let θ radian be an angle in standard position. By definition, arc length $S = r\,\theta$ (where θ is in radians) . For the unit circle, $r = 1$.

and $S = (1)\ \theta$

$S = \theta$ (where θ is in radians)

Since for any real number S , θ (in radians) = S we can replace θ by S in the trigonometric definitions to obtain

$\sin\theta = \sin S$; $\cos\theta = \cos S$; $\tan\theta = \tan S$; $\csc\theta = \csc S$; $\sec\theta = \sec S$; $\cot\theta = \cot S$

From above and in future if S is any real number, then, for example $\sin S$ may mean either the sine of the real number S or the geometric sine of S (where S in radians). Therefore if S is written without any units it is understood that S is the real number S or the angle S in radians. However if we want to indicate the measure of the angle in degrees we must explicitly indicate the degree measure.

The trigonometric function of the real number S = the trigonometric function of the angle S in radians and the above definitions become

$\sin S = \sin S$ radians

$\cos S = \cos S$ radians

$\tan S = \tan S$ radians

$\csc S = \csc S$ radians

$\sec S = \sec S$ radians

$\cot S = \cot S$ radians

Note above that no conversions are needed if θ is in radians.
Because the circle was used in obtaining these functions, they are sometimes referred to **as circular functions**

The use of the unit circle to approximate values of sin S and cos S for $0 \le S \le 2\pi$

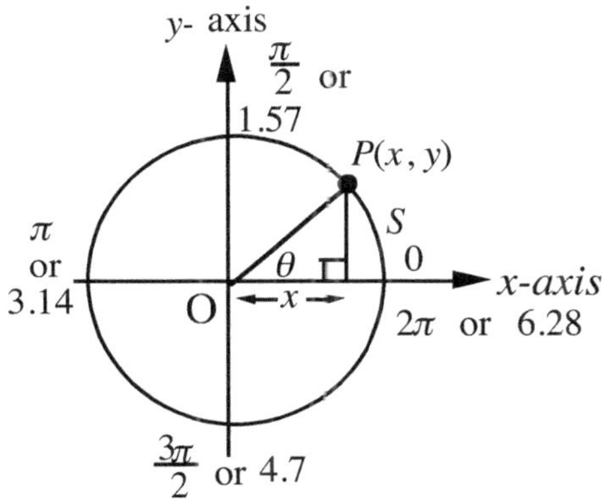

Consider a point $P(x, y)$ on the unit circle, where $x = f(s)$ and $y = f(s)$.

By definition, we have

$$\sin S = \frac{y}{r}$$

With $r = 1$, $\sin S = y$.

Similarly, $\cos \theta = \frac{x}{r}$ and

$$\cos S = x$$

$$\tan S = \frac{y}{x}$$

To find cos S we locate the value of S on the circumference of the unit circle and then read the x-coordinate of this point .(We may draw the perpendicular from this point to the x axis.)

Similarly, to find $\sin S$ we locate S on the circumference and then read the corresponding y-value.
(We may draw a horizontals to intersect the y-axis)

Reference number:

Let us recall the concept of the reference angle, where $0 \le \theta \le 2\pi$ and θ was in degrees or in radians. We have a similar case in dealing with real numbers but in this case, we have a **reference number** S which we used above in finding the trigonometric values for the cosine and sine functions.

The reference angle in this case is the acute angle in radians between the terminal side and the x-axis. Note that this acute angle, S_{ref}, is between 0 and 1.57 radians or between 0 and $\frac{\pi}{2}$.

Note below that $1.57 = \frac{\pi}{2} = 90°$; $3.14 = \pi = 180°$; $6.28 = 2\pi = 360°$

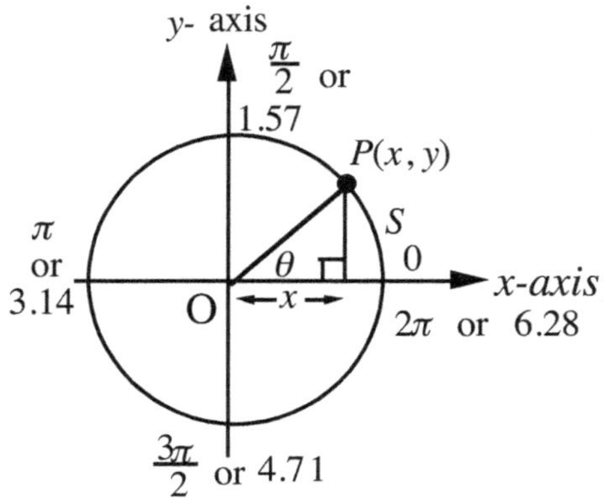

Example 1

Find the reference angle and the reference number for $S = 1.05$

Solution

Step 1: Sketch the angle 1.05 radian

Step 2: Determine the reference angle which in the above case is the same as the given angle.
The given angle is in the 1st quadrant. (It is acute and therefore less than 1.57 radians)
The reference angle, $S_{ref} = 1.05$.
The reference number is 1.05.

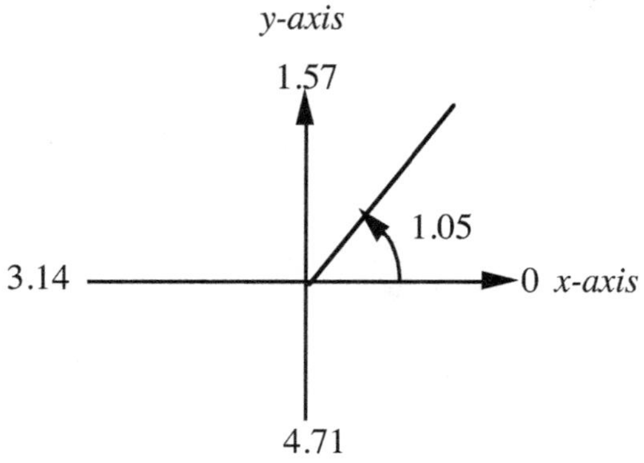

Example 2 Find the reference angle and reference number for $S = 2.09$

Solution

Step 1: Sketch (draw) the angle $S = 2.07$ radian

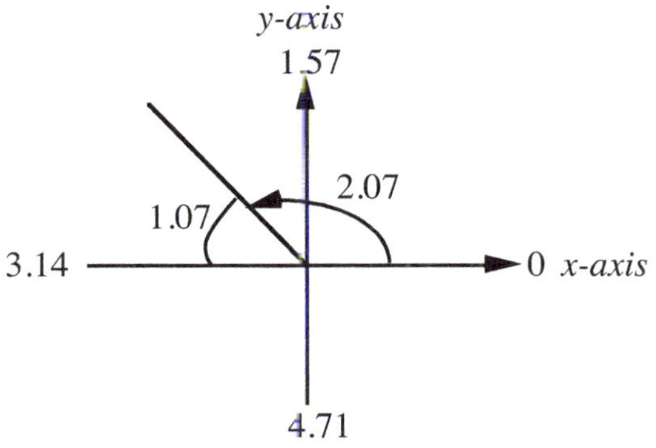

Step 2: The reference angle in this case is found by the difference between 3.14 and 2.07, noting that the reference angle is acute (less than 1.57 radians) and it is always between the terminal side and the horizontal axis. $\theta_{ref} = 3.14 - 2.09 = 1.07$ rads

The reference angle for 2.09 radians is 1.07 rad. The reference number is 1.07.

Also, note above that $\theta_{ref} = \pi - 2.09$ (since $\pi = 3.14$)

Example 3 Find the reference angle and reference number for 4.28.

Solution

Step 1: Draw the angle 4.28 in radians.

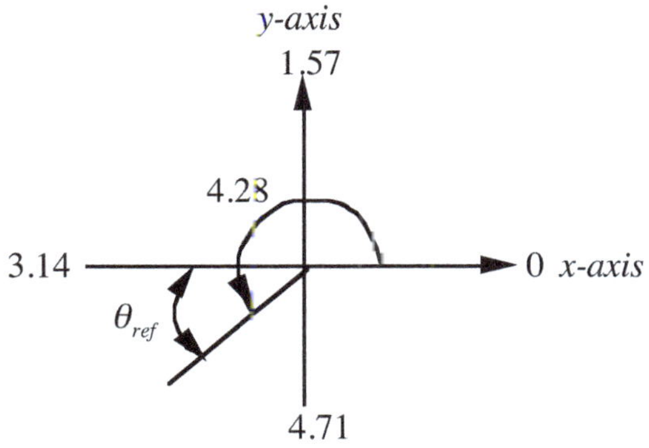

Step 2: Find the reference angle by difference

$\theta_{ref} = (4.28 - \pi)$

$\theta_{ref} = (4.28 - 3.14)$ rad (Note that $\pi = 3.14$ approximately.)

$= 1.14$ rad.

The reference angle for 4.28 radians is 1.14 radians

The reference number is 1.14.

Example 4: Find the reference angle and the reference number if $\theta = 5.15$ radians

Solution
Step 1: Sketch $\theta = 5.15$ radians

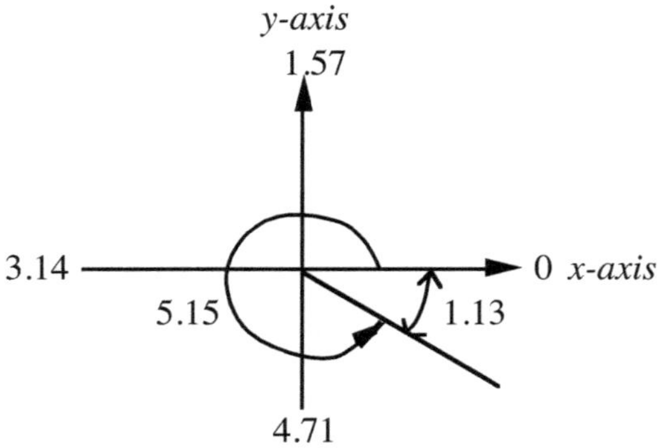

Step 2: $\theta_{ref} = (2\pi - 5.15)$ radians
$= (2(3.14) - 5.15)$ radians
$= (6.28 - 5.15)$ radians
$= 1.13$ radians

The reference angle for 5.15 radians is 1.13 radians.
The reference number is 1.13.

Example 5 Find the reference angle for $\theta = 10.42$.

Solution Refer to both Figures below.

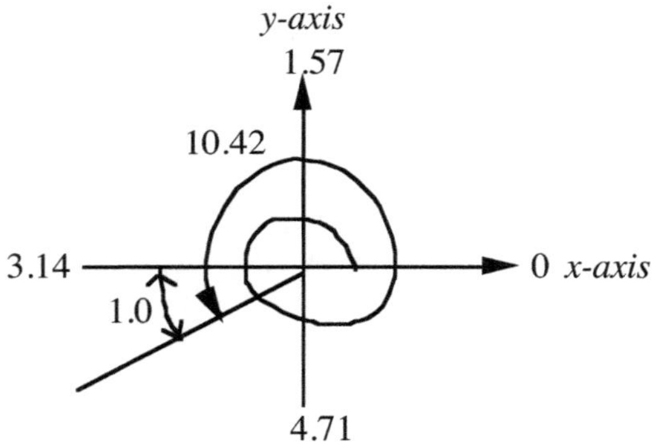

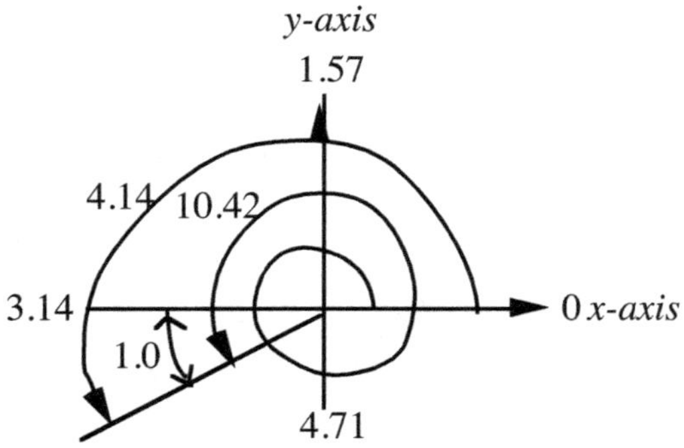

After the first revolution, there are $10.42 - 6.28 = 4.14$ radians left over.

The reference angle $= 4.14 - 3.14 = 1$ rad.
The reference number is 1

Example 6 Find the reference angle in radians and reference number if $\theta = -4.96$

Solution

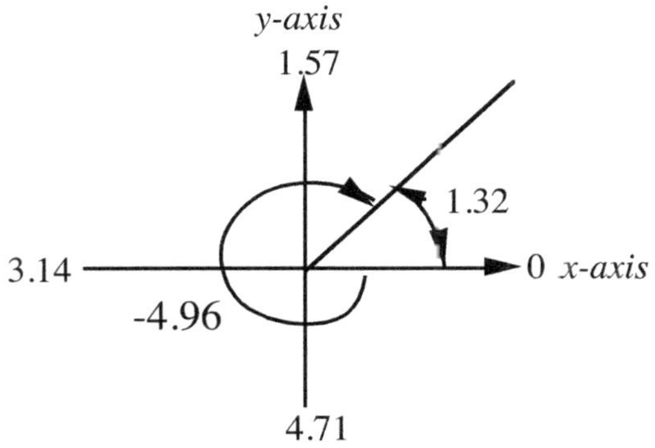

Step 1: Draw $\theta = -4.96$ in radians (Draw it clockwise.)

Step 2: The reference angle is is 6.28 - 4..96
$\theta_{ref} = 1.32$

The reference number = 1.32

Lesson 27 Exercises

Find the reference angle and the reference number for the following:
$(a)\ \theta = 5.4$ radians; $(b)\ 11.30$; $(c)\ -4.92$; $(d)\ -8.28$; $(e)\ 1.63$ radians.

Answers: (a) 0.88 radians, 0.88; (b) 1.26 radians, 1.26; (c) 1.36 radians, 1.36;
(d) 1.14 radians, 1.14; (e) 1.51 radians, 1.51.

Lesson 28

Trig Functional Values of Angles and of Real Numbers

Finding the trigonometric functional value of any angle in radians

Procedure:

Step 1: Sketch the given angle in radians

Step 2: Find the reference angle as described previously (See page168)

Step 3: From tables or calculator find the value of the trigonometric function.

Step 4: The value from Step 3 is always positive and we must determine the sign of the function according to the quadrant in which the terminal side of the angle lies (see page 123 for how to determine the signs of the functions in various quadrants.)

For some values of θ, using the calculator alone (as in Step 3) would be sufficient.

The following example parallels the Example on page123, and noting that

$1.57 = \frac{\pi}{2} = 90°$; $3.14 = \pi = 180°$; $6.28 = 2\pi = 360°$

Example Find sin θ, cos θ; tan θ if (1) if $\theta = 3$; (2) $\theta = 6.8$; (3) $\theta = -4.96$.

Solution

(1) Step 1: Draw $\theta = 3$

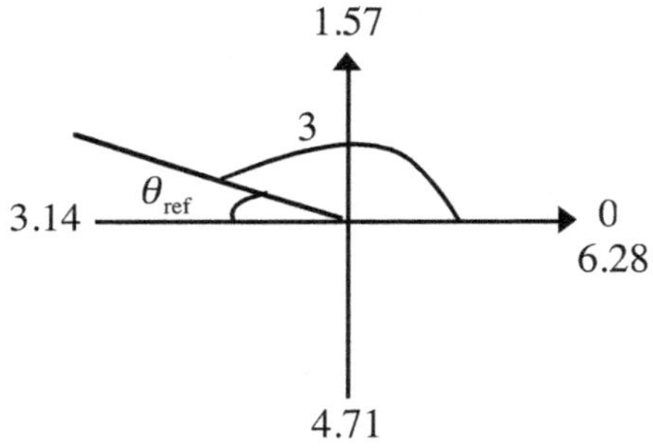

Step 2: Find the reference angle, θ_{ref}.

$\theta_{ref} = 3.14 - 3 = .14$

$\theta_{ref} = .14$

Step 3: sin 3 = 0.14 (sin 0.14 = 0.14<---- This result for the sine is interesting. In the future you will learn that for very small values of θ, $\sin\theta = \theta$, a fact we will apply when we cover the derivative of the sine function in calculus)

cos 3 = −0.99; tan 3 = −0.14

(2) For $\theta = 6.8$,

$\theta_{ref} = 6.8 - 6.28 = 0.52$ and

sin 6.8 = 0.49; cos 6.8 = 0.87; tan 6.8 = 0.57

(3) For $\theta = -4.96$,

$\theta_{ref} = 6.28 - 4.96 = 1.32$

$\theta_{ref} = 1.32$

sin (−4.96) = 0.97; cos (−4.96) = 0.25; tan (−4.96) = 4.0.

Trigonometric functional value of a real number S

Given the real number S, to find its trigonometric functional value, we assume that S is an angle in radians , find the reference angle and then find the value of the corresponding geometrical function of S in radians. (from tables, using a calculator, or from memory).

The domain of trigonometric functions consists of angles but the domain of circular functions consists of real numbers.

Relationship between θ (standard-position angle in radians) and θ_{ref}

1. If θ is in the **first quadrant**, then $\theta_{ref} = \theta$

2. If θ is in the **second quadrant**, then $\theta_{ref} = 3.14 - \theta$ or equivalently $\theta = 3.14 - \theta_{ref}$

3. If θ is in the **third quadrant**, then $\theta_{ref} = \theta - 3.14$ or equivalently $\theta = \theta_{ref} + 3.14$

4. If θ is in the **fourth quadrant**, then $\theta_{ref} = 6.28 - \theta = \theta$ or equivalently $\theta = 6.28 - \theta_{ref}$

Note that θ and θ_{ref} are always in the **same** quadrant. Also note that $\pi \approx 3.14$

Lesson 28 Exercises

A Find (a) $\sin\theta$; (b) $\cos\theta$ for the following:

(1) $\theta = 5.4$ radians; (2) $\theta = 11.30$ radians; (3) $\theta = -4.92$; (4) $\theta = -8.28$ radians; (5) $\theta = 1.63$ radians

Answers: 1. (a) -0.77, (b) 0.63; 2. (*a*) -0.95, (b) 0.30; 3. (*a*) 0.98, 0.21;
4. (a) -0.91, (*b*) -0.41; (5) (a) 1.0, (b) -0.06.

B Find θ in radians if $0 \le \theta \le 6.28$ given that:

(1) $\sin\theta = -.77$; (2) $\cos\theta = -0.06$.

Answers: (1) $\theta = 5.4$ rad, 4.02 rad; (2) $\theta = 1.63$ rad, 4.65 rad.

CHAPTER 11

Graphs of Trigonometric Functions

Lesson 29: **Introduction; Labeling the Coordinate Axes;Illustration of the Periodicity of Trigonometric Functions**

Lesson 30: **Sketching the Graph of** $y = \sin x$

Lesson 31: **Sketching the Graph of** $y = \cos x$

Lesson 32: **Graphs of Discontinuous Trigonometric Functions:** $y = \tan x$; $y = \csc x$; $y = \sec x$; $y = \cot x$

Lesson 33: **Direct Procedure for Sketching the Graphs of** $y - k = a \sin (bx - h)$, and $y - k = a \cos (bx - h)$

Lesson 29

Introduction; Labeling the Coordinate Axes; Illustration of the Periodicity of Trigonometric Functions

Introduction

Having mastered the sketching of polynomial and rational functions, we should **not** develop any phobias when we learn how to sketch trigonometric graphs. The approach and principles are similar to those outlined for graphing polynomial and rational functions.

As in any sketching in an x–y coordinate plane, what we need to do is to obtain important points (ordered pairs) on each curve, plot these points, and connect them to obtain the characteristic shape.

In the main, we should have the following objectives in mind when drawing trigonometric graphs.

1. From which starting point, through which points and to which endpoint do we draw the curve

2. The unit interval for the coordinate axes.

3. The number of the basic curves to draw: This depends on the domain and period of the function.

4. For each basic curve we need **five important points** (critical points) 2 end points (a starting point and a stopping point), 1 mid-point and two quarter-points .
Each point is an ordered pair (x, y). The five points include points at which the curve crosses (intersects) the x- and y-axes; and the highest point and the lowest point on the curve.
The highest and the lowest points are determined by a constant called the amplitude of the function.
Examples: The amplitude of $y = 3\sin x$ is 3 . The amplitude of $y = 2\sin x$ is 2; The amplitude of $y = \sin x$ is 1. The amplitude of $y = \frac{1}{2}\cos x$ is $\frac{1}{2}$.

There are a number of approaches for sketching the graphs of trigonometric functions. Each approach depends on the complexity of the given function.

Case 1: Sketching the graphs of the basic functions $y = \sin x$. $y = \cos x$, $y = \tan x$, $y = \csc x$, $y = \sec x$, $y = \cot x$. We will first cover the basic curves (one cycle), the sine, cosine, and the tangent functions as well as the basic curves for the cosecant, secant and cotangent functions. We will then cover drawing the graphs given the general equations of the sine and cosine functions. See Lesson 33.

Case 2 Sketching the graphs of functions such as $y = 3\sin x$, $y = \sin x + 2$, $y = \sin 3x$ (See Lesson 30)

Case 3: Sketching the graphs of the more general forms such as $y - 2 = 3\sin(4x + \pi)$, $y - 2 = 3\sin(4x + 1)$; $y - 2 = 3\cos(4x + \pi)$; $y - 2 = 3\cos(4x + 1)$; $y - 2 = 3\tan(4x + \pi)$; $y - 2 = 3\tan(4x + 1)$, See Lesson 33.

Case 4: Sketching the graphs of the inverse trigonometric functions such as $y = \text{Arcsin}\, x$, $y = \arccos x$,, $y = \text{Arctan}\, x$, $y = \text{arcsec}\, x$, $y = arc\csc x$, $y = \text{Arccot}\, x$.

Labeling the Coordinate Axes of Trigonometric Graphs

Sometimes, the difficulty students may have in graphing trigonometric functions is in labeling the coordinate-axes . We will therefore devote sometime to learn how to label these axes.

Labeling the number line both forwards and backwards

Suppose we want to label the number line or the x-axis with the unit interval $\frac{1}{2}$, then the end of the first interval is labeled $1 \bullet \frac{1}{2}$ or $\frac{1}{2}$, the end of the next interval is labeled $2 \bullet \frac{1}{2}$ or 1; the end of the next interval is labeled $3 \bullet \frac{1}{2}$ or $\frac{3}{2}$; the end of the fourth interval is labeled $4 \bullet \frac{1}{2}$ or 2. Notice the sequence of the factors 1, 2, 3, and 4 in determining multiples of the unit interval.
Note that in labeling the line, the convention is that all fractions must be reduced to lowest terms. For example, $4 \bullet \frac{1}{2}$ is normally labeled 2; $4 \bullet \frac{\pi}{2}$ is normally be labeled 2π; and similarly $3 \bullet \frac{1}{2}$ and $3 \bullet \frac{\pi}{2}$ are normally labeled $\frac{3}{2}$ and $\frac{3\pi}{2}$ respectively.

(a) **Using fractions**

Using fractions: (x-axis)

-2	$-\frac{3}{2}$	-1	$-\frac{1}{2}$	0	$\frac{1}{2}$	1	$\frac{3}{2}$	2	$\frac{5}{2}$
or	or	or	or		or	or	or	or	or
$-4 \bullet \frac{1}{2}$	$-3 \bullet \frac{1}{2}$	$-2 \bullet \frac{1}{2}$	$-1 \bullet \frac{1}{2}$		$1 \bullet \frac{1}{2}$	$2 \bullet \frac{1}{2}$	$3 \bullet \frac{1}{2}$	$4 \bullet \frac{1}{2}$	$5 \bullet \frac{1}{2}$

(b) **Using degrees**

If we use degrees for labeling, with the " unit interval $90°$, we obtain the figure below:

Using degrees (x-axis)

$-360°$	$-270°$	$-180°$	$-90°$	0	$90°$	$180°$	$270°$	$360°$	$450°$
or	or	or	or		or	or	or	or	or
$-4 \bullet 90°$	$-3 \bullet 90°$	$-2 \bullet 90°$	$-1 \bullet 90°$		$1 \bullet 90°$	$2 \bullet 90°$	$3 \bullet 90°$	$4 \bullet 90°$	$5 \bullet 90°$

(c) **Using π units**

If we use π units for labeling, with the " unit interval" $\frac{\pi}{2}$, we obtain the figure below:

Using π units: (x-axis)

-2π	$-\frac{3\pi}{2}$	$-\pi$	$-\frac{\pi}{2}$	0	$\frac{\pi}{2}$	π	$\frac{3\pi}{2}$	2π	$\frac{5\pi}{2}$
or	or	or	or		or	or	or	or	or
$-4 \bullet \frac{\pi}{2}$	$-3 \bullet \frac{\pi}{2}$	$-2 \bullet \frac{\pi}{2}$	$-1 \bullet \frac{\pi}{2}$		$1 \bullet \frac{\pi}{2}$	$2 \bullet \frac{\pi}{2}$	$3 \bullet \frac{\pi}{2}$	$4 \bullet \frac{\pi}{2}$	$5 \bullet \frac{\pi}{2}$

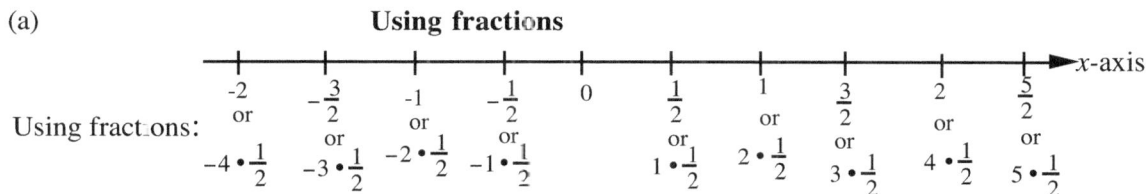

(d) **Using Decimals**

If we use decimals for labeling, with the "unit interval" 1.57, we obtain the figure below:

										x-axis
Using decimal units:	-6.28	-4.71	-3.14	-1.57	0	1.57	3.14	4.71	6.28	7.85
	or	or	or	or		or	or	or	or	or
	-4(1.57)	-3(1.57)	-2(1.57)	-1(1.57)		1(1.57)	2(1.57)	3(1.57)	4(1.57)	5(1.57)
Also:	-2π	$-\frac{3\pi}{2}$	$-\pi$	$-\frac{\pi}{2}$		$\frac{\pi}{2}$	π	$\frac{3\pi}{2}$	2π	$\frac{5\pi}{2}$

Note also that $1.57 = \frac{\pi}{2} = 90°$; $3.14 = \pi = 180°$; $6.28 = 2\pi = 360°$

Illustrating the periodicity of trigonometric functions

One common property the graphs of trigonometric functions have is the property of periodicity (regular repetition of the same y-values for different x-values.).We will use the graph of the sine function to illustrate this repetitive property. The other trigonometric functions also have repetitive properties.

The graph of $y = \sin x$

The graph of $y = \sin x$ (x in radians) is drawn below, using a table of values (or a set of ordered pairs). We will learn how to graph this function later, but for the meantime, let us familiarize ourselves with the basic properties. Observe the shape of the curve and its relationship with the axes (x- and y-axes). The period of $y = \sin x$ is 2π. Thus, the curve drawn will **repeat itself every** 2π units (Note: $\pi \approx 3.14$, and therefore $2\pi \approx 2(3.14) \approx 6.28$). Algebraically, from time to time, we obtain the same y-values when we substitute different x-values in $y = \sin x$.

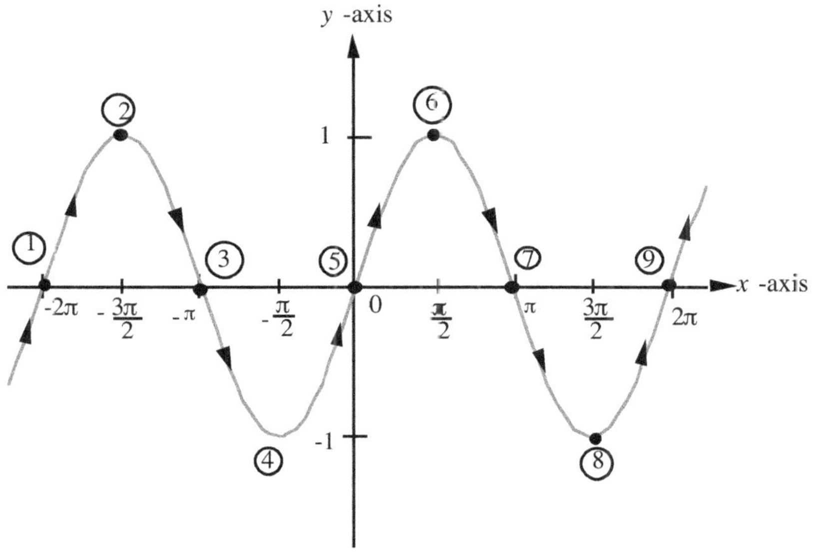

Figure

Graphical meaning of the phrase "**repeats itself**"

In the above figure, observe the curve at point **1**, with the coordinates $x = -2\pi$, $y = 0$, and note the arrow and its direction (the arrow is used for illustrative purposes only and we do not show the arrows when we draw these graphs). At point **1,** the curve is increasing (going up) and crossing the x- axis. At point **5** (with $x = 0$, $y = 0$) the curve does the same thing as it did at point **1**, namely it is increasing (going up) and crossing the x-axis again. Thus, at point **5**, the curve repeats what it did at point **1**, namely, crossing the x-axis and going up. The distance between point **1** and point **5** is the period of this function. In the above case, this period is 2π. Hence we say that the curve repeats itself every 2π (or 6.28 units). Similarly, at point **9**, the curve is increasing (going up) and crossing the x-axis. Thus, the curve repeats itself again at point **9**. Again, the distance between the points **5** and **9** is 2π units, the period of this function. Observe that repetition as explained above involves two simultaneous (at the same time) events, namely, crossing the x-axis and increasing (going up). Instead of starting from point **1** or from point **5**, we could also start from point **3**. Suppose, we start from point **3**, then observe that at this point, the curve is crossing the x-axis and decreasing (going down). At point **7**, the curve repeats the same two events occurring at point **3**, namely, crossing the x-axis and decreasing (going down). Thus, starting from point **3**, the curve repeats itself at point **7.** The distance between point **3** and point **7** is 2π units. Similarly, if the curve starts at point **2** (turning point), it will repeat the pattern (starting from the turning point and decreasing) at point **6.**

If the curve starts at point **4**, it will repeat at point **8**.

Therefore, to determine the period from the graph, we pick a starting point on the curve, observe two simultaneous events at this point and then observe the next time the same "simultaneous events" take place and then determine the distance between these two nearest "simultaneous- events" points.

Another property of the graph of the sine function is the **amplitude**. The amplitude determines the highest and the lowest points on the graph of this function. In the above figure, the amplitude is the vertical distance from any of the points marked **2**, **4**, **6**, or **8** to the x-axis.

Examples of amplitudes: The amplitude of $y = 3\sin x$ is 3 . The amplitude of $y = 2\sin x$ is 2; The amplitude of $y = \sin x$ is 1. The amplitude of $y = \frac{1}{2}\cos x$ is $\frac{1}{2}$.

In the next section, we will now learn how to sketch the graph of $y = \sin x$ for $0 \le x \le 2\pi$.

Lesson 29 Exercises

1. Use the following unit intervals to label the number line from 0° to 360°:
(a) 30°; (b) 40° , (c) 45°

2. Use the following unit intervals to label the number line from 0 to 2π: (a) $\frac{\pi}{3}$; (b) $\frac{\pi}{4}$.

3. Use the following unit interval to label the number line from 0 to 6: $\mathbf{\frac{1}{4}}$

4. Use the following unit interval to label the number line from 0 to 6.28: **1.57**

1.(a)

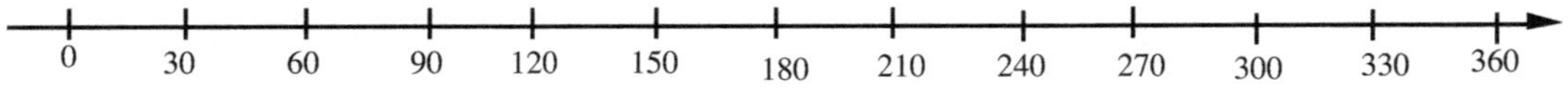

1.(b)

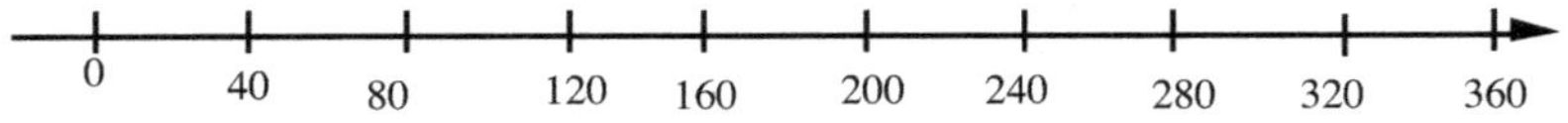

1.(c)

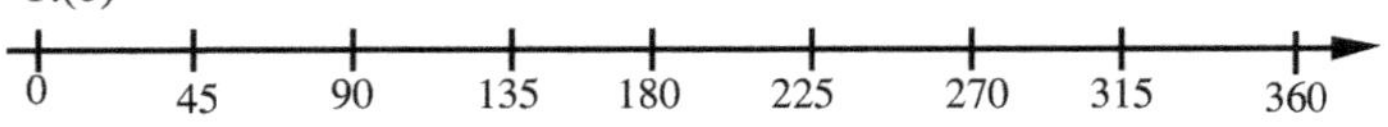

2.(a)

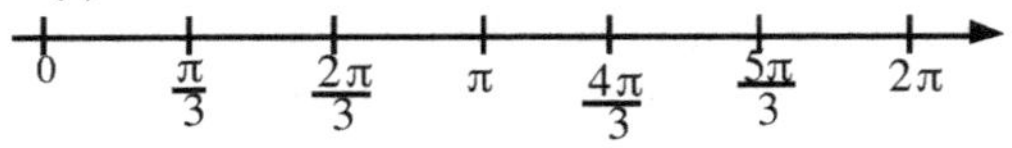

2.(b)

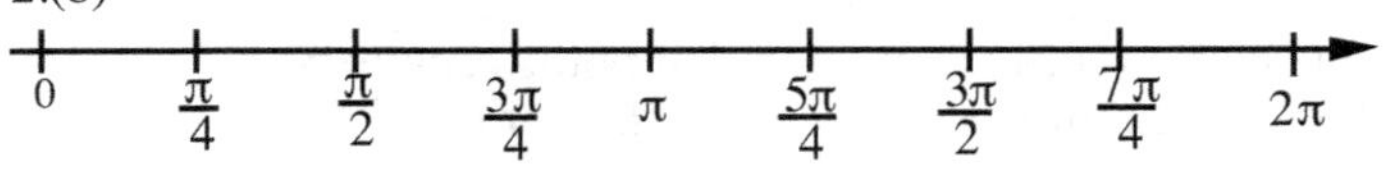

3.

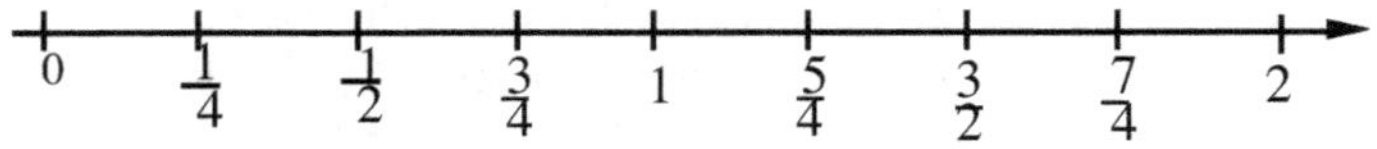

4.

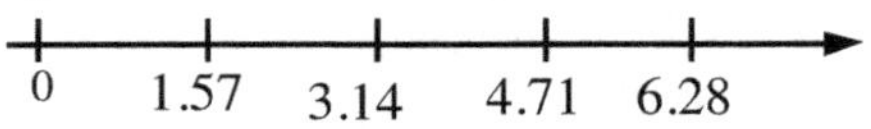

Lesson 30

Sketching the Graph of $y = \sin x$

To sketch the graph of $y = \sin x$ we must know the following basic properties of this curve:

1. Its **period** is 2π. Its **amplitude** which is 1.

2. Five important points (**critical points):** 2 end points (a starting point and a stopping point), 1 mid-point and two quarter-points are needed. Each point is an ordered pair (x, y). The five points include the x-intercepts, the y-intercept; and the amplitude. The amplitude determines the highest and the lowest points on the graph of this function

Example 1 Sketch the graph of $y = \sin x$ for $0 \le x \le 2\pi$.

Step 1: We construct a table containing these important points, using the formula $y = \sin x$

x	0	$\frac{\pi}{2}$	π	$\frac{3\pi}{2}$	2π
$y = \sin x$	0	1	0	-1	0

Step 2: From the origin and on the x-axis, mark-off a convenient interval and label its end-point 2π, the period of $y = \sin x$. Divide this interval into four equal intervals and label them with $\frac{\pi}{2}$, π, and $\frac{3\pi}{2}$, using Figure (c) p, 459 as a guide.

Step 3: On the y-axis, choosing a convenient interval, mark and label the points $y = 1$ and $y = -1$.

Step 4: Plot the points $(0,0)$, $(\frac{\pi}{2},1)$, $(\pi,0)$, $(\frac{3\pi}{2},-1)$ and $(2\pi,0)$. (Points on $y = \sin x$)

Step 5: Connect the points from Step 4 by a smooth curve to obtain the characteristic shape of $y = \sin x$. (Figure)

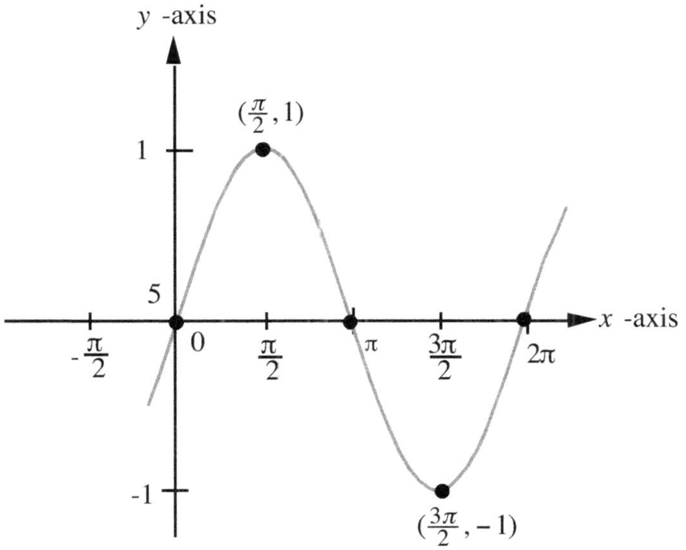

Figure: The graph of $y = \sin x$

Practice sketching this graph a number of times, noting its features, so that this graph (picture) becomes part of your "trigonometric vocabulary", just like the multiplication tables.

Example 2 Sketch the graph of $y = \sin x$ for $-2\pi \le x \le 2\pi$

Solution: The steps are the same as those for Example 1 above, except that we extend the graph to the left of the *y*-axis. .

Step 1: From the origin and on the *x*-axis, mark-off a convenient interval and label its end-point 2π, the period of $y = \sin x$.

Step 2: Divide this interval from 0 to 2π into four equal intervals and label them with $0, \frac{\pi}{2}, \pi$, and $\frac{3\pi}{2}$ (See also Figure (c), page 175, as a guide.)

Step 3: Similarly, to the left of the *y*-axis, on the *x*-axis, mark-off another interval 2π units, and label this endpoint -2π. Divide this interval into four equal intervals, and label this interval on the negative *x*-axis with the negatives of the *x*-values from Step 2.

Step 4: On the *y*-axis, using a convenient interval, mark-off and label the points $y = 1$ and $y = -1$.

Step 5: Plot the points $(0,0)$, $(\frac{\pi}{2},1)$, $(\pi,0)$, $(\frac{3\pi}{2},-1)$, $(2\pi,0)$ and also plot $(-\frac{\pi}{2},-1)$, $(-\pi,0)$, $(-\frac{3\pi}{2},1)$ and $(-2\pi,0)$.

Step 7: Connect the plotted points, from left to right, by a smooth curve to obtain the characteristic shape of $y = \sin x$. (two cycles) (Figure)

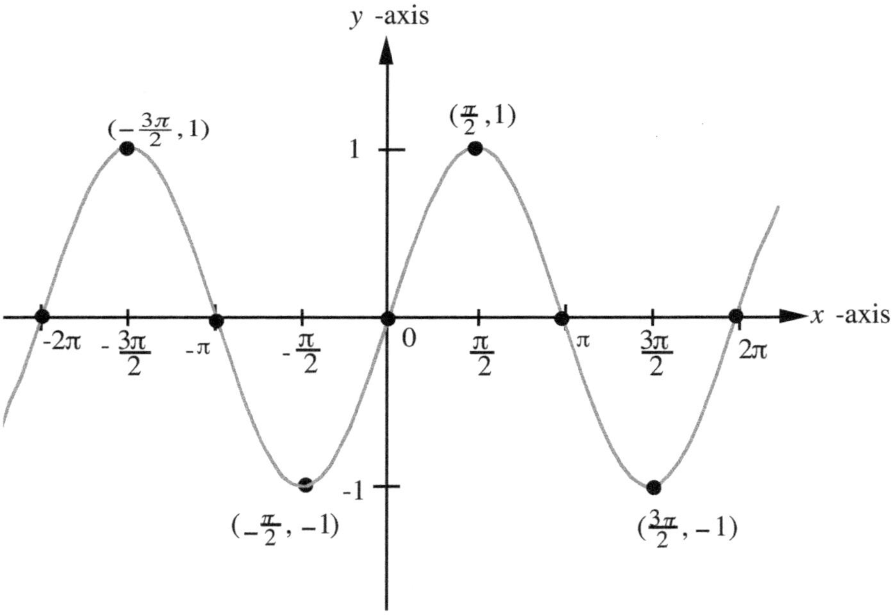

Figure : The graph of $y = \sin x$ for $-2\pi \le x \le 2\pi$

Example 3 Given $y = \sin x$, draw the graph of $y = \sin x + 2$

Rewrite $y = \sin x + 2$ as $y - 2 = \sin x$ and apply the transformation method. See **Figure 3.**

Here, $k = 2, h = 0, a = 1, b = 1$ (also review p. 75-76)

$y = \sin x$		$y - 2 = \sin x$		
x_0	y_0	$x_n = x_0$	$y_n = y_0(a) + k$	(x_n, y_n)
0	0	0	$0(1) + 2 = 2$	$(0,2)$
$\frac{\pi}{2}$	1	$\frac{\pi}{2}$	$1(1) + 2 = 1 + 2 = 3$	$(\frac{\pi}{2},3)$
π	0	π	$0(1) + 2 = 0 + 2 = 2$	$(\pi, 2)$
$\frac{3\pi}{2}$	-1	$\frac{3\pi}{2}$	$-1(1) + 2 = -1 + 2 = 1$	$(\frac{3\pi}{2},1)$
2π	0	2π	$0(1) + 2 = 0 + 2 = 2$	$(2\pi, 2)$

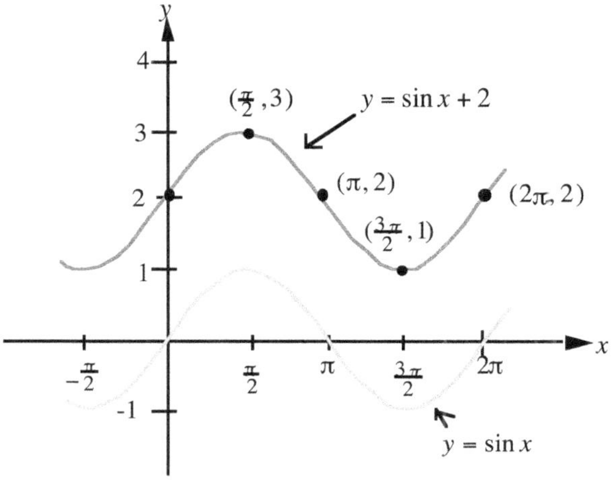

Figure 3: The graph of $y = \sin x + 2$

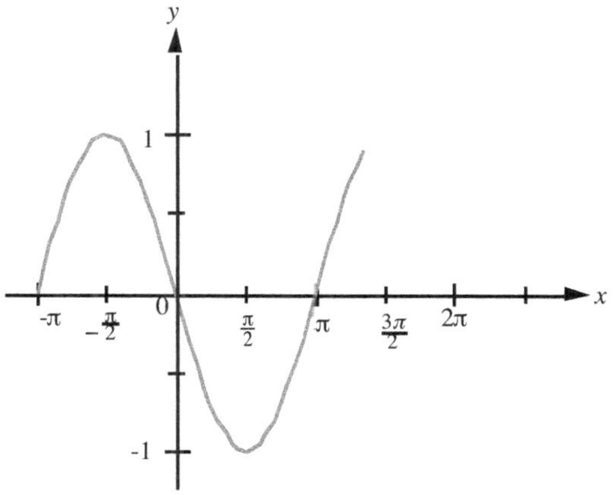

Figure 4: Graph of $y = \sin(x + \pi)$

Example 4 : Sketch the graph of $y = \sin(x + \pi)$

Solution We can obtain the graph of $y = \sin(x + \pi)$ from the graph of $y = \sin x$ by moving each critical point π units to the left, keeping the y-coordinates unchanged. More formally, we apply the transformation method. See **Figure 4.** $y = \sin(x + \pi)$. Here, $k = 0, h = -\pi, a = 1, b = 1$ (also review p.75-76)

$y = \sin x$		$y = \sin(x + \pi)$		
x_0	y_0	$x_n = \frac{x_0}{b} + h$	$y_n = y_0$	(x_n, y_n)
0	0	$\frac{0}{1} - \pi = 0 - \pi = -\pi$	0	$(-\pi,0)$
$\frac{\pi}{2}$	1	$\frac{\pi}{2 \cdot 1} - \pi = \frac{\pi}{2} - \pi = -\frac{\pi}{2}$	1	$(-\frac{\pi}{2},1)$
π	0	$\frac{\pi}{1} - \pi = 0$	0	$(0,0)$
$\frac{3\pi}{2}$	-1	$\frac{3\pi}{2 \cdot 1} - \pi = \frac{3\pi}{2} - \pi = \frac{\pi}{2}$	-1	$(\frac{\pi}{2},-1)$
2π	0	$\frac{2\pi}{1} - \pi = 2\pi - \pi = \pi$	0	$(\pi,0)$

Example 5 Sketch the graph of $y = 5\sin(2x - 3)$

Solution See figure below.
We can apply the transformation method as in the previous examples
Rewrite $y = 5\sin(2x - 3)$ as $y = 5\sin[2(x - \frac{3}{2})]$ and apply the transformation method.

Here, $k = 0$, $h = \frac{3}{2}$, $a = 5$, $b = 2$. Construct a table of values and graph . **Figure 5**

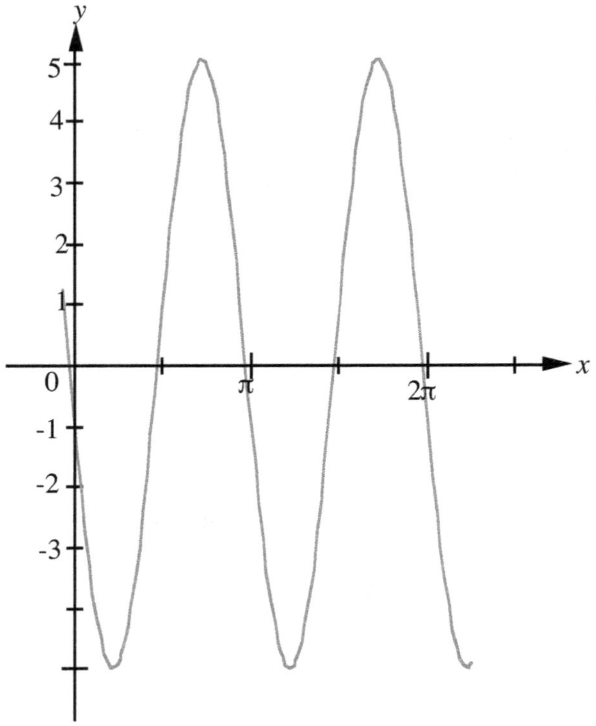

Figure : The Graph of $y = 5\sin(2x - 3)$

Now, having more or less mastered the sketching of the sine curves, let us go on to learn to sketch t he cosine, tangent, secant and the cosecant curves. In Lesson 33, we will learn a direct method for sketching the graphs of trigonometric functions such as $y - k = a\sin(bx - h)$ and $y - k = a\cos(bx - h)$. However always keep in mind the transformation method as well as the general method for sketching the graphs of functions.
We will use the same principles that we used in sketching sine curves. The only thing new we have to know are the properties of each curve.
We will begin by first drawing the graph of $y = \cos x$ using values from memory, tables or calculator.

Lesson 30 Exercises

Sketch the graphs of the following:

1. $y = \sin x$ for $0 \le x \le 2\pi$; **2.** $y = -\sin x$ for $0 \le x \le 2\pi$;; **3.** $y = 3\sin x$ for $0 \le x \le 2\pi$; ;

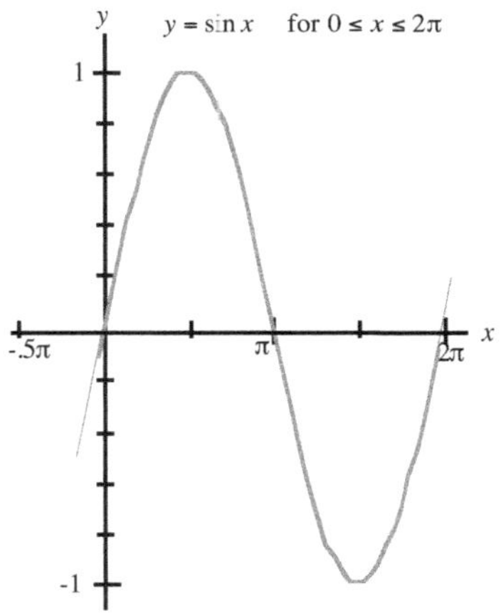

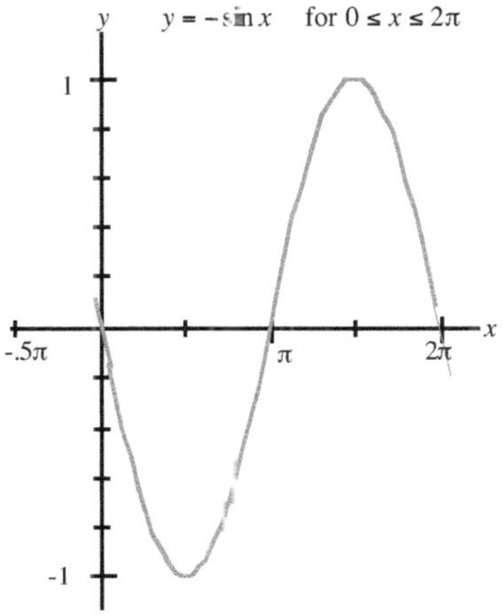

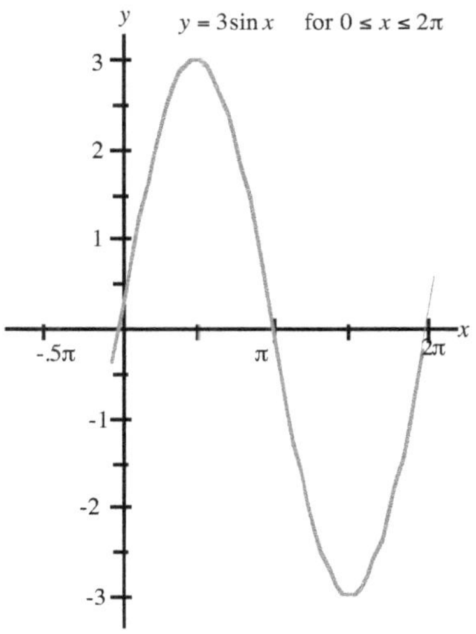

Lesson 31
Sketching the Graph of $y = \cos x$

Properties of the cosine curve $y = \cos x$

(1) The period of $y = \cos x$ is 2π, the same period as that of $y = \sin x$.
(2) The amplitude is I.

Observe in **Figure** that if the curve begins from the point (0, 1) marked **2**, on the y-axis, then (in going from left to right) we observe that, at the point $(2\pi, 1)$, marked **5**, the curve repeats. Thus the curve is again at its maximum point at **5** as it was at **2**. The period is the distance between point **2** and point **5,** and it is 2π, the period of the cosine curve. However, if we had started drawing the curve from the point $(-\frac{\pi}{2}, 0)$ marked **1,** then the next time the curve repeats. would be at point **4**. The period in this case will be the distance between **1** and **4**. This period is also 2π, the same period as before.

In fact, from trigonometric identities, $\cos x = \sin(x + \frac{\pi}{2})$

This means that given the sketch of the $y = \sin x$, to obtain the sketch of $y = \cos x$, we shift the sine curve horizontally to the left by $\frac{\pi}{2}$ units. (We shift each important point on $y = \sin x$).

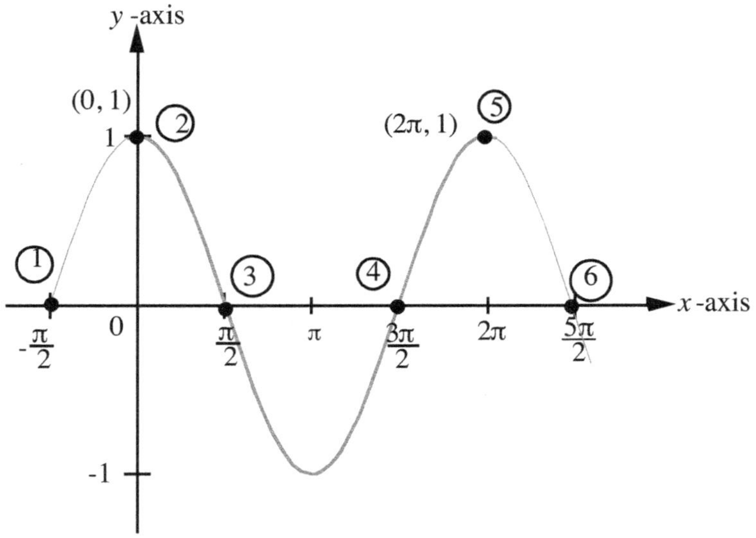

Figure : The graph of $y = \cos x$ for $-\frac{\pi}{2} \le x \le \frac{5\pi}{2}$

Observe the shape of the cosine curve and its relationship with the x- and y- axes. Also observe the similarities and differences between the sine and the cosine curves.

How to sketch the graph of $y = \cos x$

To sketch the graph of $y = \cos x$ we must know the following basic properties of this curve:

1. Its **period** is 2π; and its **amplitude** is 1.

2. Five important points (**critical points**) 2 end points (a starting point and a stopping point), 1 mid-point and two quarter-points are needed. Each point is an ordered pair (x, y). The five points include the x-intercepts, the y-intercept; and the amplitude. The amplitude determines the highest and the lowest points on the graph of this function

Example 6 Sketch the graph of $y = \cos x$ for $0 \le x \le 2\pi$.

Step 1: We construct a table containing these important points, using the formula $y = \cos x$

Table for $y = \cos x$

x	0	$\frac{\pi}{2}$	π	$\frac{3\pi}{2}$	2π
$y = \cos x$	1	0	-1	0	1

Step 2: From the origin and on the x-axis, mark-off a convenient interval and label its end-point 2π, the period of $y = \cos x$. Divide this interval into four equal intervals and label them with $\frac{\pi}{2}$, π, and $\frac{3\pi}{2}$,, using Figure .. p.466 as a guide.

Step 4: On the y-axis, using a convenient interval, mark-off and label the points $y = 1$ and $y = -1$.

Step 5: Plot the points $(0, 1)$, $(\frac{\pi}{2}, 0)$, $(\pi, -1)$, $(\frac{3\pi}{2}, 0)$ and $(2\pi, 1)$. (obtained from the above table)

Step 6: Connect the points from Step 5 by a smooth curve to obtain the characteristic shape of $y = \cos x$. (Figure)

Practice sketching this graph a number of times, noting its features, so that this graph (picture) becomes part of your "trigonometric vocabulary", just like the multiplication tables.

Lesson 31 Exercises

Sketch the graphs of the following:

1. $y = \cos x$ for $0 \le x \le 2\pi$; **2.** $y = -\cos x$ for $0 \le x \le 2\pi$; **3.** $y = 3\cos x$ for $0 \le x \le 2\pi$;

4. $y = -\cos x$ for $0 \le x \le 2\pi$;

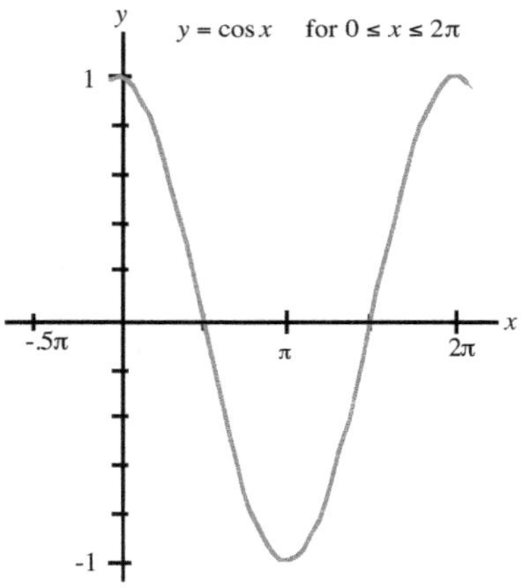

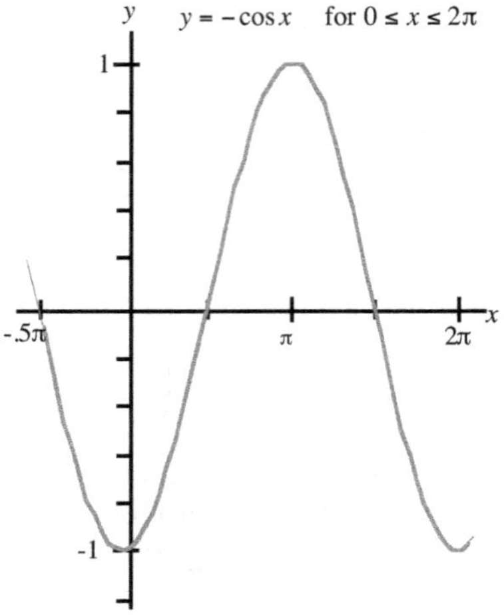

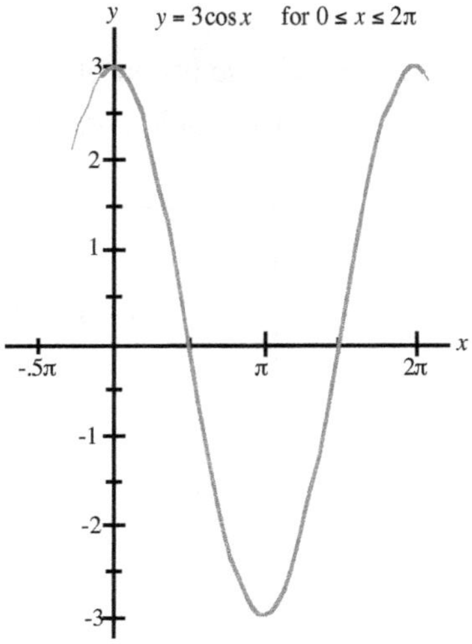

Lesson 32

Graphs of Discontinuous Trigonometric Functions: $y = \tan x$; $y = \csc x$; $y = \sec x$; $y = \cot x$

The approach used in sketching the graphs $y = \tan x$ and the reciprocal functions, $y = \csc x$, $y = \sec x$ and $y = \cot x$, is similar to that used in sketching the graphs of rational functions

The Graph of the Tangent Function, $y = \tan x$

Here, we will use the same principles that we used in sketching the sine and cosine curves. However, we must first learn the properties of the tangent curve.

Properties of $y = \tan x$

(1) The period of $y = \tan x$ is π (Thus, the curve repeats itself every π units)

(2) The x-intercepts are at where $\sin x = 0$ and $\cos x \neq 0$, since for x-intercepts, $y = \tan x = \frac{\sin x}{\cos x} = 0$..

(3) The y-intercept is at where $x = 0$. (Here, $\sin x = 0$, $\cos x = 1$)

(4) The curve is not defined where $\cos x$ is zero: That is it is not defined at $x = -\frac{\pi}{2}, \frac{\pi}{2}, \frac{3\pi}{2}, \frac{5\pi}{2}$ etc. The curve is discontinuous at these points and the curve is asymptotic to the vertical lines $x = -\frac{\pi}{2}, \frac{\pi}{2}, \frac{3\pi}{2}, \frac{5\pi}{2}$ etc

(4) The tangent curve has points of inflection at all x-intercepts.

Table for $y = \tan x$

x	$-\frac{\pi}{2}$	$-\frac{\pi}{4}$	0	$\frac{\pi}{4}$	$\frac{\pi}{2}$
$y = \tan x$	undefined	-1	0	1	undefined

Note that $\tan x$ and $\sec x$ have the same vertical asymptotes $\tan x = \frac{\sin x}{\cos x}$. and $\sec x = \frac{1}{\cos x}$
Note also that the tangent curve is discontinuous (broken up) at the x-intercepts (zeros) of the cosine curve.

How to Sketch the Graph of $y = \tan x$

Example 7 Sketch the graph of $y = \tan x$ for $-\frac{\pi}{2} \le x \le \frac{\pi}{2}$.

Solution

Step 1: Determine the x-intercept. The x-intercept is at $(0, 0)$.

Step 2: Determine the y-intercept. The y-intercept is at $(0, 0)$.

Step 3: Find the vertical asymptotes: The vertical asymptotes are the lines $x = -\frac{\pi}{2}$, and $x = \frac{\pi}{2}$.

(We may use the cosine curve as a guide to help identify the vertical asymptotes since the tangent ($\tan x = \frac{\sin x}{\cos x}$) is not defined at where the cosine is 0)

Step 4: Determine the quarter points for $x = -\frac{\pi}{4}$, and $\frac{\pi}{4}$,

The quarter points are at $(-\frac{\pi}{4}, -1)$ and $(\frac{\pi}{4}, 1)$.

Step 5: Plot the points from Steps 1, 2, and 4, above.

Step 6: Using broken or dotted lines, draw the vertical asymptotes $x = -\frac{\pi}{2}$, and $x = \frac{\pi}{2}$.

Step 7: Connect the points in Step 5 by a smooth solid curve noting that the curve does not meet its asymptotes but rather approaches them smoothly as the curve and its asymptotes are extended indefinitely. (Figure below)

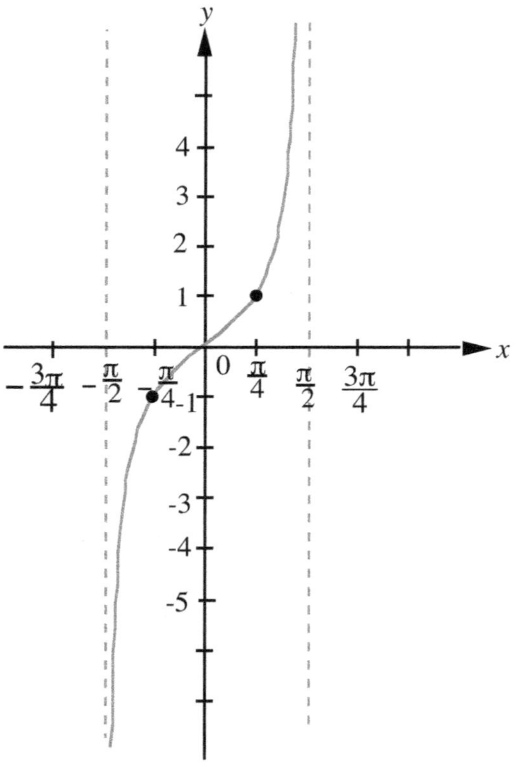

The graph of $y = \tan x$ for $-\frac{\pi}{2} \le x \le \frac{\pi}{2}$

Example 8 Sketch the graph of $y = \tan x$ for $-\pi \le x \le 2\pi$

Solution See also Example 7, above.

Step 1: Determine the x-intercepts. The x-intercepts are at $(-\pi, 0)$, $(0, 0)$, $(\pi, 0)$ and $(2\pi, 0)$

Step 2: Determine the y-intercept. The y-intercept is at $(0, 0)$.

Step 3: Find the vertical asymptotes: The vertical asymptotes are the lines $x = -\frac{\pi}{2}$, $x = \frac{\pi}{2}$, and $x = \frac{3\pi}{2}$. (We can use the cosine curve as a guide to help identify the vertical asymptotes since the tangent ($\tan x = \frac{\sin x}{\cos x}$) is not defined at where the cosine is 0)

Step 4: Determine the quarter points:

The quarter points are: $(-\frac{3\pi}{4}, 1)$, $(-\frac{\pi}{4}, -1)$, $(\frac{\pi}{4}, 1)$, $(\frac{3\pi}{4}, -1)$, $(\frac{5\pi}{4}, 1)$, $(\frac{7\pi}{4}, -1)$.

Step 5: Plot the points from Steps 1, 2, and 4,

Step 6: Using broken or dotted lined, draw the vertical asymptotes from Step 3.

Step 7: Using solid curves draw the tangent curves (referring to Example 7) as shown in Figure.

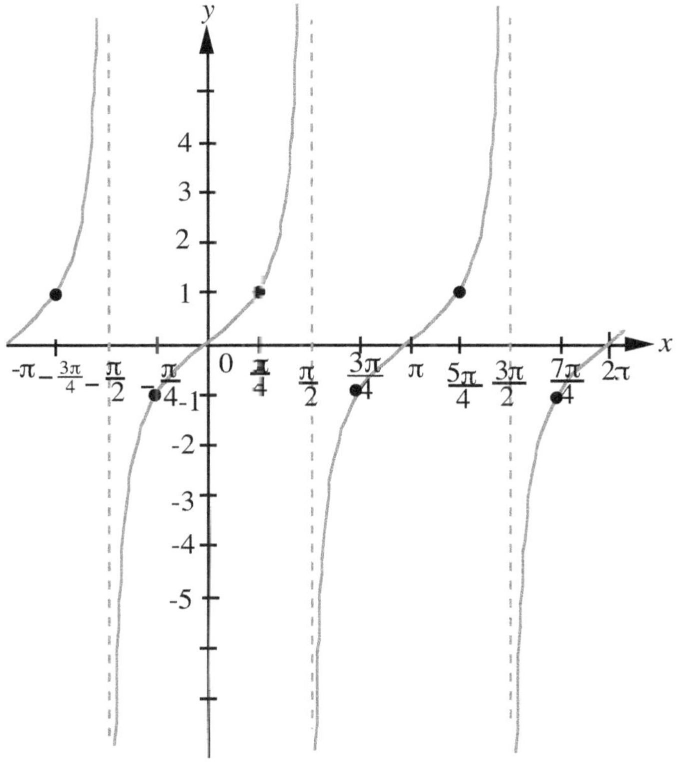

Figure : The graph of $y = \tan x$ $-\pi \le x \le 2\pi$

The Graph of $y = \csc x$

By definition, $\csc x = \frac{1}{\sin x}$ and so $\csc x$ is not defined where $\sin x = 0$. The cosecant curve is discontinuous (broken up) at the x-intercepts of the sine curve, and has vertical asymptotes at these intercepts. In addition, the concavities of the cosecant curve are the reversed concavities of the sine curve. The maximum (absolute maximum) points of the sine curve are the minimum (relative minimum) points of the cosecant curve. The period of $\csc x$ is 2π (same as that of $\sin x$). See also the description of the graphs of reciprocal functions on page 46-51.

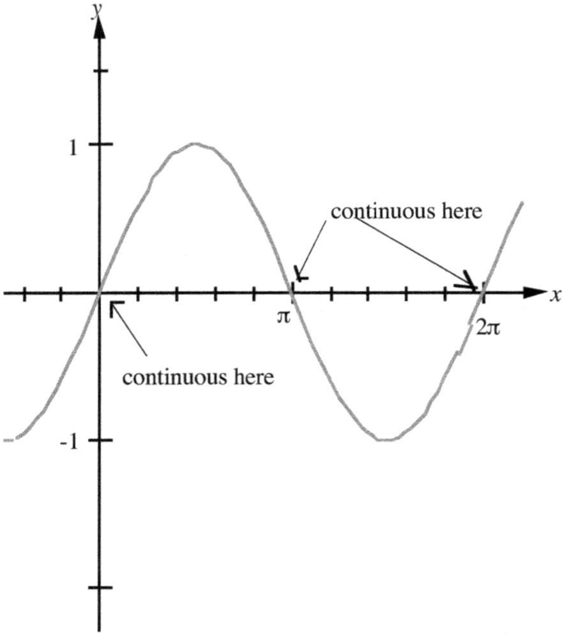

(k) The graph of $y = \sin x$.

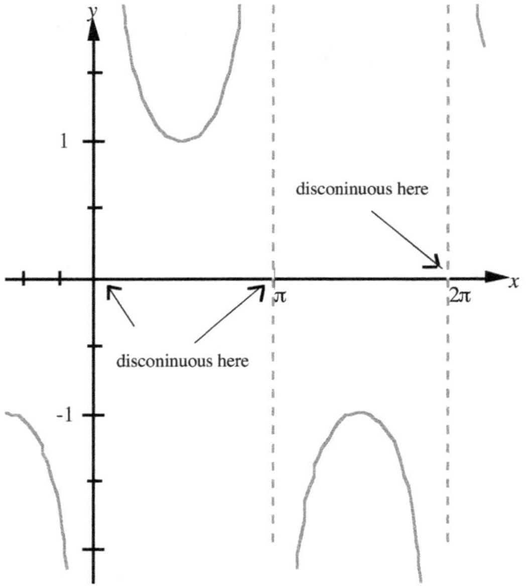

(l) The graph of $y = \frac{1}{\sin x} = \csc x$

How to Sketch the Graph of $y = \csc x$; $0 \le x \le 2\pi$

(See also page 46 for the graphs of functions and their reciprocals)

Example Sketch the graph of $y = \csc x$ for $0 \le x \le 2\pi$.

Note: There are no x-intercepts and no y-intercepts.

Step 1: Using dotted lines, sketch the graph of the corresponding sine curve.

Step 2: Draw vertical asymptotes at the x-intercepts of the sine curve.

Step 3: Using solid curved lines, draw the cosecant curve by visually breaking up the sine curve at its x-intercepts, and reversing the concavity each portion of broken curve to obtain the cosecant curve as shown in Figure.

Note: You dot not have to sketch the sine curve before sketching the cosecant curve. Using the sine curve is only a mnemonic device, since by now we are very familiar with the sine curve.

Table for the graph

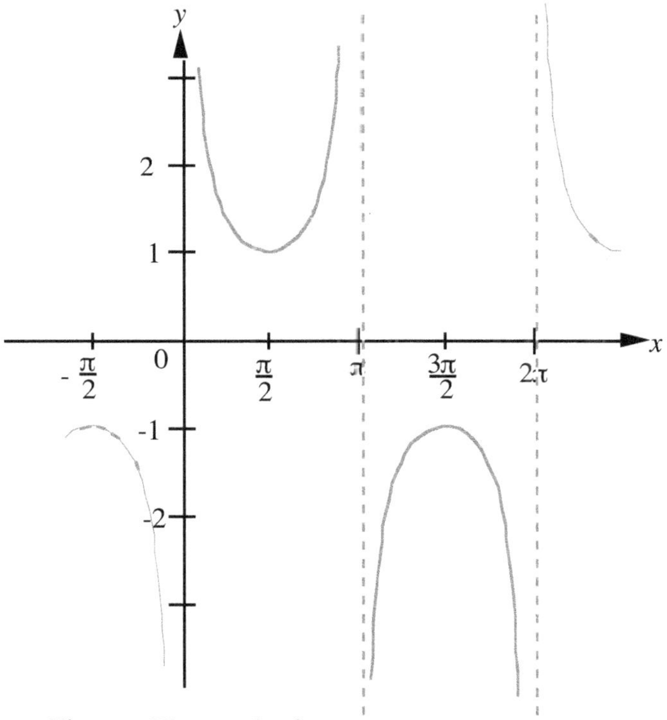

Figure : The graph of $\csc x$

The graph of $y = \sec x$

By definition, $\sec x = \frac{1}{\cos x}$ and so $\sec x$ is not defined where $\cos x = 0$. The secant curve is discontinuous (broken up) at where $\cos x = 0$ and the graph of $y = \sec x$ is broken up at the x-intercepts of the cosine curve, and has vertical asymptotes at these intercepts. Thus we can sketch $y = \sec x$ by first sketching the cosine curve (since by now we are used to sketching the cosine curve) and noting the following: The secant curve has no absolute maximum or minimum but has relative maximum and minimum. The concavity of the secant curve is the reversed concavity of the cosine curve. The maximum point of the cosine curve becomes the minimum point of the secant curve and the minimum point of the cosine curve becomes the maximum (relative maximum) point of the secant curve. The period of $\sec x$ is 2π (same as that of $\cos x$)

How to sketch the graph of $y = \sec x$

(See also page 46 for the graphs of functions and their reciprocals)

We can sketch the secant curve in a similar way as we sketched the cosecant curve except that we can obtain it from the cosine curve.

Step 1: Using dotted lines, sketch the graph of the corresponding cosine function.

Step 2: Draw vertical asymptotes, at the x-intercepts of the cosine curve.

Step 3: Using solid curved lines, draw the secant curve by visually breaking up the cosine curve at its x-intercepts, and reversing the concavity each portion of broken curve to obtain the secant curve.

Note: You dot not have to sketch the cosine curve before sketching the secant curve. Using the cosine curve is only a mnemonic device, since by now we are very familiar with the cosine curve.

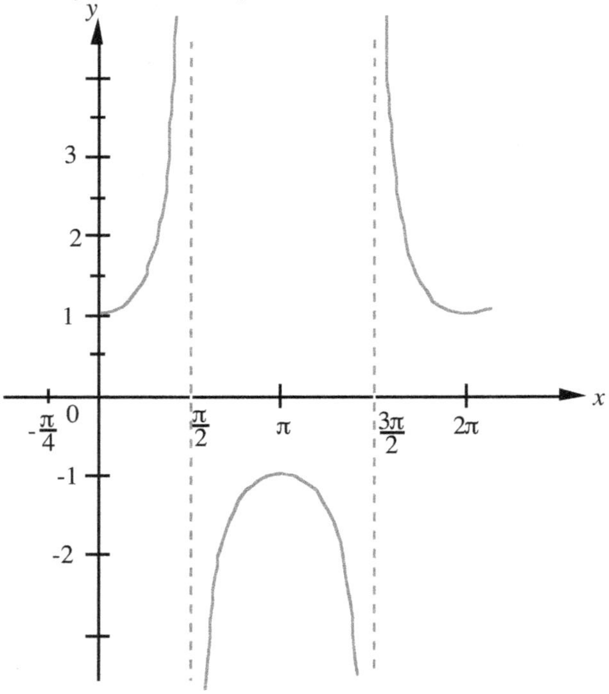

Figure : The graph of $y = \sec x$

The graph of $y = \cot x$

By definition, $\cot x = \frac{1}{\tan x} = \frac{\cos x}{\sin x}$. Thus $\cot x$ is not defined at the x-intercepts of $\tan x$ or $\sin x$, (since $\tan x = \frac{\sin x}{\cos x}$). The graph $y = \cot x$ has vertical asymptotes at the x-intercepts of $\tan x$ or $\sin x$.

How to sketch the graph of $y = \cot x$

Step 1: Using broken lines, draw the vertical asymptotes at the intercepts above.

Step 2: Noting that the cotangent curve and the tangent curve have the same shape except that the concavities of the cotangent curve are the reversed concavities of the tangent curve. Using solid curves draw the cotangent curve as shown in figure.

Table for the graph

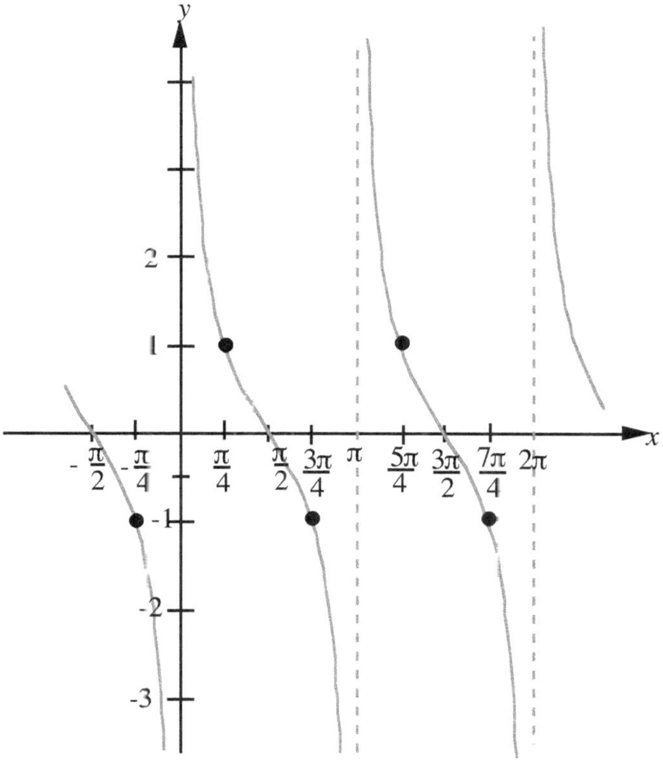

Figure :The graph of $y = \cot x$

The author's previous description (page 46, repeated below) of the relationship between the graphs of continuous functions and their reciprocals applies also to $y = \tan x$ and $y = \cot x$ if we consider the graph of $y = \tan x$ as a collection of continuous pieces of a curve and apply the description to each piece.

"Whenever a given curve becomes infinitely discontinuous at a point, the given curve breaks up into two pieces at this point; an asymptote is formed at this point, and each piece reverses its concavity and orientates itself such that the end of each piece smoothly becomes asymptotic to the asymptote so formed. "

Lesson 32 Exercises

Sketch the graphs of $y = 2\tan x$ for $0 \le x \le \pi$

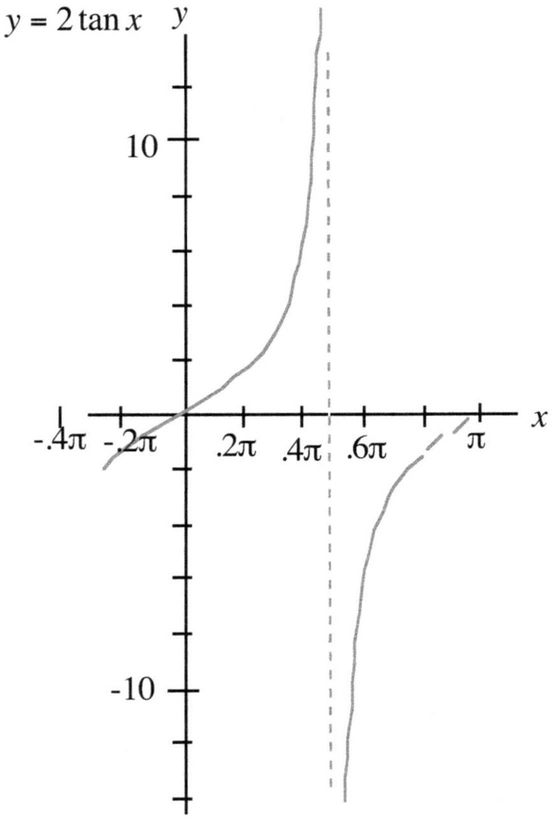

Lesson 33

Direct Procedure for Sketching the Graphs of

$y - k = a \sin(bx - h)$, and $y - k = a \cos(bx - h)$

Note below that in addition to the direct methods, we can use the transformation method to sketch the graphs of the more general functions. See method 2 of Example 2

General Equations of the Sine and Cosine Curves

Generally the equation of the sine curve is given by

$$\boldsymbol{y - k = a \sin(bx - h)} \qquad (1) \quad \text{<-- general form}$$

and that of the cosine is given by

$$\boldsymbol{y - k = a \cos(bx - h)} \qquad (2) \quad \text{<-- general form}$$

where (in each equation) k is the vertical shift, a is the amplitude, b is the constant used in determining the period; b is also the number of cycles in a given interval:

The period, $p = \frac{2\pi}{b}$.

If $a > 0$, the graph has the regular shape, but if $a < 0$, the graph has the inverted shape.

When the constant, b, of the right-hand side of $y - k = a\sin(bx - h)$ and $y - k = a\cos(bx - h)$ is factored out, we obtain, respectively,

$$\boldsymbol{y - k = a \sin b(x - \tfrac{h}{b})} \quad \text{<----------- standard form}$$

and

$$\boldsymbol{y - k = a \cos b(x - \tfrac{h}{b})} \quad \text{<----------- standard form}$$

in which $\frac{h}{b}$ is the horizontal shift or the phase shift or phase constant.

We may obtain $\frac{h}{b}$ from (2) by setting $bx - h = 0$ and solving for x.

If $h = 0$ (no horizontal shift from origin), $k = 0$ (no vertical shift from the origin), $a = 1$, and $b = 1$, we obtain

$$y = \sin x \qquad (3)$$
$$y = \cos x \qquad (4)$$

Equations (3) and (4) are the simplest forms of the equations of the sine and cosine curves, respectively. Its period is $\frac{2\pi}{1}$ or 2π..

The graph starts from $x = h$ and ends at $x = 2\pi + h$. The minimum point is at $y = K - A$, and the maximum point, $y = K + A$

Note:

In the main, we should have the following objectives in mind when drawing the trigonometric curves.

1. From which starting point, through which points and to which final point do we draw the curve
2. The unit interval for the coordinate axes.
3. The number of the basic curves to draw: This depends on the domain and period.
4. For each basic curve we need five important points (critical points) 2 end points (a starting point and a stopping point), 1 mid-point and two quarter-points . Each point is an ordered pair (x, y).
   The five points include points at which the curve crosses (intersects) the x- and y-axes; and how far up and how far down (its amplitude) in the y-direction does it go.

Procedure for sketching the graph of the general equation, $y - k = a \sin(bx - h)$ and

Step 1: Find the period p.

$$p = \frac{2\pi}{b}$$

Step 2: Find the beginning-point x-coordinate, x_1, of the first cycle.

Set $bx - h = 0$ and solve for x. to obtain $x = \frac{h}{b}$. Let this x-value be x_1

Step 3: Find the end-point x-coordinate, x_5, of the first cycle.

Set $bx - h = 2\pi$ and solve for x to obtain $x = \frac{2\pi + h}{b}$ Let this x-value be x_5

(We may combine Steps 1 and 2, by solving the domain inequality: $0 \le bx - h \le 2\pi$ to obtain

$$\frac{h}{b} \le x \le \frac{2\pi + h}{b}$$

Step 4: Determine the domain (x-values) to use to obtain ordered pairs for the points..
Divide the period p by 4 to obtain "the unit interval" for the x-axis.

Add $\frac{p}{4}$ successfully to x_1 until x_5 is reached (to obtain the x-values for the next Step)

Step 5: Using these x-values (from Step 4) calculate the corresponding y-values by substituting these values in turns in the functional equation to obtain ordered pairs or a table of values for x and y.

Step 6: Find the amplitude, The amplitude is $|a|$

For this step, we can also solve the range inequality $-a \le y - k \le a$ to obtain

$k - a \le y \le k + a$ (**Note:** For recall $-a \le y - k \le a$ is equivalent to $|y - k| \le a$)

Step 7: Plot the ordered pairs from Step 5.

Step 8 : Determine the concavities of the basic curve
If $a > 0$ the curve is regular (has the regular concavity). If $a < 0$, the curve is inverted. (Each portion of the regular curve is reflected about the x-axis))

Step 9: Determine the number of cycles for the specified interval.
The number of cycles $= b$ $(0 \le x \le 2\pi)$. If only one cycle is asked for, then draw it for only one cycle (period).

Step 10: Draw the remaining cycles by repetition and following Step 4 to Step 8.

Note: Skipping Step 5

If you do not want to calculate y but use only the following information to graph: the x-values, $|a|$, p, $a > 0$ or $a < 0$; then note the following:

1. There are five x-values on the x-axis, There are four equal intervals between the first x-value and the fifth x-value.

2. In starting to draw the curve note that **if** the first point is **on** the x-axis (a zero), the next point , (the second point) is **not** on the x-axis (a minimum or maximum point); the third point is **on** the x-axis (a zero) the fourth point is **not** on the x-axis (a minimum or maximum point) , and the fifth point is **on** the x-axis (a zero).. However, **if** the first point is **not** on the x-axis (**not** a zero), the second point is **on** the x-axis (a zero), the third point is **not** on the x-axis (not a zero), the fourth point is **on** the x-axis, and the fifth point is **not** on the x-axis. Using this alternation in the location of points on or off the x-axis, together with the amplitude, we can draw the curves. We will begin with an example in which $k = 0$.

Example 1: Sketch the graph of $y = 3\sin(4x + \pi)$

Solution

Method 1

Step 1: Find the period p: $b = 4$, and $p = \frac{2\pi}{b} = \frac{2\pi}{4} = \frac{\pi}{2}$

Step 2: Find the beginning point x-coordinate, x_1, of the first cycle.

Set $4x + \pi = 0$ and solve for x. to obtain $x = -\frac{\pi}{4}$. Let $x_1 = -\frac{\pi}{4}$.

Step 3: Find the endpoint x-coordinate, x_5, of the first cycle:

Set $4x + \pi = 2\pi$., and solve to obtain $x = \frac{\pi}{4}$. Let $x_5 = \frac{\pi}{4}$.

(We may combine Steps 2 and 3, by solving the domain inequality: $0 \le 4x + \pi \le 2\pi$ to obtain

$-\frac{\pi}{4} \le x \le \frac{\pi}{4}$ (The first cycle of the curve is between $-\frac{\pi}{4}$ and $\frac{\pi}{4}$.

Step 4: Determine the domain (x-values) to use to obtain ordered pairs for the points.

Divide the period p by 4 to obtain "the quarter -interval", $\frac{p}{4}$, for the x-axis.

Add $\frac{p}{4}$ successfully to x_1 until x_5 is reached (to obtain the x-values for the next Step)

$\frac{p}{4} = \frac{\pi}{2} \bullet \frac{1}{4}$ or $\frac{1}{4}$ of $\left(\frac{\pi}{2}\right)$ $\quad (p = \frac{\pi}{2})$

$= \frac{\pi}{8} < - - - - - \frac{p}{4}$

$x_1 = -\frac{\pi}{4}$ (from Step 2)

$x_2 = x_1 + \frac{\pi}{8}$

$= -\frac{\pi}{4} + \frac{\pi}{8}$

$= -\frac{\pi}{8}$

$x_3 = x_2 + \frac{\pi}{8}$

$= -\frac{\pi}{8} + \frac{\pi}{8}$

$= \mathbf{0}$

$x_4 = x_3 + \frac{\pi}{8}$

$= 0 + \frac{\pi}{8}$

$= \frac{\pi}{8}$

$x_5 = \frac{\pi}{8} + \frac{\pi}{8}$

$= \frac{2\pi}{8}$

$= \frac{\pi}{4}$ (same value as in Step 3)

Summarizing, the x-coordinates are $-\frac{\pi}{4}$, $-\frac{\pi}{8}$, 0, $\frac{\pi}{8}$, and $\frac{\pi}{4}$.

Step 5: Label the x-axis (using $\frac{p}{4} = \frac{\pi}{8}$ as the unit interval). See Fig. 1, p.200.

Step 6: Find the y-coordinates of the minimum and the maximum points.

The amplitude is $|a| = |-3| = 3$.

For this step, we may also write the range inequality $-3 \le y \le 3$ (noting that $k = 0$)

The y--coordinates of the minimum and the maximum points are -3 and 3 respectively.

Step 7: Using the x-coordinates (from Step 4) calculate the corresponding y-coordinates by substituting these values in turns in the functional equation to obtain ordered pairs or a table of values for x and y. The ordered pairs are:

$(-\frac{\pi}{4}, 0)$; $(-\frac{\pi}{8}, 3)$; $(0, 0)$; $(\frac{\pi}{8}, -3)$; $(\frac{\pi}{4}, 0)$.

For the calculations of the y-coordinates, see **Box 1** below.

Step 8: Label the y-axis.

Step 9: Plot the ordered pairs from Step 7.

Step 10**:** Determine the concavities of the basic curve
Since $a > 0$, the curve has the regular concavity.

Step 11 Connect the points plotted, noting the concavities and the characteristic shape of the sine curve.

Step 12: Determine the number of cycles for the specified interval.
The number of cycles = b. If asked for only one cycle, draw the curve for one cycle.

Step 13: Draw the remaining cycles by extension and following Step 4 to Step 11..

Step 14: The number of cycles to draw = b, and in the present problem, $b = 4$, and therefore we draw 3 more cycles.

Box 1: Calculations for Step 7

$$x_1 = -\frac{\pi}{4},\ y = 3\sin[4(-\tfrac{\pi}{4}) + \pi] = 3\sin[-\pi + \pi] = 3\sin[0] = 0 \quad (\sin 0 = 0) \quad \Rightarrow (-\tfrac{\pi}{4}, 0)$$

$$x_2 = -\frac{\pi}{8},\ y = 3\sin[4(-\tfrac{\pi}{8}) + \pi] = 3\sin[-\tfrac{\pi}{2} + \pi] = 3\sin[\tfrac{\pi}{2}] = 3(1) = 3 \quad (\sin\tfrac{\pi}{2} = 1) \quad \Rightarrow (-\tfrac{\pi}{8}, 3)$$

$$x_3 = 0,\ y = 3\sin[4(0) + \pi] = 3\sin[0 + \pi] = 3\sin[\pi] = 3(0) = 0 \quad (\sin\pi = 0) \quad \Rightarrow (0, 0)$$

$$x_4 = \frac{\pi}{8},\ y = 3\sin[4(\tfrac{\pi}{8}) + \pi] = 3\sin[\tfrac{\pi}{2} + \pi] = 3\sin[\tfrac{3\pi}{2}] = 3(-1) = -3 \quad \Rightarrow (\tfrac{\pi}{8}, -3)$$

$$x_5 = \frac{\pi}{4},\ y = 3\sin[4(\tfrac{\pi}{4}) + \pi] = 3\sin[\pi + \pi] = 3\sin[2\pi] = 3(0) = \mathbf{0} \quad \Rightarrow (\tfrac{\pi}{4}, 0)$$

Note : $\sin[2\pi] = 0$

Example 1: Sketch the graph of $y = 3\sin(4x + \pi)$

Method 2 (You may also use method 2 (transformation) of Example 2)

Step 1: Find the period p. $b = 4$, and $p = \frac{2\pi}{b} = \frac{2\pi}{4} = \frac{\pi}{2}$

Step 2: Find the beginning-point x-coordinate, x_1, of the first cycle.

Set $4x + \pi = 0$ and solve for x.: Let $x_1 = -\frac{\pi}{4}$

Step 3: Find the end-point x-coordinate, x_5, of the first cycle:

Set $4x + \pi = 2\pi$. and solve for x to obtain $x = \frac{\pi}{4}$. Let $x_5 = \frac{\pi}{4}$.

(We may combine Steps 2 and 3, by solving the domain inequality: $0 \le 4x + \pi \le 2\pi$ to obtain

$-\frac{\pi}{4} \le x \le \frac{\pi}{4}$ (The first cycle of the curve is between $-\frac{\pi}{4}$ and $\frac{\pi}{4}$.)

There are 4 cycles, since $b = 4$.

Step 4: Determine the domain (x-values) to use to obtain points (for the first cycle).

Divide the period p by 4 to obtain "the quarter -interval", $\frac{p}{4}$, for the x-axis.

Add $\frac{p}{4}$ successfully to x_1 until x_5 is reached (to obtain the x-values for the next Step)

$\frac{p}{4} = \frac{\pi}{2} \bullet \frac{1}{4}$ or $\frac{1}{4}$of $\left(\frac{\pi}{2}\right)$ $\quad (p = \frac{\pi}{2})$

$= \frac{\pi}{8} < - - - - - \frac{p}{4}$

$x_1 = -\frac{\pi}{4}$ (from Step 2)

$x_2 = x_1 + \frac{\pi}{8}$

$= -\frac{\pi}{4} + \frac{\pi}{8}$

$= -\frac{\pi}{8}$

$x_3 = x_2 + \frac{\pi}{8}$

$= -\frac{\pi}{8} + \frac{\pi}{8}$

$= 0$

$x_4 = x_3 + \frac{\pi}{8}$

$= 0 + \frac{\pi}{8}$

$= \frac{\pi}{8}$

$x_5 = \frac{\pi}{8} + \frac{\pi}{8}$

$= \frac{2\pi}{8}$

$= \frac{\pi}{4}$ (same value as in Step 3)

Summarizing, the x-coordinates are $-\frac{\pi}{4}$, $-\frac{\pi}{8}$, 0, $\frac{\pi}{8}$, and $\frac{\pi}{4}$

Step 5: Label the x-axis (using $\frac{p}{4} = \frac{\pi}{8}$ as the unit interval)

Step 6: Determine the corresponding y-coordinates. Instead of calculating the y-coordinates by substituting the x-coordinates in the functional equation, we shall **alternatively** find the y-coordinates by using the mnemonic device "YoMaxYoMinYo (pronounced yo-max-yo-min-yo) or MaxYoMinYoMax where if $k = 0$, Yo = 0, Max = $|a| = 3$, and Min = $-|a| = -3$. Therefore, the y-coordinates are 0, 3, 0, -3 and 0.

Pairing these y-values with the corresponding x-values, $-\frac{\pi}{4}$, $-\frac{\pi}{8}$, 0, $\frac{\pi}{8}$, $\frac{\pi}{4}$ (from Step 4),

we obtain the ordered pairs $(-\frac{\pi}{4}, 0)$; $(-\frac{\pi}{8}, 3)$; $(0, 0)$; $(\frac{\pi}{8}, -3)$; $(\frac{\pi}{4}, 0)$. for the graph.

Step 7: Label the y-axis

Step 8: Plot the ordered pairs from Step 6.

Step 9: Determine the concavity of the basic curve. Since $a > 0$, the curve has the regular concavity.

Step 10: Connect the five points plotted noting the concavity and applying the following:

If the first point is **on** the x-axis (a zero), the next point (the second point) is **not** on the x-axis (a minimum or maximum point); the third point is **on** the x-axis (a zero); the fourth point is **not** on the x-axis (a minimum or maximum point); and the fifth point is **on** the x-axis (a zero). However, **if** the first point is **not** on the x-axis (**not** a zero), the second point is **on** the x-axis (a zero); the third point is **not** on the x-axis (not a zero), the fourth point is **on** the x-axis, and the fifth point is **not** on the x-axis. Using this alternation in the location of points on or off the x-axis, together with the amplitude, we can draw the curves.

Step 11: The number of cycles to draw = 4, since $b = 4$, and therefore we would draw 3 more cycles by repetition.

Step 12: Draw the remaining cycles by extension and following Step 4 to Step 10.

Figure 1 shows the first cycle, Figure 2 shows all the cycles between 0 and 2π, using a different scale. Since $b = 4$, four cycles are drawn.

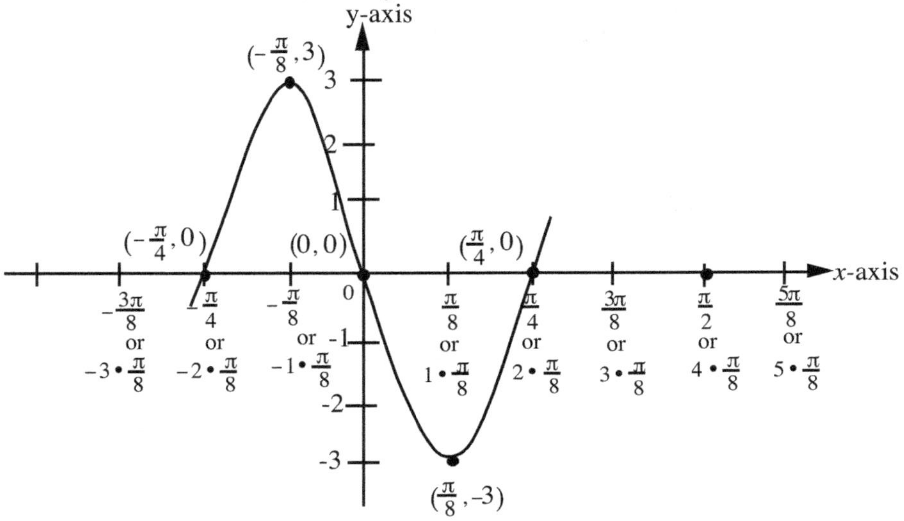

Figure 1

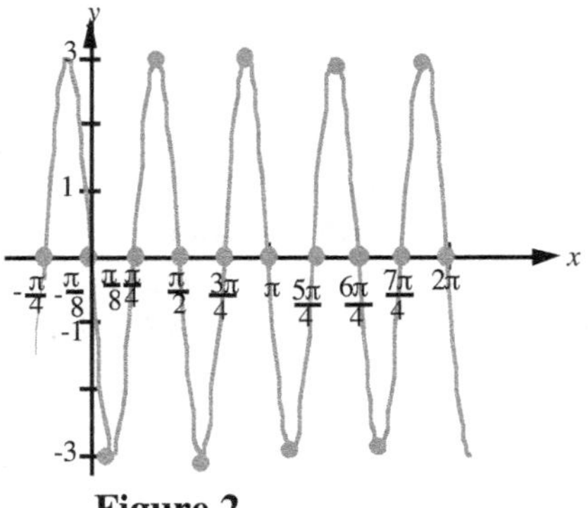

Figure 2

Example 1b Sketch the graph of $y = 3\sin(4x + \pi)$ $\quad 0 \le x \le 2\pi$

The sketch is the same as that of Example 1a, except that only that part of the curve between 0 and 2π is shown. (You may also use method 2 of Example 2)

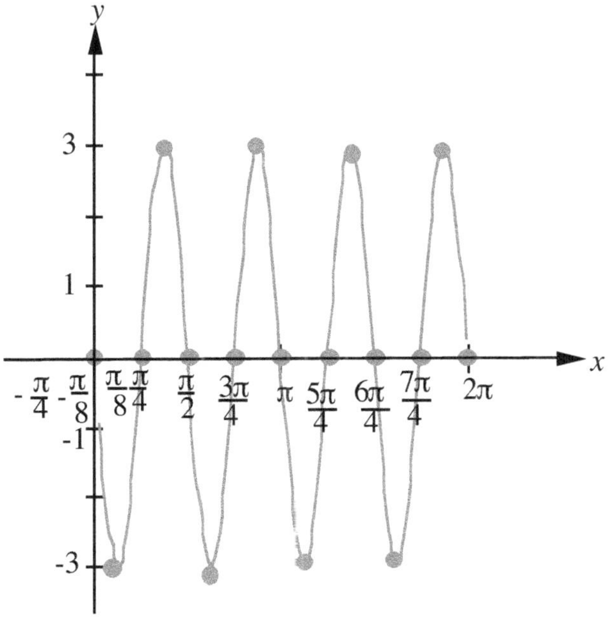

Example 2: Sketch the graph of $y - 1 = 3\sin(4x + \pi)$

We cover three methods

Method 1

This Example is the same as Example 1, except that here, k = 1. . Each point on the graph in Example 1 is shifted 1 unit up, (a vertical shift). Steps 1 to 4 are the same. The difference in Step 5 is the 1. Redo the calculations for Step 5, and observe later that we could obtain the ordered pairs in Step 5 in Example 2 by increasing the y-value of each ordered pair in Example 1 by 1. We repeat the steps in Example 1 with the appropriate changes for Example 2..

Step 1: Find the period p. $b = 4$, and $p = \frac{2\pi}{b} = \frac{2\pi}{4} = \frac{\pi}{\mathbf{2}}$

Step 2: Find the beginning point x-coordinate, x_1 of the first cycle

Set $4x + \pi = 0$ and solve for x.: Let $x_1 = -\frac{\pi}{4}$

Step 3: Find the end-point x-coordinate, x_5, of the first cycle of the curve:

Set $4x + \pi = 2\pi$.and solve for x Let $x_5 = \frac{\pi}{4}$

(We may combine Steps 2 and 3, by solving the domain inequality: $0 \le 4x + \pi \le 2\pi$ to obtain

$-\frac{\pi}{4} \le x \le \frac{\pi}{4}$ (The first cycle of the curve is between $-\frac{\pi}{4}$ and $\frac{\pi}{4}$.

Step 4: Determine the domain (x-values) to use to obtain ordered pairs for the points..

Divide the period p by 4 to obtain "the quarter -interval", $\frac{p}{4}$, for the x-axis.

Add $\frac{p}{4}$ successfully to x_1 until x_5 is reached (to obtain the x-values for the next Step)

$\frac{p}{4} = \frac{\pi}{2} \bullet \frac{1}{4}$ or $\frac{1}{4}$ of $\left(\frac{\pi}{2}\right)$ $\quad (p = \frac{\pi}{2})$

$= \frac{\pi}{8} < - - - - \frac{p}{4}$

$x_1 = -\frac{\pi}{4}$ (from Step 2)

$x_2 = x_1 + \frac{\pi}{8} = -\frac{\pi}{4} + \frac{\pi}{8} = -\frac{\pi}{8}$

$x_3 = x_2 + \frac{\pi}{8} = -\frac{\pi}{8} + \frac{\pi}{8} = 0$

$x_4 = x_3 + \frac{\pi}{8} = 0 + \frac{\pi}{8} = \frac{\pi}{8}$

$x_5 = \frac{\pi}{8} + \frac{\pi}{8} = \frac{2\pi}{8} = \frac{\pi}{4}$

(same value in Step 3)

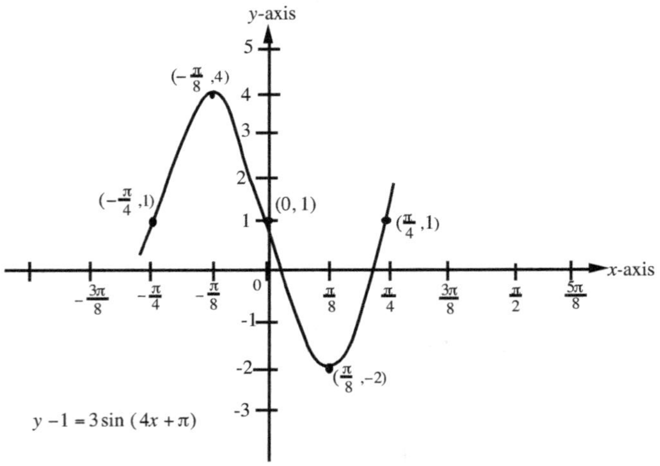

Figure 3

Summarizing, the x-coordinates are $-\frac{\pi}{4}$, $-\frac{\pi}{8}$, 0, $\frac{\pi}{8}$, and $\frac{\pi}{4}$

Step 5: Label the x-axis (using $\frac{p}{4} = \frac{\pi}{8}$ as the unit interval)

Step 6: Using these x-coordinates (from Step 4) calculate the corresponding y-coordinates by substituting these values in turns in the functional equation to obtain ordered pairs or a table of values for x and y. The ordered pairs are $(-\frac{\pi}{4}, 1)$; $(-\frac{\pi}{8}, 4)$; $(0, 1)$; $(\frac{\pi}{8}, -2)$; $(\frac{\pi}{4}, 1)$.
For the calculations of the y-coordinates see Box 2 below.

Step 7: Find the y-coordinates of the minimum and maximum points.
The amplitude is $|a| = 3, k = 1$ We can also solve the range inequality $-3 \le y - 1 \le 3$ (from $-a \le y - k \le a$) to obtain $-2 \le y \le 4$.
The y-coordinate of the minimum point is -2 and that of the maximum point is 4.

Step 8: Label the y-axis

Step 9: Plot the ordered pairs from Step 6.

Step 10: Determine the concavities of the basic curve
Since $a > 0$, the curve has the regular concavity.

Step 11: Connect the points plotted, noting the concavities and the characteristic shape of the sine curve.

Step 12: Determine the number of cycles for the specified interval.
The number of cycles = b. If only one cycle is required, draw the curve for one cycle.

Step 13: Draw the remaining cycles by extension and following Step 4 to Step 10.
See **Figure 3**, above.

Box 2: Calculations for Step 6

$$x_1 = -\frac{\pi}{4},\ y = 3\sin[4(-\tfrac{\pi}{4}) + \pi] + 1 = 3\sin[-\pi + \pi] + 1 = 3\sin[0] + 1 \quad (\sin 0 = 0) \quad = 1 \ \Rightarrow (-\tfrac{\pi}{4}, 1)$$

$$x_2 = -\frac{\pi}{8},\ y = 3\sin[4(-\tfrac{\pi}{8}) + \pi] + 1 = 3\sin[-\tfrac{\pi}{2} + \pi] + 1 = 3\sin[\tfrac{\pi}{2}] + 1 = 3(1) \quad (\sin\tfrac{\pi}{2} = 1) \quad = 4 \ \Rightarrow (-\tfrac{\pi}{8}, 4)$$

$$x_3 = 0,\ y = 3\sin[4(0) + \pi] + 1 = 3\sin[0 + \pi] + 1 = 3\sin[\pi] + 1 = 3(0) + 1 \quad (\sin\pi = 0) \quad = 1 \ \Rightarrow (0,1)$$

$$x_4 = \frac{\pi}{8},\ y = 3\sin[4(\tfrac{\pi}{8}) + \pi] + 1 = 3\sin[\tfrac{\pi}{2} + \pi] + 1 = 3\sin[\tfrac{3\pi}{2}] + 1 = 3(-1) + 1 = -2 \ \Rightarrow (\tfrac{\pi}{8}, -2)$$

$$x_5 = \frac{\pi}{4},\ y = 3\sin[4(\tfrac{\pi}{4}) + \pi] + 1 = 3\sin[\pi + \pi] + 1 = 3\sin[2\pi] + 1 = 3(0) + 1 = \mathbf{1} \ \Rightarrow (\tfrac{\pi}{4}, 1).$$

Example 2

Method 2 : (Transformation of functions method) (also review p.75-76)

Given $y = \sin x$, draw the graph of $y - 1 = 3\sin(4x + \pi)$

Rewrite $y - 1 = 3\sin(4x + \pi)$ as $y - 1 = 3\sin[4(x + \frac{\pi}{4})]$. Here, $k = 1$, $h = -\frac{\pi}{4}$, $a = 3$, $b = 4$

$y = \sin x$		$y - 1 = 3\sin[4(x + \frac{\pi}{4})]$		
x_0	y_0	$x_n = \frac{x_0}{b} + h$	$y_n = y_0(a) + k$	(x_n, y_n)
0	0	$\frac{0}{4} - \frac{\pi}{4} = 0 + -\frac{\pi}{4} = -\frac{\pi}{4}$	$0(3) + 1 = 0 + 1 = 1$	$(-\frac{\pi}{4}, 1)$
$\frac{\pi}{2}$	1	$\frac{\pi}{2 \bullet 4} - \frac{\pi}{4} = \frac{\pi}{8} - \frac{\pi}{4} = -\frac{\pi}{8}$	$1(3) + 1 = 3 + 1 = 4$	$(-\frac{\pi}{8}, 4)$
π	0	$\frac{\pi}{4} - \frac{\pi}{4} = 0$	$0(3) + 1 = 0 + 1 = 1$	$(0,1)$
$\frac{3\pi}{2}$	-1	$\frac{3\pi}{2 \bullet 4} - \frac{\pi}{4} = \frac{3\pi}{8} - \frac{\pi}{4} = \frac{\pi}{8}$	$-1(3) + 1 = -3 + 1 = -2$	$(\frac{\pi}{8}, -2)$
2π	0	$\frac{2\pi}{4} - \frac{\pi}{4} = \frac{\pi}{2} - \frac{\pi}{4} = \frac{\pi}{4}$	$0(3) + 1 = 0 + 1 = 1$	$(\frac{\pi}{4}, 1)$
$-\frac{\pi}{2}$	-1	$-\frac{\pi}{2 \bullet 4} - \frac{\pi}{4} = -\frac{\pi}{8} - \frac{\pi}{4} = -\frac{3\pi}{8}$	$-1(3) + 1 = -3 + 1 = -2$	$(-\frac{3\pi}{8}, -2)$
$-\pi$	0	$-\frac{\pi}{4} - \frac{\pi}{4} = -\frac{\pi}{2}$	$0(3) + 1 = 0 + 1 = 1$	$(-\frac{\pi}{2}, 1)$
$-\frac{3\pi}{2}$	1	$-\frac{3\pi}{2 \bullet 4} - \frac{\pi}{4} = -\frac{3\pi}{8} - \frac{\pi}{4} = -\frac{5\pi}{8}$	$1(3) + 1 = 3 + 1 = 4$	$(-\frac{5\pi}{8}, 4)$
-2π	0	$-\frac{2\pi}{4} - \frac{\pi}{4} = -\frac{3\pi}{4}$	$0(3) + 1 = 0 + 1 = 1$	$(-\frac{3\pi}{4}, 1)$

Step 1: Quickly prepare a table of values for x and y, using the basic $\sin x$ equation and convenient x-values.

Step 2: Using the table of values from Step 1, determine the new coordinates x_n and y_n.

Step 3: Plot the new points and connect them by a solid curve (See Figure 3, previous page) or next page.

In Example 2 above, compare method 1 and method 2, the transformation method. Also compare the calculations in step 6 of Method 1 and the calculations in Method 2. Observe the compactness of the transformation method compared to method 1. The first five rows of method 2 cover all the calculations in method 1. A word to the wise is enough.

Example 2: Sketch the graph of $y - 1 = 3\sin(4x + \pi)$

Method 3

This Example is the same as Example 1, except that here $k = 1$. . Each point on the graph in Example 1 is shifted 1 unit up, (a vertical shift). Steps 1 to 4 are the same. We repeat the steps with the appropriate changes.

Step 1: Find the period p. $b = 4$, and $p = \frac{2\pi}{b} = \frac{2\pi}{4} = \frac{\pi}{\mathbf{2}}$

Step 2: Find the beginning point x-coordinate, x_1, of the first cycle.

Set $4x + \pi = 0$ and solve for x.: Let $x_1 = -\frac{\pi}{4}$

Step 3: Find the end-point x-coordinate of the first cycle:

Set $4x + \pi = 2\pi$. and solve for x to obtain $x = \frac{\pi}{4}$. Let $x_5 = \frac{\pi}{4}$.

(We may combine Steps 2 and 3, by solving the domain inequality: $0 \le 4x + \pi \le 2\pi$ to obtain

$-\frac{\pi}{4} \le x \le \frac{\pi}{4}$ (The first cycle of the curve is between $-\frac{\pi}{4}$ and $\frac{\pi}{4}$.

Four 4 cycles are to be drawn. since $b = 4$.

Step 4: Determine the domain (x-values) to use to obtain points for the (for the first cycle).

Divide the period p by 4 to obtain "the quarter -interval", $\frac{p}{4}$, for the x-axis.

Add $\frac{p}{4}$ successfully to x_1 until x_5 is reached (to obtain the x-values for the next Step)

$$\frac{p}{4} = \frac{\pi}{2} \bullet \frac{1}{4} \text{ or } \frac{1}{4}\text{of}\left(\frac{\pi}{2}\right) \qquad (p = \frac{\pi}{2})$$

$$= \frac{\pi}{8} < - - - - \frac{p}{4}$$

$x_1 = -\frac{\pi}{\mathbf{4}}$ (from Step 2)

$x_2 = x_1 + \frac{\pi}{8}$

$= -\frac{\pi}{4} + \frac{\pi}{8}$

$= -\frac{\pi}{\mathbf{8}}$

$x_3 = x_2 + \frac{\pi}{8}$

$= -\frac{\pi}{8} + \frac{\pi}{8}$

$= \mathbf{0}$

$x_4 = x_3 + \frac{\pi}{8}$

$= 0 + \frac{\pi}{8}$

$= \frac{\pi}{\mathbf{8}}$

$x_5 = \frac{\pi}{8} + \frac{\pi}{8}$

$= \frac{2\pi}{8}$

$= \frac{\pi}{\mathbf{4}}$ (same value as in Step 3)

Summarizing, the x-coordinates are $-\frac{\pi}{4}$, $-\frac{\pi}{8}$, 0, $\frac{\pi}{8}$, and $\frac{\pi}{4}$.

Step 5: Label the x-axis (using $\frac{p}{4} = \frac{\pi}{8}$ as the unit interval)

Step 6: Determine the corresponding y-coordinates. We can obtain the y-coordinates by substituting the x-values in the functional equation, $y - 1 = 3\sin(4x + \pi)$.

However, we can **alternatively** find these values by using the mnemonic device "YoMaxYoMinYo (pronounced yo-max-yo-min-yo) or MaxYoMinYoMax, where if $k = 1$, Yo $= k = 1$, Max $= k + |a| = 1 + 3 = 4$, Min $= = 1 - 3 = -2$.
The corresponding y-coordinates are 1, 4, 1, -2 and 1. Pairing these y-values with the corresponding x-values, $-\frac{\pi}{4}, -\frac{\pi}{8}, 0, \frac{\pi}{8}, \frac{\pi}{4}$ (from Step 5), we obtain the ordered pairs $(-\frac{\pi}{4}, 1)$; $(-\frac{\pi}{8}, 4)$; $(0, 1)$; $(\frac{\pi}{8}, -2)$; $(\frac{\pi}{4}, 1)$. for the graph.

Step 7: Label the y-axis.

Step 8: Plot the ordered pairs from Step 6.

Step 9**:** Determine the concavity of the basic curve. Since $a > 0$, the curve has the regular concavity.

Step 10: Connect the five points plotted, noting the concavity, the characteristic shape of the sine curve, and applying the following:

Step 11: The number of cycles to draw = 4, since $b = 4$; and therefore, we draw 3 more cycles by repetition.

Step 12: Draw the remaining cycles by extension and following Step 4 to Step 10.
Figure shows the first cycle Figure shows all the cycles between 0 and 2π, using a different scale. Since $b = 4$, Four cycles are drawn.

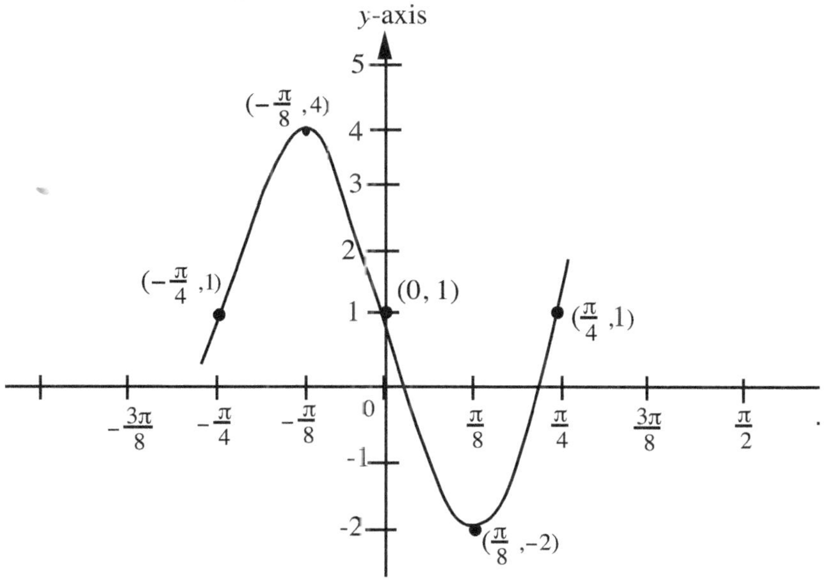

Figure: Graph of $y - 1 = 3\sin(4x + \pi)$

Sketching the graph of general cosine curve $y - k = a \cos (bx - h)$

Example 3: To draw the graph of $y = 4 \cos (2x - \pi)$ or $y - 1 = 3 \cos (4x + \pi)$ imitate the method of the last two examples for the sine curves, $y = 3 \sin (4x + \pi)$ and $y - 1 = 3 \sin (4x + \pi)$. In particular, imitate the **transformation method**.

The graph of $y = 4 \cos (2x - \pi)$ is shown in Figure below

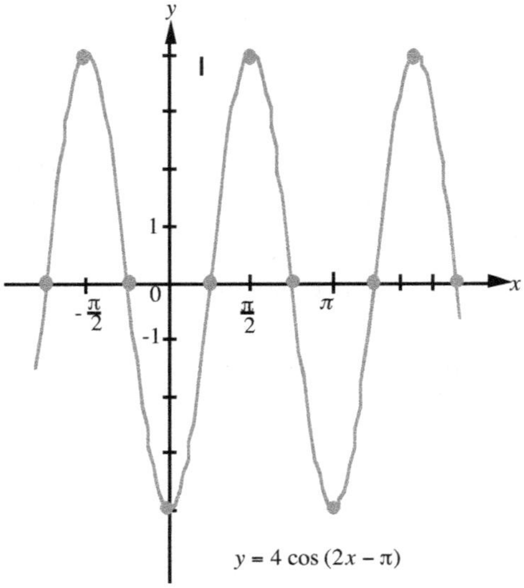

More Examples on Trigonometric Graphs

Example 4 Sketch the graph of $y - 1 = 3 \sin (4x + \pi)$ $\quad 0 \le x \le 2\pi$

Solution

The graph is the same as that of Example 2, except that we extend the graph to 2π and delete the part of the graph for which $x < 0$. Imitate the **transformation method**, and make sure x_n goes up to 2π.

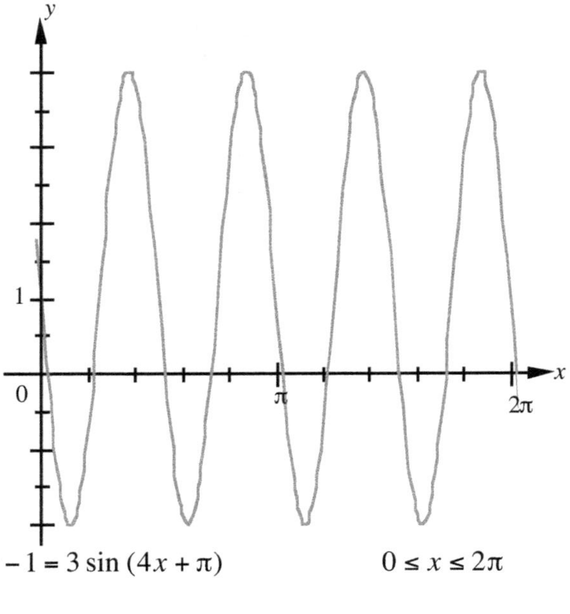

Graph of $y - 1 = 3 \sin (4x + \pi)$ $\quad 0 \le x \le 2\pi$

Example 5 Sketch the graph of $y = -\sin x$ $(0 \le x \le 2\pi)$.

Solution Imitate the **transformation method,** .and make sure the x_n's satisfy $0 \le x \le 2\pi$.

The graph is the reflection of $y = \sin x$ about the x-axis.

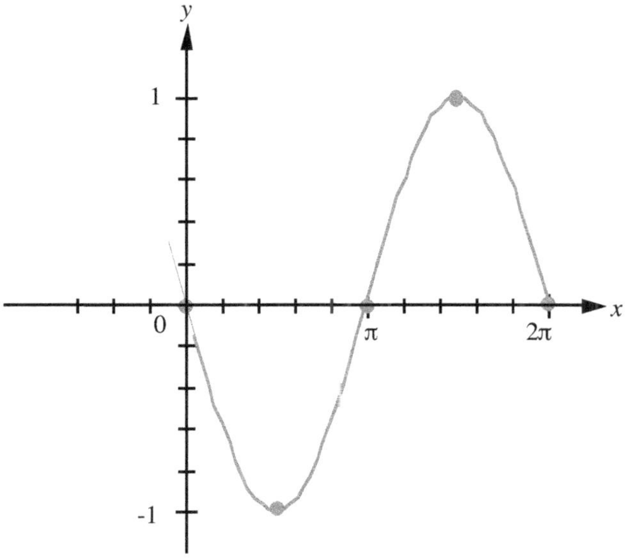

$y = -\sin x \quad (0 \le x \le 2\pi)$.

Example 6 Sketch the graph of $y = -\sin x$ $(-\pi \le x \le 2\pi)$.

Solution: Imitate the **transformation method,** and make sure the x_n's satisfy $-\pi \le x \le 2\pi$ ......

The sketch is similar to that of Example 5 above, except that we extend the curve to -π.

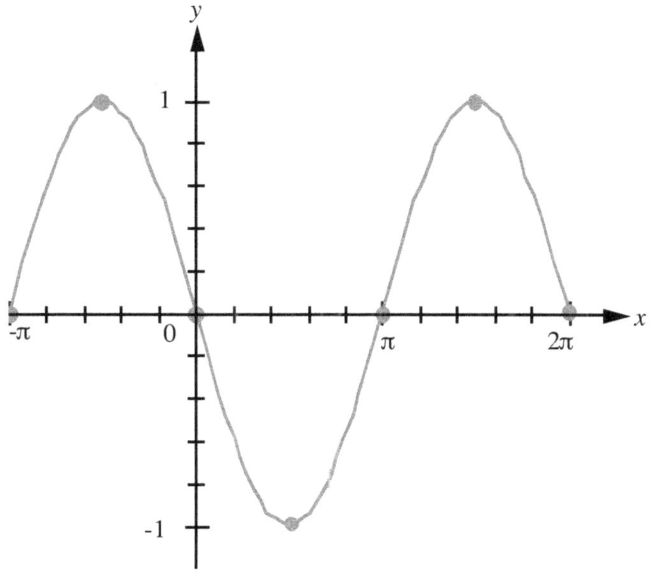

$y = -\sin x \quad (-\pi \le x \le 2\pi)$.

Example 7 Sketch the graph of $y = \sin x \quad (0 \le x \le 3\pi)$

Solution In this case, we sketch one and half sine curve starting from the origin and terminating at the point $(3\pi, 0)$.

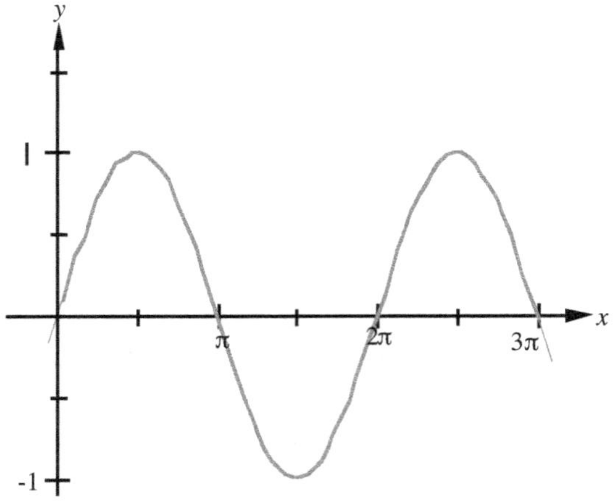

$y = \sin x \qquad 0 \le x \le 3\pi$

Example 8 Sketch the graph of $y = 3 \sin x \qquad (0 \le x \le 2\pi)$

Given $y = a \sin x$, a is called the amplitude. The amplitude indicates how far up (maximum value or highest point on this curve) or how far down (minimum value or lowest point) the curve goes from the x-axis. The maximum y-value for $y = \sin x$ is 1 and the minimum value is -1. The maximum y-value for $y = 3 \sin x$ is 3 and the minimum y-value is -3.

The graph of $y = 3 \sin x$ and $y = \sin x$ have the same period, 2π.

To sketch $y = 3 \sin x$, repeat Step 1- Step 5 in Example 7 above, but replace $(\frac{\pi}{2}, 1)$ by $(\frac{\pi}{2}, 3)$ and replace $(\frac{3\pi}{2}, -1)$ by $(\frac{3\pi}{2}, -3)$.

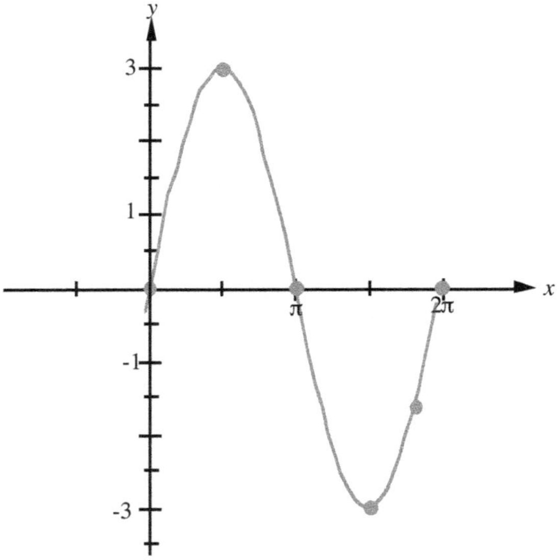

Figure $y = 3 \sin x \qquad (0 \le x \le 2\pi)$

Example 9 Sketch the graph of $y = \sin 4x$ (see p.199-205 for the general case)

Method 1

Step 1: $b = 4$ the period, $p = \frac{2\pi}{4} = \frac{\pi}{2}$

Since the period is $\frac{\pi}{2}$ the graph the curve must repeat itself every $\frac{\pi}{2}$ units.

Step 2: Find the starting point: Set $4x = 0$ and solve to obtain $x = 0$.

Step 3: Find the endpoint: Set $4x = 2\pi$ and solve for x to obtain $x = \frac{\pi}{2}$.

Step 4: Determine the domain: The unit interval $u = \frac{1}{4}(p) = \frac{1}{4}(\frac{\pi}{2}) = \frac{\pi}{8}$.

$$x_1 = 0$$

$$x_2 = 0 + \frac{\pi}{8} = \frac{\pi}{8}$$

$$x_3 = \frac{\pi}{8} + \frac{\pi}{8} = \frac{2\pi}{8} \text{ or } \frac{\pi}{4}$$

$$x_4 = \frac{2\pi}{8} + \frac{\pi}{8} = \frac{3\pi}{8}$$

$$x_5 = \frac{3\pi}{8} + \frac{\pi}{8} = \frac{4\pi}{8} \text{ or } \frac{\pi}{2}$$

Step 5: The amplitude of $y = \sin 4x$ is 1 (i.e. $a = 1$).

The range is $-1 \le y \le 1$. The minimum point is at $y = -1$ and the maximum point is at $y = 1$.

Step 6: Label the x-axis with the x-values from Step 4 .

Step 7: Label the y-axis with the results from Step 5.

Step 8: Draw a complete sine curve starting from the origin $(0, 0)$) so that the curve repeats at the mark, $\frac{\pi}{2}$. (Review the properties of the sine curve).

However, we have not finished sketching yet, because the interval is $0 \le x \le 2\pi$. So we must repeat the sketch until we have drawn enough sine curves to complete the interval up to the 2π mark.

So, from the point $B(\frac{\pi}{2}, 0)$ we draw three more additional sine (cycles) curves (using the properties of the sine curve). For the interval $0 \le x \le 2\pi$, the number of cycles is b..

For $y = \sin 4x$, $0 \le x \le 2\pi$ the number of cycles is 4.

$y = \sin 4x$:

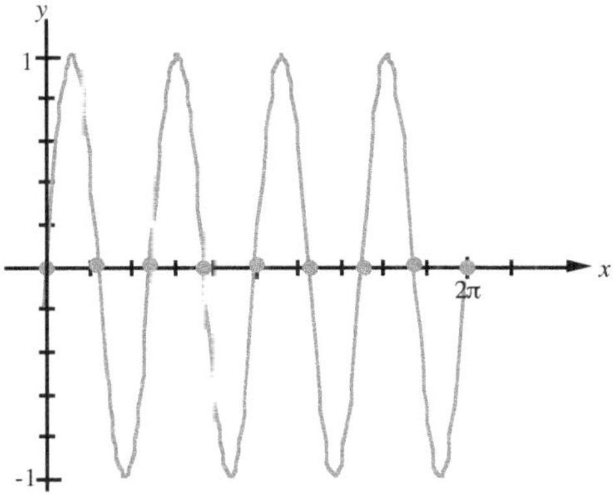

Method 2: Apply the transformation method

Example 10 Sketch the graph of $y = 3\sin 4x$ $\quad 0 \le x \le 2\pi$

Method 1

Step 1: The amplitude is 3

Step 2: Find the period: the period, $p = \frac{2\pi}{4} = \frac{\pi}{2}$

Step 3: Number of cycles = 4

The graph is shown below.

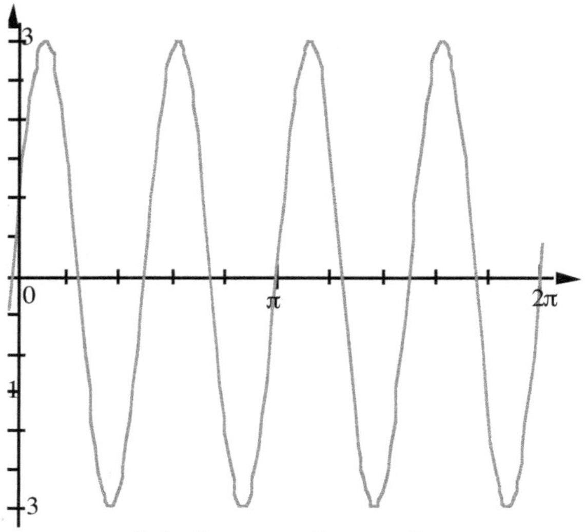

$y = 3\sin 4x$ $\quad 0 \le x \le 2\pi$

Method 2: Apply the transformation method

Example 11 Sketch $y = \sin x + 2$ $\quad 0 \le x \le 2\pi$

In standard form $y - 2 = \sin x$ Apply the transformation method)

Each point on the graph of $y = \sin x$ is shifted up 3 units, keeping the x-values unchanged.

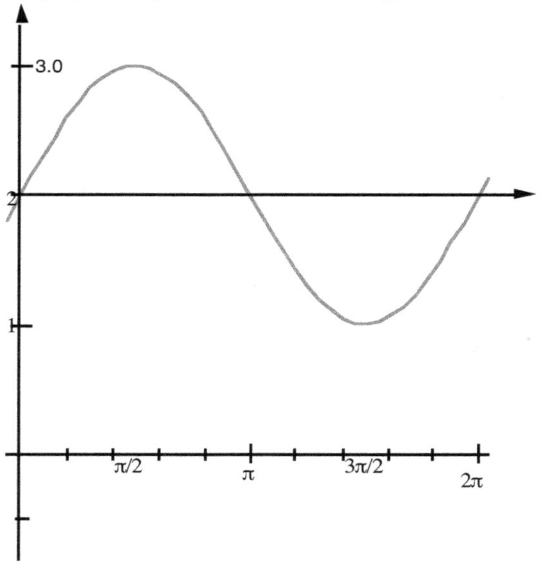

Example 12 Sketch the graph of $y - 2 = \sin(x + 3)$

Method 1

The period is $p = \frac{2\pi}{1} = 2\pi$

Domain: $0 \le x + 3 \le 6.28$

$0 - 3 \le x \le 6.28 - 3$

$-3 \le x \le 3.28$

The graph of the given equation is the graph of $y = \sin x$ with each x-coordinate shifted 3 units to the left and each y-coordinate shifted up 2 units, simultaneously.

Use decimal units to label the x-axis.

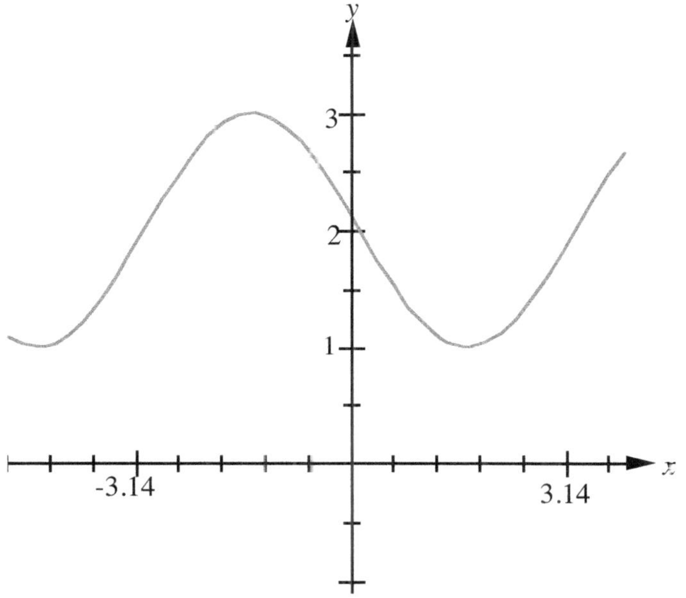

$y - 2 = \sin(x + 3)$

Method 2: Apply the transformation method

Sketching graphs of functions by addition of *y*-coordinates (ordinates)

Example Sketch the graph of $f(x) = \sin x - \cos x$,

Let the functions $f(x)$, $g(x)$ and $h(x)$ have the same domain and that $f(x) = g(x) + h(x)$ (1)

To sketch $f(x)$, we first sketch $g(x)$ and $h(x)$ separately on the same set of rectangular axes, then (at each) for each critical or important x-value, the corresponding y-values of $g(x)$ and $h(x)$) are added together. The sum so obtained gives the (new) y-value of $f(x)$. These summed y-values are plotted. The new points are then connected by a smooth curve. Note above that we add only the y-values keeping the x-values unchanged.

Example 13: Sketch the graph of $y = \sin x + \cos x$

Let $y_1 = \sin x$ and $y_2 = \cos x$

Step 1: Sketch the graph of $y_1 = \sin x$

Step 2: Sketch the graph of $y_2 = \cos x$ on the same axes as $y_1 = \sin x$

Step 3: For each x-coordinate of important or critical points add the corresponding y_1 and y_2 to obtain $y = y_1 + y_2$, and ordered pairs (x, y)

Step 4: Plot the ordered pairs from Step 3, noting that the x-coordinates remain the same.

Step 5: Connect the new points with a smooth curve.

Note that y_1 and y_2 are added algebraically, and we must take into consideration the signs (positive or negative) when we add of y_1 and y_2.

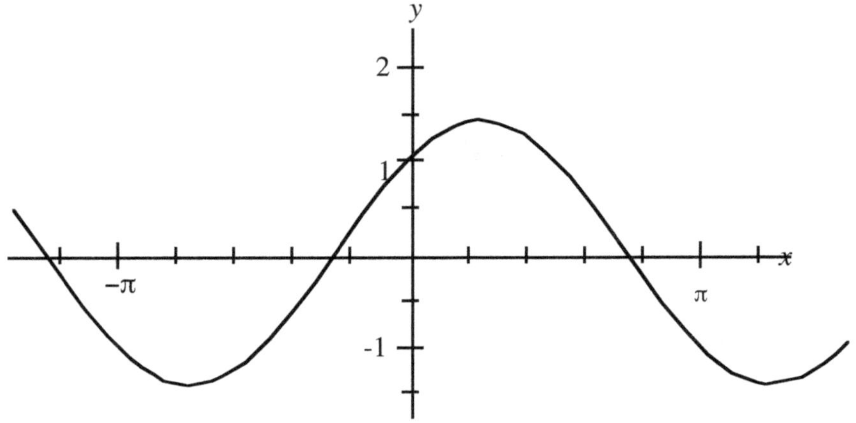

Example 14 Sketch the graph of $y = \sec x$ for $2\pi \le x \le 3\pi$

Solution See graph below

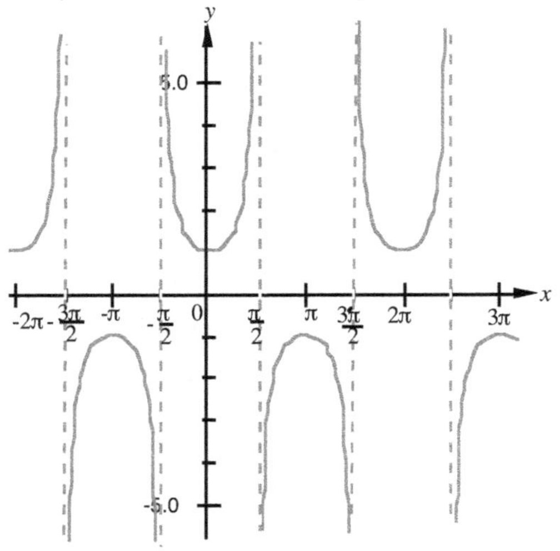

Example 15: Sketch the graph of $y = 2\tan(x + \frac{\pi}{3})$

Method 1

Solution Before proceeding, see page 187 to review the basic properties of the tangent curve.

Step 1: Find the period p. $b = 1$, and $p = \frac{\pi}{b} = \frac{\pi}{1} = \pi$

Step 2: Find the beginning-point x-coordinate, x_1, of the first cycle which locates a vertical asymptote.

Set $x + \frac{\pi}{3} = -\frac{\pi}{2}$, and solve for x to obtain $x_1 = -\frac{5\pi}{6}$ <---This is a vertical asymptote.

Step 3: Find the end-point x-coordinate, x_5, of the first cycle which locates a vertical asymptote.

Set $x + \frac{\pi}{3} = \frac{\pi}{2}$ and solve for x to obtain $x_5 = \frac{\pi}{6}$.<---This is a vertical asymptote.

(We may combine Steps 2 and 3, by solving the domain inequality: $-\frac{\pi}{2} \le x + \frac{\pi}{3} \le \frac{\pi}{2}$ to obtain

$-\frac{5\pi}{6} \le x \le \frac{\pi}{6}$ (The first cycle of the curve is between the asymptotes $x = -\frac{5\pi}{6}$ and $x = \frac{\pi}{6}$)

Step 4: Find the x-intercepts: Set $y = 0$ in $y = 2\tan(x + \frac{\pi}{3})$

Then $0 = 2\tan(x + \frac{\pi}{3})$ or $\tan(x + \frac{\pi}{3}) = 0$

$x + \frac{\pi}{3} = 0$ or π. When $x + \frac{\pi}{3} = 0$, $x = -\frac{\pi}{3}$. When $x + \frac{\pi}{3} = \pi$, $x = \frac{2\pi}{3}$

Since for the cycle of interest, $-\frac{5\pi}{6} \le x \le \frac{\pi}{6}$, we consider only $x = -\frac{\pi}{3}$.

The x-intercept is at $\left(-\frac{\pi}{3}, 0\right)$

Step 5: Find the y-intercept. Let $x = 0$ in $y = 2\tan(x + \frac{\pi}{3})$

Then $y = 2\tan(0 + \frac{\pi}{3})$ and $y = 2\tan\frac{\pi}{3}$ or $y = 2(1.732)$ ($\tan\frac{\pi}{3} = 1.732$)

$y = 3.46$. The y-intercept is at $(0, 3.5)$.

Step 6: Determine the domain (x-values) to use to obtain points (for the first cycle).

Divide the period p by 4 to obtain "the quarter -interval", $\frac{p}{4}$, for the x-axis.

Add $\frac{p}{4}$ successfully to x_1 until x_5 is reached (to obtain the x-values for the next Step)

$$\frac{p}{4} = \frac{\pi}{1} \bullet \frac{1}{4} \text{ or } \frac{1}{4} \text{of} \left(\frac{\pi}{1}\right) \qquad (p = \pi)$$

$$= \frac{\pi}{4} < - - - - \frac{p}{4}$$

$x_1 = -\frac{5\pi}{6}$ (from Step 2, Also a vertical asymptote)

$x_2 = -\frac{5\pi}{6} + \frac{\pi}{4}$

$= -\frac{7\pi}{12}$

$x_3 = x_2 + \frac{\pi}{8} = -\frac{7\pi}{12} + \frac{\pi}{4} = -\frac{4\pi}{12}$

$x_4 = -\frac{4\pi}{12} + \frac{\pi}{4}$

$= -\frac{\pi}{12}$

$x_5 = -\frac{\pi}{12} + \frac{\pi}{4}$

$= \frac{2\pi}{12}$ or $\frac{\pi}{6}$ (From Step 3. A vertical asymptote)

Summarizing, the x-coordinates are $-\frac{5\pi}{6}$ or $-\frac{10\pi}{12}$, $-\frac{7\pi}{12}$, $-\frac{4\pi}{12}$, $-\frac{\pi}{12}$, and $\frac{2\pi}{12}$, noting that the first and the last values locate vertical asymptotes.

Step 7: Label the x-axis with the summary from Step 6.

Step 8: Determine the corresponding y-coordinates by substituting the x-coordinates $-\frac{7\pi}{12}$, $-\frac{4\pi}{12}$, $-\frac{\pi}{12}$ in the equation $y = 2\tan(x + \frac{\pi}{3})$ to obtain -2, 0, and 2 respectively. (For calculations, see Box 3, below)

Step 9: Pair these y-values with $-\frac{7\pi}{12}$, $-\frac{4\pi}{12}$, $-\frac{\pi}{12}$ respectively to obtain

$$\left(-\tfrac{7\pi}{12}, -2\right), \left(-\tfrac{4\pi}{12}, 0\right), \left(-\tfrac{\pi}{12}, 2\right)$$

Step 10: Label the y-axis

Step 11: Plot the ordered pairs from Step 9.

Step 12: Draw the vertical asymptotes $x_1 = -\frac{5\pi}{6}$ and $x_5 = \frac{\pi}{6}$

Step 13: Connect the three points plotted noting the concavity , the characteristic shape of the tangent curve and extend the curve so that each end is asymptotic to its asymptote.

Box 3: Calculations for Step 8

$$\left.\begin{aligned} x_2 = -\tfrac{7\pi}{12},\ y &= 2\tan(-\tfrac{7\pi}{12} + \tfrac{\pi}{3}) \\ &= 2\tan(-\tfrac{3\pi}{12}) \\ &= 2(-\tan\tfrac{3\pi}{12}) = 2(-1) = -2 \end{aligned}\right\} \Rightarrow (-\tfrac{7\pi}{12}, -2)$$

$$\left.\begin{aligned} x_3 = -\tfrac{4\pi}{12},\ y &= 2\tan(-\tfrac{4\pi}{12} + \tfrac{\pi}{3}) \\ &= 2\tan(0) \\ &= 0 \end{aligned}\right\} \Rightarrow (-\tfrac{4\pi}{12}, 0)$$

$$\left.\begin{aligned} x_4 = -\tfrac{\pi}{12},\ y &= 2\tan(-\tfrac{\pi}{12} + \tfrac{\pi}{3}) \\ &= 2\tan(\tfrac{3\pi}{12}) = 2(1) = 2 \end{aligned}\right\} \Rightarrow (-\tfrac{\pi}{12}, 2)$$

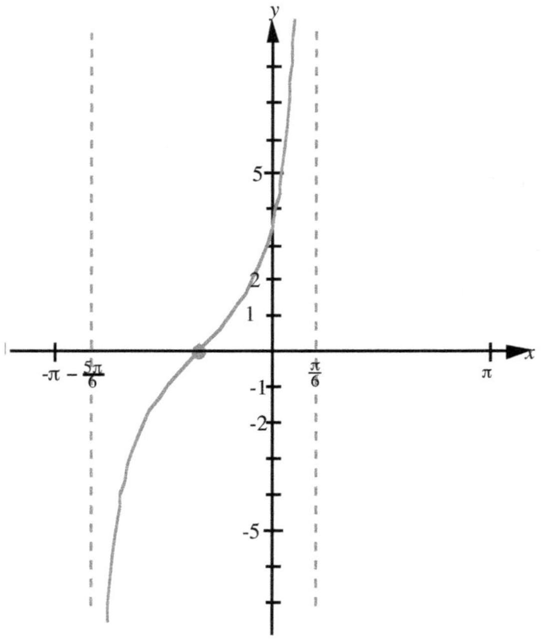

Method 2: Apply the transformation method (imitate page 203.

Lesson 33 Exercises

Sketch the graph of each of the following:

1. $y = 4\sin(2x + \pi)$; **2.** $y = -2\sin(3x + 4)$; **3.** $y - 1 = 3\cos(x - \frac{\pi}{4})$;

4. $y = 3\tan\left(x - \frac{\pi}{4}\right)$ for $0 \le x \le 2\pi$; **5.** $y = -\cos x$ for $0 \le x \le 2\pi$.

Answers:

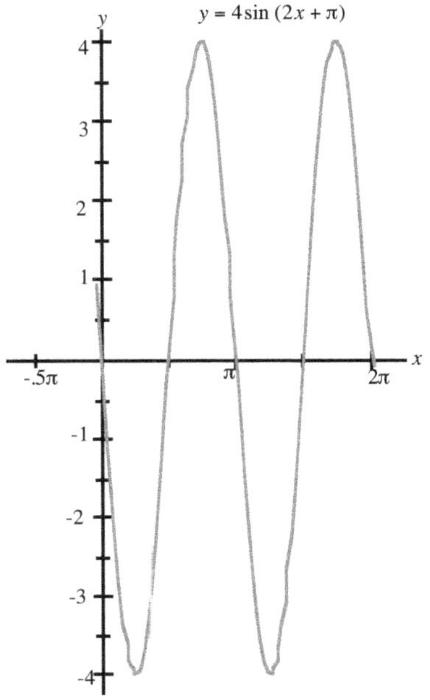

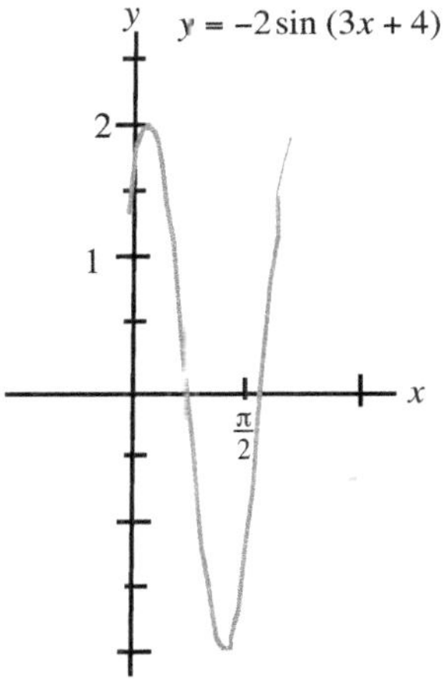

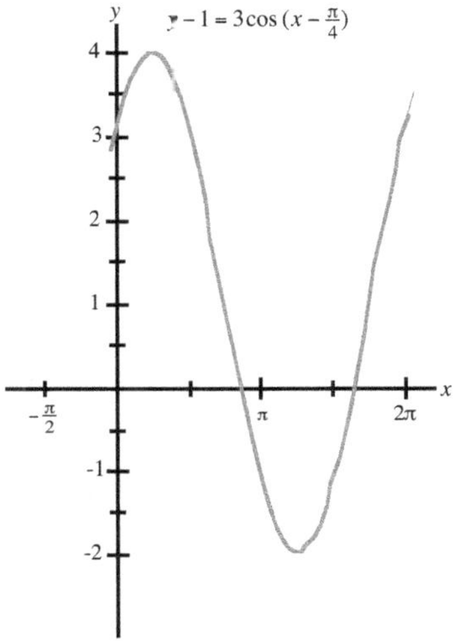

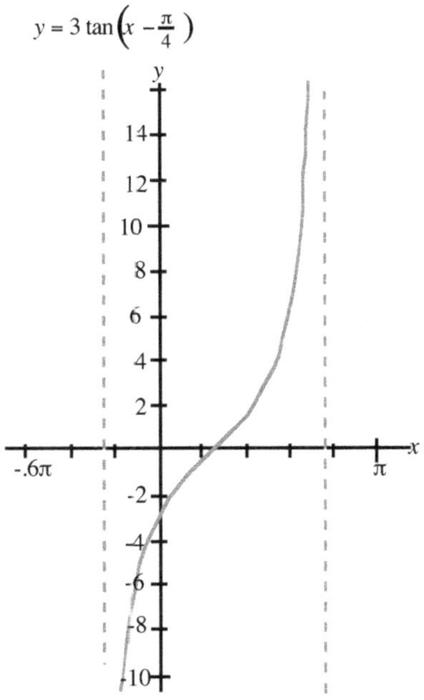

CHAPTER 12

From Trigonometric Functions to
Inverse Trigonometric Functions

("Exchange is no Robbery")

Lesson 34: **Introduction; Definitions; Operations Involving Inverse Trigonometric Functions**
Lesson 35: **How to Sketch the Graph of** $y = \text{Arcsin}\, x$
Lesson 36: **How to Sketch the Graph of** $y = \text{Arccos}\, x$
Lesson 37: **How to Sketch the Graph of** $y = \text{Arctan}\, x$

Lesson 34

Introduction; Definitions; Operations Involving Inverse Trigonometric Functions

We use inverse trigonometric functions in solving trigonometric equations. In calculus, we use inverse trigonometric functions to obtain formulas for various results involving algebraic functions.

In Chapter 6, we learned how to find the functional values (y-values), given the measures of the angles. There, we obtained the functional values using trigonometric tables, the electronic calculator or from memory. We also learned how to find the measures of angles given the functional values.
With inverse trigonometric functions, we are also given the functional value, say y, and we are asked to find an angle or a number, say x. The process of finding y and the process of finding x are inverse operations (in much the same way as multiplication and division are inverse operations).

In Chapter 2, we also learned how to determine the inverses of algebraic functions, given the various functional forms. If the function was specified by a set form, we obtained its inverse by interchanging the first and second components of each ordered pair. With trigonometric functions, interchanging the first and second components of each ordered pair of numbers generally produce relations which are **not** functions because of the periodicity of the trigonometric functions.
Consider the **trigonometric relation** $y = \sin x$, which is specified by the set of ordered pairs,

$A = \{(-2\pi, 0), (-\frac{3\pi}{2}, 1), (-\pi, 0), (-\frac{\pi}{2}, -1), (0, 0), (\frac{\pi}{2}, 1), (\pi, 0), (\frac{3\pi}{2}, -1), (2\pi, 0)\}$.

This relation is also a **function** since no two distinct ordered pairs have the same first component. (A function is a set of ordered pairs of numbers in which no two distinct ordered pairs have the same first component).
On interchanging the first and second components of each ordered pair, we obtain the inverse r elation, $B = \{(0, -2\pi), (1, -\frac{3\pi}{2}), (0, -\pi), (-1, -\frac{\pi}{2}), (0, 0), (1, \frac{\pi}{2}), (0, \pi), (-1, \frac{3\pi}{2})\ (0, 2\pi)\}$.

This new **relation** is **not a function** since the first, the third, the fifth, the seventh and the ninth ordered pairs have the same first component, namely 0, or also, the second, and the sixth ordered pairs have the same first component, namely 1. We observe that the original set, A, does **not** represent a **one-to-one function** and therefore, the inverse relation is not a function. The graph in the figure below fails the horizontal line test.

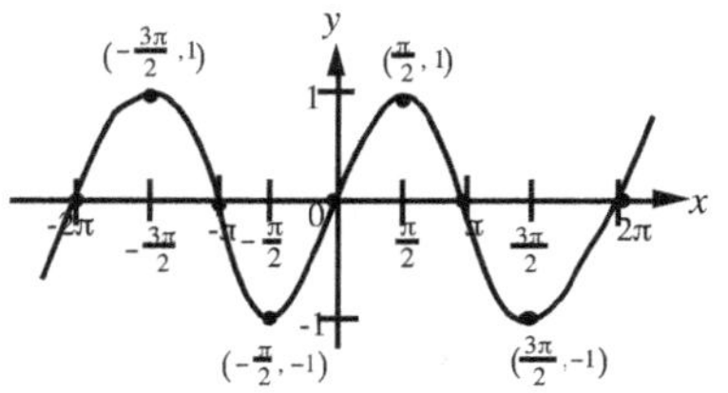

To make the set B a function, we can eliminate some of the ordered pairs so that no two distinct ordered pairs have the same first component. If we consider the ordered pair $(0, 0)$ and choose ordered pairs to its right and ordered pairs to its left so that no two distinct ordered pairs have the same first component, we obtain the following: $C = \{(-1, -\frac{\pi}{2}), (0, 0), (1, \frac{\pi}{2})\}$..This set C is a **function** since all the first components are different from one another. We observe also that in the original function given by set A, if we keep the fourth, the fifth and the sixth ordered pairs and drop the other pairs, we obtain $\{(-\frac{\pi}{2}, -1), (0, 0), (\frac{\pi}{2}, 1)\}$ which is a one-to-one function and therefore, its inverse relation is also a function.

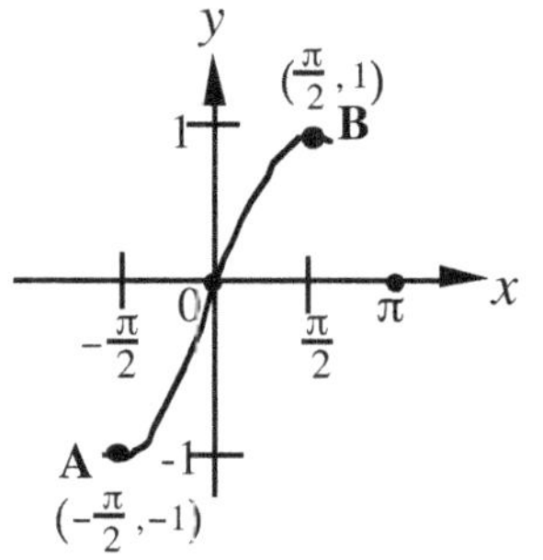

Fig 1. $y = \sin x$ for $-\frac{\pi}{2} \le x \le \frac{\pi}{2}$
$-1 \le y \le 1$

Principal branch

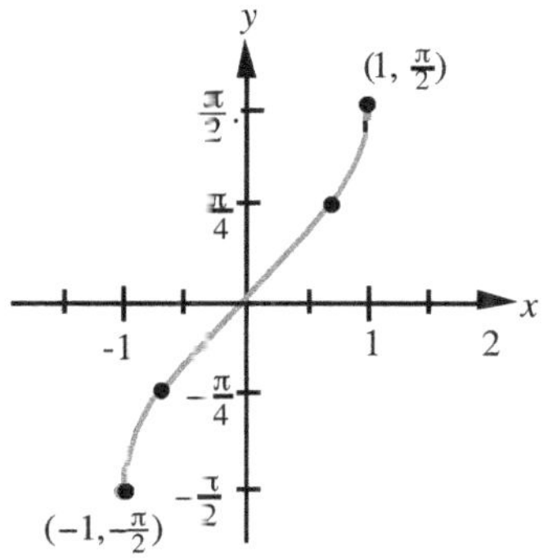

Fig. 2 $x = \sin y, -\frac{\pi}{2} \le y \le \frac{\pi}{2}$.
$-1 \le x \le 1$

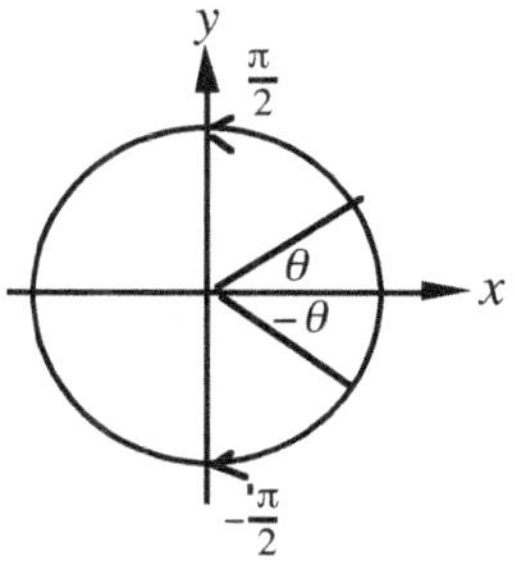

Fig. 3: For $-\frac{\pi}{2} \le y \le \frac{\pi}{2}$.

From above, we notice that the inverse relation for the sine function, $y = \sin x$, is generally **not** a function. Similarly, the inverse relations for $y = \cos x$, $y = \sec x$, $y = \csc x$, $y = \tan x$, and $y = \tan x$ are generally **not** functions.

Some of these inverse relations are many-valued and consist of a collection of single-valued functions known as branches. Since it is more convenient to work with functions, we will make these inverse relations become functions by restricting their ranges. By placing restrictions on the domains of those original functions which are **not** one-to-one, we obtain a particular branch called the **principal branch** of each inverse relation. Each branch **chosen** is an inverse function and we capitalize the first character of the functional name or its abbreviation. (Note that some authors do not adhere to this **capitalization** and in which case the principal branch and its range are to be understood).

Inverse Sine Function

Definition Given, the equation, $y = \sin x$ $\quad -\frac{\pi}{2} \le x \le \frac{\pi}{2}$, what is its inverse?

Since the inverse is found by **interchanging** the roles of x and y,

Domain ↓ Range ↓

the inverse of $y = \sin x$ for $-\frac{\pi}{2} \le x \le \frac{\pi}{2}$, $\quad -1 \le y \le 1$ is

$x = \sin y$ for $-\frac{\pi}{2} \le y \le \frac{\pi}{2}$, $\quad -1 \le x \le 1$

Range Domain

(Observe that the domain and range are interchanged.)

If we solve $x = \sin y$ for y, and use the notation, $\mathrm{Sin}^{-1}x$, we obtain

$y = \mathrm{Sin}^{-1}x$ (Recall previously that the inverse of $f(x)$ was symbolized $f^{-1}(x)$)

Another notation for this inverse is $y = \mathrm{Arcsin}\, x$. We may use either $y = \mathrm{Sin}^{-1}x$ or $y = \mathrm{Arcsin}\, x$.

Example 1 If $\frac{\pi}{6} = \text{Arcsin}\left(\frac{1}{2}\right)$ or $\text{Sin}^{-1}\left(\frac{1}{2}\right)$, then $\sin\frac{\pi}{6} = \frac{1}{2}$

(is read " if $\frac{\pi}{6}$ is the angle or number whose sine is $\frac{1}{2}$.")
(Memorize this example to help remember the definitions)
$y = \text{Arcsin } x$ (y is the angle or the number whose sine is x.

Inverse Cosine Function

Definition Given, the equation, $y = \cos x$ $\quad 0 \le x \le \pi$, what is its inverse?

The inverse is found by interchanging the roles of x and y.

Domain ↓ Range ↓

The inverse of $y = \cos x$ for $0 \le x \le \pi$, $-1 \le y \le 1$ is
$x = \cos y$ for $0 \le y \le \pi$, $-1 \le x \le 1$

Range Domain

If we solve $x = \cos y$ for y, and use the notation, $\text{Cos}^{-1}x$, we obtain $y = \text{Cos}^{-1}x$ or $y = \text{Arccos } x$

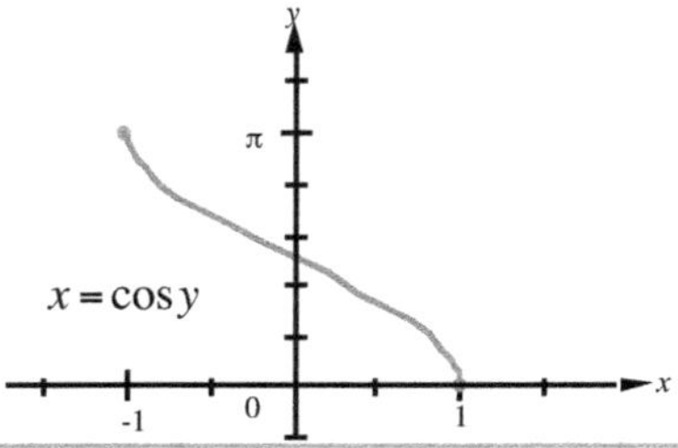

Inverse Tangent Function

Definition Given, the equation, $y = \tan x$, $-\frac{\pi}{2} < x < \frac{\pi}{2}$, what is its inverse?

The inverse is found by interchanging the roles of x and y.

Domain ↓ Range ↓

The inverse of $y = \tan x$ for $-\frac{\pi}{2} < x < \frac{\pi}{2}$, $-\infty < y < \infty$ is
$x = \tan y$ for $-\frac{\pi}{2} < y < \frac{\pi}{2}$, $-\infty < x < \infty$

Range Domain

If we solve $x = \tan y$ for y, and use the notation, $\text{Tan}^{-1}x$, we obtain $y = \text{Tan}^{-1}x$ or $y = \text{Arctan } x$

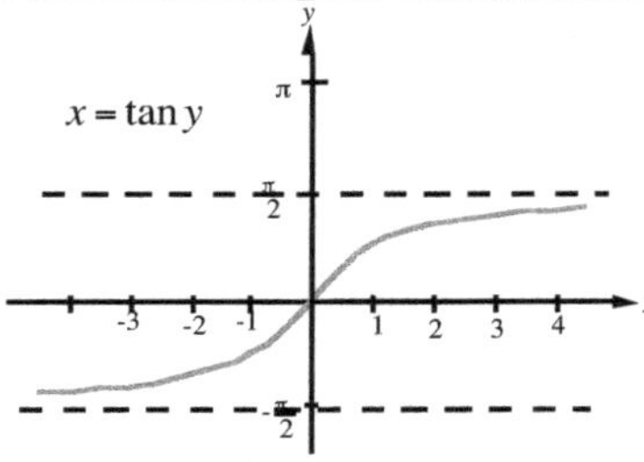

Other Inverse Trigonometric Functions

The inverse functions $\text{Csc}^{-1}x$, $\text{Sec}^{-1}x$, and $\text{Cot}^{-1}x$ can be expressed in terms of $\text{Sin}^{-1}x$, $\text{Cos}^{-1}x$, and $\text{Tan}^{-1}x$ respectively as follows:

$\text{Csc}^{-1}x = \text{Sin}^{-1}\left(\frac{1}{x}\right)$ for $|x| \ge 1$; $\text{Sec}^{-1}x = \text{Cos}^{-1}\left(\frac{1}{x}\right)$ for $|x| \ge 1$; and $\text{Cot}^{-1}x = \text{Tan}^{-1}\left(\frac{1}{x}\right)$.

For example, $\text{Csc}^{-1}2 = \text{Sin}^{-1}\left(\frac{1}{2}\right) = \frac{\pi}{6}$. That is, to find $\text{Csc}^{-1}2$, first find the reciprocal of 2, which is $\frac{1}{2}$ and then find $\text{Sin}^{-1}\left(\frac{1}{2}\right)$, which is $\frac{\pi}{6}$. This approach gives only the reference angle and you have to place the angle in standard position for the original reciprocal inverse function according to the quadrant for its definition. For example, if x is negative and we use $\text{Cot}^{-1}x = \text{Tan}^{-1}\left(\frac{1}{x}\right)$, we must place the reference angle in the second quadrant for $\text{Cot}^{-1}x$, and write the standard position angle.

Choice of the ranges and quadrants for the inverse functions (See also p. 222)

If x is positive, $\text{Sin}^{-1}x$ $\text{Cos}^{-1}x$, $\text{Tan}^{-1}x$, $\text{Csc}^{-1}x$, $\text{Sec}^{-1}x$, and $\text{Cot}^{-1}x$ are all in the first quadrant, or appropriately quadrantal. However, if x is negative, $\text{Sin}^{-1}x$,$\text{Csc}^{-1}x$, or $\text{Tan}^{-1}x$ are in the fourth quadrant or appropriately quadrantal and they are expressed as negative angles; while if x is negative, $\text{Cos}^{-1}x$, $\text{Sec}^{-1}x$, and $\text{Cot}^{-1}x$ are in the second quadrant or appropriately quadrantal.

Note the difference between the quadrants for, say, the angle, $\text{Sin}^{-1}x$ (Fig. 3, previous page) and the quadrants for the graph of $y = \text{Sin}^{-1}x$ (Fig. 2, previous page)..The graph consists of the points (x,y).

We summarize the definitions of the **inverse trigonometric functions** below. The restrictions made to obtain the inverse functions are arbitrary, and other branches may be used. The above branches are the ones that are usually used.

Function	Domain	Range	Inverse	Notation	Domain	Range
$y = \sin x$,	$-\frac{\pi}{2} \le x \le \frac{\pi}{2}$,	$-1 \le y \le 1$	$x = \sin y$,	$y = \text{Arc}\sin x$ or $\text{Sin}^{-1}x$;	$-1 \le x \le 1$;	$-\frac{\pi}{2} \le y \le \frac{\pi}{2}$
$y = \cos x$,	$0 \le x \le \pi$,	$-1 \le y \le 1$	$x = \cos y$,	$y = \text{Arc}\cos x$ or $\text{Cos}^{-1}x$;	$-1 \le x \le 1$;	$0 \le y \le \pi$
$y = \tan x$,	$-\frac{\pi}{2} < x < \frac{\pi}{2}$,	$-\infty < y < \infty$	$x = \tan y$,	$y = \text{Arc}\tan x$ or $\text{Tan}^{-1}x$;	$-\infty < x < \infty$;	$-\frac{\pi}{2} < y < \frac{\pi}{2}$
$y = \cot x$,	$0 < x < \pi$,	$-\infty < y < \infty$	$x = \cot y$,	$y = \text{Arc}\cot x$ or $\text{Cot}^{-1}x$;	$-\infty < x < \infty$;	$0 < y < \pi$
(Also: $\text{Tan}^{-1}x + \text{Cot}^{-1}x = \frac{\pi}{2}$)				or $y = \text{Cot}^{-1}x = \text{Tan}^{-1}\left(\frac{1}{x}\right) = \frac{\pi}{2} - \text{Tan}^{-1}x$;		$0 < y < \pi$
$y = \sec x$,	$0 \le x \le \pi,\ x \ne \frac{\pi}{2}$,	$\lvert y\rvert \ge 1$	$x = \sec y$,	$y = \text{Arc}\sec x$ or $\text{Sec}^{-1}x$;	$\lvert x\rvert \ge 1$,	$0 \le y \le \pi,\ y \ne \frac{\pi}{2}$ Also,
				$y = \text{Sec}^{-1}x = \text{Cos}^{-1}\left(\frac{1}{x}\right)$;	$\lvert x\rvert \ge 1$	$0 \le y \le \pi,\ y \ne \frac{\pi}{2}$
$y = \csc x$,	$-\frac{\pi}{2} \le x \le \frac{\pi}{2},\ x \ne 0$,	$\lvert y\rvert \ge 1$	$x = \csc y$,	$y = \text{Arc}\csc x$ or Csc^{-1};	$\lvert x\rvert \ge 1$,	$-\frac{\pi}{2} \le y \le \frac{\pi}{2},\ y \ne 0$
(Also: $\text{Sec}^{-1}x + \text{Csc}^{-1}x = \frac{\pi}{2}$)				Also, $y = \text{Csc}^{-1}x = \text{Sin}^{-1}\left(\frac{1}{x}\right)$;	$\lvert x\rvert \ge 1$,	$-\frac{\pi}{2} \le y \le \frac{\pi}{2},\ y \ne 0$

Inverse Operations

Finding the **sine** of an angle or a number and finding the inverse sine or **arcsine** are inverse processes of each other in much the same way that **multiplication and division** are inverse operations of each other. Two operations which are inverses of each other undo (reverse the action of) each other. Thus the function $\text{Arcsin}\,x$ reverses the action of $\sin x$, and vice versa.

Example 2 Find Arcsin (0.866). (Note that Arcsin (.866) is an angle or a number.)

Step 1: Let $y = \text{Arcsin}\ (0.866)$ $\quad -\frac{\pi}{2} \le y \le \frac{\pi}{2}$. (We will find y, the angle whose sine is .866)

Two conditions must be satisfied: 1. $\sin y = 0.866$); **2.** $-\frac{\pi}{2} \le y \le \frac{\pi}{2}$.

Step 2: Find the reference angle for the angle whose sine 0.866

From memory or calculator (or tables), the reference angle is $\frac{\pi}{3}$. (Note that $\frac{\pi}{3} = 60°$)

Step 3: Find the quadrant in which the sine function is positive.
The sine function is positive in the **first** and second quadrants.

Since by definition, $\text{Arcsin}\,x$ must be in either the **first** or the fourth quadrant, $(-\frac{\pi}{2} \le y \le \frac{\pi}{2})$,

$\text{Arcsin}\,x$ is in the **first** quadrant where it is a positive angle.

Therefore, $\text{Arcsin}\ (0.866) = \frac{\pi}{3}$. (Note that $-\frac{\pi}{2} \le \frac{\pi}{3} \le \frac{\pi}{2}$)

Example 3 Find Arcsin (-0.5). (Note that Arcsin (- 0.5) is an angle or a number.)

Step 1: Let $y = \text{Arcsin}\ (-0.5)$ $\quad -\frac{\pi}{2} \le y \le \frac{\pi}{2}$. (We will find y, the angle whose sine is -0.5)

Step 2: Find the reference angle for the angle whose sine is - 0.5.

From tables or memory (or calculator), the reference angle is $\frac{\pi}{6}$.

Step 3: Find the quadrant in which the sine function is negative. The sine function is negative in the third and **fourth** quadrants. Since $\text{Arcsin}\,x$ must be in either the first or the **fourth** quadrant,

$(-\frac{\pi}{2} \le y \le \frac{\pi}{2})$, $\text{Arcsin}\,x$ is in the fourth quadrant where it is a negative angle.

Therefore, $\text{Arcsin}\ (-0.5) = -\frac{\pi}{6}$. Summarily, $\text{Sin}^{-1}(-x) = -\text{Sin}^{-1}x$

Example 4 Find $\text{Cot}^{-1}(-2.7)$ using a calculator.

Method 1: Step 1: Find $\frac{1}{2.7} = 0.37037037$ (Summarily, $\text{Cot}^{-1}(-2.7) = \pi - \text{Cot}^{-1}(2.7) = \pi - \text{Tan}^{-1}(\frac{1}{2.7})$)

Step 2: Find $\text{Tan}^{-1}\left(\frac{1}{2.7}\right) = \text{Tan}^{-1}(0.37037037) = 0.35$ <--- Reference angle in radians

The reference angle for $\text{Cot}^{-1}(-2.7) = 0.35$. By definition, when x is negative, $\text{Cot}^{-1}x$ is in the second quadrant and the terminal side of the angle is in the second quadrant.

Therefore, $\text{Cot}^{-1}(-2.7) = \pi - 0.35 \boxed{= 2.79}$.

Method 2: Applying $\text{Cot}^{-1}x + \text{Tan}^{-1}x = \frac{\pi}{2}$ and $\text{Tan}^{-1}(-x) = -\text{Tan}^{-1}x$, $-\infty < x < \infty$, see also p.222..

$$\text{Cot}^{-1}(-2.7) = \frac{\pi}{2} - \text{Tan}^{-1}(-2.7) = \frac{\pi}{2} - [-\text{Tan}^{-1}(2.7)] \boxed{= \frac{\pi}{2} + \text{Tan}^{-1}(2.7)} = 1.57 + 1.216 \boxed{= 2.79}$$

General Composition of Trigonometric and Inverse Trigonometric Functions

Consider the following question at the beginning level of trigonometry.

" If $\sin x = \frac{1}{2}$, and the terminal side of x is in the first quadrant, find $\cot x$ ".

However, at the present level, we might pose the same question as:

" find $\cot\left(\text{Sin}^{-1}\frac{1}{2}\right)$ " <--Find the cotangent of the angle between $-\frac{\pi}{2}$ and $\frac{\pi}{2}$ whose sine is $\frac{1}{2}$.

Example 5 Find the value of $\cot\left(\text{Sin}^{-1}\left(\frac{1}{2}\right)\right)$.

Solution: Same as if $\sin\theta = \frac{1}{2}$, and θ is in the first quadrant, find $\cot\theta$.

(or If $\sin\theta = \frac{1}{2}$ and $-\frac{\pi}{2} \le \theta \le \frac{\pi}{2}$, find $\cot\theta$.) (Note $\sin\theta = \sin\frac{\pi}{6} = \frac{1}{2}$, $\theta = \frac{\pi}{6}$)

Step 1: Draw a right triangle, and let $\text{Sin}^{-1}\left(\frac{1}{2}\right) = \theta$

Then $\sin\theta = \frac{\text{opposite side}}{\text{hypotenuse}} = \frac{1}{2}$, so that the side opposite θ is 1, and the hypotenuse is 2.

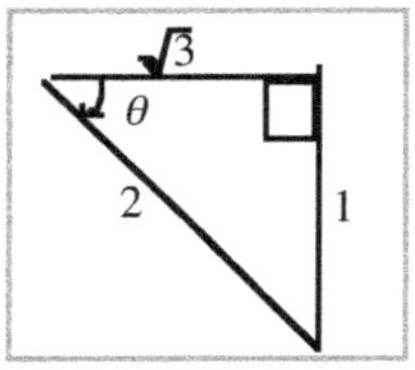

Step 2 Using the Pythagorean theorem, the side adjacent to θ is found to be $\sqrt{3}$.

Step 3: Since $\sin\theta = +\frac{1}{2}$, $\text{Sin}^{-1}\left(\frac{1}{2}\right)$ is in the first quadrant

Step 4: $\cot\theta = \frac{\text{adjacent side}}{\text{opposite side}} = \frac{\sqrt{3}}{1} = \sqrt{3}$

.Step 5: Determine the sign of $\cot\theta$. Since $\text{Sin}^{-1}\left(\frac{1}{2}\right)$ is in the first quadrant, where $\cot\theta$ is also positive, $\cot\left(\text{Sin}^{-1}\left(\frac{1}{2}\right)\right) = \sqrt{3}$.

Example 6 Find the value of $\cot\left(\text{Sin}^{-1}\left(-\frac{1}{2}\right)\right)$. Same as if $\sin\theta = -\frac{1}{2}$, and θ is in the fourth quadrant, find $\cot\theta$.(or If $\sin\theta = -\frac{1}{2}$ and $-\frac{\pi}{2} \le \theta \le \frac{\pi}{2}$, find $\cot\theta$) See also p.222.

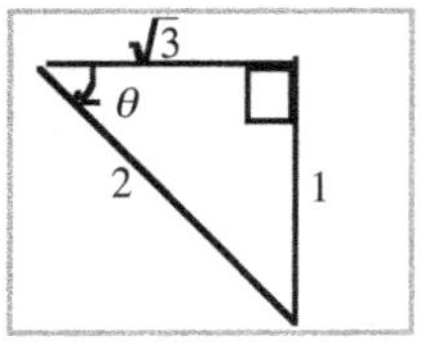

Step 1: Draw a right triangle, and let $\text{Sin}^{-1}\left(\frac{1}{2}\right) = \theta$. Then $\sin\theta = \frac{\text{opposite side}}{\text{hypotenuse}} = \frac{1}{2}$, so that the side opposite θ is 1, and the hypotenuse is 2.

Step 2 Using the Pythagorean theorem, the side adjacent to θ is found to be $\sqrt{3}$

Step 3: $\sin\theta = -\frac{1}{2}$ (negative), $\text{Sin}^{-1}\left(-\frac{1}{2}\right)$ is in the fourth quadrant, See also p.222.

Step 4: $\cot\theta = \dfrac{\text{adjacent side}}{\text{opposite side}} = \dfrac{\sqrt{3}}{1} = \sqrt{3}$.

Step 5: Determine the sign of $\cot\theta$. Since $\text{Sin}^{-1}\left(-\frac{1}{2}\right)$ is in the fourth quadrant, where the cotangent is negative, $\cot\left(\text{Sin}^{-1}\left(-\frac{1}{2}\right)\right) = -\sqrt{3}$.

Special Cases of Composition of Trigonometric Functions

Example The function $y = \text{Sin}^{-1}x$ and the restricted sine function $y = \sin x$ are inverses of each other, and as such, each reverses the action of the other,

Definitions

	Note
$\text{Sin}^{-1}(\sin x) = x;\quad -\frac{\pi}{2} \le x \le \frac{\pi}{2}$ <-domain of the inside function $\sin x$	$\pi \approx 3.14$
$\sin(\text{Sin}^{-1}x) = x;\quad -1 \le x \le 1$ <-domain of the inside function $\sin^{-1}x$	$\frac{\pi}{2} \approx \frac{3.142}{2} = 1.57$
$\cos(\text{Cos}^{-1}x) = x;\quad -1 \le x \le 1$ <-domain of the inside function $\cos^{-1}x$	
$\text{Cos}^{-1}(\cos x) = x;\quad 0 \le x \le \pi$ <-domain of the inside function $\cos x$	
$\tan(\text{Tan}^{-1}x) = x;\quad -\infty < x < \infty$ <-domain of the inside function $\tan^{-1}x$	
$\text{Tan}^{-1}(\tan x) = x;\quad -\frac{\pi}{2} < x < \frac{\pi}{2}$ <-domain of the inside function $\tan x$	

Examples:

Case 1

Example 1 Evaluate: $\text{Sin}^{-1}(\sin\frac{\pi}{6})$ (A)

Step 1: Evaluate the inside function. From memory or tables $\sin\frac{\pi}{6} = \frac{1}{2}$

Step 2: Evaluate $\text{Sin}^{-1}\left(\frac{1}{2}\right)$

Step 3: Find the angle in radians or number between $-\frac{\pi}{2}$ and $\frac{\pi}{2}$ whose sine is $\frac{1}{2}$

From memory or tables the angle is $\frac{\pi}{6}$.

Therefore $\text{Sin}^{-1}(\sin\frac{\pi}{6}) = \frac{\pi}{6}$ (B)

Observe that we could have obtained the angle $\frac{\pi}{6}$ by inspection and applying the definition,

$$\sin^{-1}(\sin x) = x;\ \text{if } -\frac{\pi}{2} \le x \le \frac{\pi}{2} \quad \text{or } -1.57 \le x \le 1.57 \qquad (C)$$

Here, $x = \frac{\pi}{6} \approx 0.52$ and $-\frac{\pi}{2} \le \frac{\pi}{6} \le \frac{\pi}{2}$ or $-1.57 \le 0.52 \le 1.57$

Example 2: Evaluate $\text{Sin}^{-1}(\sin\frac{\pi}{4})$

The argument $\frac{\pi}{4}$ of the inside function is between $-\frac{\pi}{2}$ and $\frac{\pi}{2}$, and satisfies the domain

$-\frac{\pi}{2} \le x \le \frac{\pi}{2}$ of the restricted sine function.

Therefore, the domain of the function is satisfied, and definition (C) above is applicable, and

$\text{Sin}^{-1}(\sin\frac{\pi}{4}) = \frac{\pi}{4}$ (The inverse sine function undoes what the sine function does)

Case 2 Example 3 Evaluate $\sin\left(\text{Sin}^{-1}\frac{1}{2}\right)$

The argument $\frac{1}{2}$ of the inside function is between -1 and 1, and satisfies the domain $-1 \le x \le 1$. Therefore, $\sin\left(\text{Sin}^{-1}\frac{1}{2}\right) = \frac{1}{2}$. (The sine function undoes what the inverse sine function does.)

Example 4 Evaluate $\sin\left(\text{Sin}^{-1}4\right)$

The argument 4 of the inside function does not satisfy the domain $-1 \le x \le 1$ of $\text{Sin}^{-1}x$. Therefore, $\sin\left(\text{Sin}^{-1}4\right)$ is not defined. There is no angle or number between $-\frac{\pi}{2}$ and $\frac{\pi}{2}$ whose sine is 4.

Case 3 Example 5 Evaluate $\text{Sin}^{-1}(\sin\frac{3\pi}{4})$

The argument $\frac{3\pi}{4}$ of the inside function is **not** between $-\frac{\pi}{2}$ and $\frac{\pi}{2}$ does not satisfy the domain $-\frac{\pi}{2} \le x \le \frac{\pi}{2}$ and therefore $\text{Sin}^{-1}(\sin x) = x$ is not applicable. However, $\sin\frac{3\pi}{4} = 0.7071$

$\text{Sin}^{-1}(0.7071) = 0.7854$ (Note: $\frac{3\pi}{4} \approx 2.3562$, and therefore, $\sin^{-1}(\sin\frac{3\pi}{4})$ is not equal to $= \frac{3\pi}{4}$)

Therefore, $\text{Sin}^{-1}(\sin\frac{3\pi}{4}) = 0.7854$. **Clarification of the ranges of inverse trig functions follows:**

Range of Inverse Trig	**Functions when $x \ge 0$**	**Range of Inverse Trig**	**Functions when $x < 0$**
$0 \le \text{Sin}^{-1}x \le \frac{\pi}{2}$ (Ist qd)	$0 < \text{Cot}^{-1}x \le \frac{\pi}{2}$ (Ist qd)	$-\frac{\pi}{2} \le \text{Sin}^{-1}x < 0$ (4th qd)	$\frac{\pi}{2} < \text{Cot}^{-1}x < \pi$ (2nd qd)
$0 \le \text{Cos}^{-1}x \le \frac{\pi}{2}$ (Ist qd)	$0 \le \text{Sec}^{-1}x < \frac{\pi}{2}$ (Ist qd)	$\frac{\pi}{2} < \text{Cos}^{-1}x \le \pi$ (2nd qd)	$\frac{\pi}{2} < \text{Sec}^{-1}x \le \pi$ (2nd qd))
$0 \le \text{Tan}^{-1}x < \frac{\pi}{2}$ (Ist qd)	$0 < \text{Csc}^{-1}x \le \frac{\pi}{2}$ (Ist qd)	$-\frac{\pi}{2} < \text{Tan}^{-1}x < 0$ (4th qd)	$-\frac{\pi}{2} \le \text{Csc}^{-1}x < 0$ (4th qd)

qd = quadrant

Inverse Cofunction Identities

1. $\text{Sin}^{-1}x + \text{Cos}^{-1}x = \frac{\pi}{2}$, $[-1,1]$; **2.** $\text{Tan}^{-1}x + \text{Cot}^{-1}x = \frac{\pi}{2}$, $(-\infty,\infty)$; **3** $\text{Sec}^{-1}x + \text{Csc}^{-1}x = \frac{\pi}{2}$, $|x| \ge 1$

Each identity is subject to the domain of definition, Application: $\text{Cot}^{-1}x = \frac{\pi}{2} - \text{Tan}^{-1}x$

For proofs of these identities, see Appendix B, p.252.

Negative angle relationships

1. $\text{Sin}^{-1}(-x) = -\text{Sin}^{-1}x$; **3.** $\text{Tan}^{-1}(-x) = -\text{Tan}^{-1}x$; **5.** $\text{Sec}^{-1}(-x) = \pi - \text{Sec}^{-1}x$;

2. $\text{Cos}^{-1}(-x) = \pi - \text{Cos}^{-1}x$; **4.** $\text{Cot}^{-1}(-x) = \pi - \text{Cot}^{-1}x$; **6.** $\text{Csc}^{-1}(-x) = -\text{Csc}^{-1}x$.

Lesson 34 Exercises

Find the following

1.. $\text{Arcsin}\frac{1}{2}$; **2.** $\text{Arctan}\sqrt{3}$; **3.** $\text{Arccos}\frac{1}{2}$

4. $\text{Arcsin}\left(-\frac{\sqrt{3}}{2}\right)$; **5.** $\text{Sin}^{-1}1$; **6.** $\text{Arccot}\,(-0.33)$

Without using tables or a calculator find :

7. $\cot\left(\text{Arcsin}\left(-\frac{1}{4}\right)\right)$; **8.** $\text{Arccos}\left(\cos\left(\frac{15\pi}{11}\right)\right)$

11. Express $\tan\left(\text{Arcsin}\,x\right)$ in terms of x.

Is each of the following true for all x:? Why?

9. $\text{Arcsin}\left(\sin x\right) = x$; **10.** $\sin\left(\text{Arcsin}\,x\right) = x$

12. Express $\cos\left(\text{Arctan}\,x\right)$ in terms of x.

13. True or false $\tan(\text{Tan}^{-1}x) = x$ for all x.

Answers : **1.** $\frac{\pi}{6}$; **2.** $\frac{\pi}{3}$; **3.** $\frac{\pi}{3}$; **4.** $-\frac{\pi}{3}$; **5.** $\frac{\pi}{2}$; **6.** 1.88; **7.** $-\sqrt{15}$; **8.** $\frac{7\pi}{11} \approx 1.999$; **9.** No : $-\frac{\pi}{2} \le x \le \frac{\pi}{2}$;

10. No : $-1 \le x \le 1$; **11.** $\frac{x}{\sqrt{1-x^2}}$; **12.** $\frac{1}{\sqrt{x^2+1}}$; **13.** True. $-\infty < x < \infty$

Note: In Problems **11** and **12,** the composition of a trigonometric function with an inverse trigonometric function is an algebraic expression.

Lesson 35

How to Sketch the Graph of $y =$ Arcsin x

The graph of $y = \sin x$ is not a one-to-one function by "the horizontal line" rule since a horizontal line drawn crosses the graph of $y = \sin x$ for $0 \le x \le 2\pi$ in more than one point and therefore, the inverse relation of $x = \sin y$ is generally not a function. In sketching the graphs of the inverse trigonometric functions, we will be guided by the domains and ranges agreed to previously in the definitions of the inverse trigonometric functions. See page 218.

To sketch the graph of $y =$ Arcsin x we will sketch the graph of the equivalent equation

$x = \sin y$ for $-\frac{\pi}{2} \le y \le \frac{\pi}{2}$, by obtaining important points and connecting them.

2. The important points include 2 end points (a starting point and a stopping point), 1 mid-point and two quarter-points are needed. The points include the x- and the y--intercepts

Example 1 Sketch the graph of $y =$ Arcsin x

Solution We will sketch the graph of the equivalent equation

$$x = \sin y \text{ for } -\frac{\pi}{2} \le y \le \frac{\pi}{2}$$

Step 1: Choose convenient y-values from $-\frac{\pi}{2}$ to $\frac{\pi}{2}$, and calculate the corresponding x-values using the equation $x = \sin y$ to obtain ordered pairs. It is more convenient to choose y and calculate x. The first time we chose y and calculated x was when we graphed logarithmic functions such as $y = \log_2 x$). We must be careful of how we list x- and y-values obtained from the calculations, since we are choosing y and calculating x. A table containing these values are shown below.

$x = \sin y$	-1	-.71	0	.71	1
y	$-\frac{\pi}{2}$	$-\frac{\pi}{4}$	0	$\frac{\pi}{4}$	$\frac{\pi}{2}$

Step 2: Plot the points $(-1,-\frac{\pi}{2})$, $(-.71,-\frac{\pi}{4})$, $(0,0)$, $(.71,\frac{\pi}{4})$, $(1,\frac{\pi}{2})$. (Points on $x = \sin y$)

Step 3: Connect the points from Step 2 by a smooth curve to obtain the characteristic shape.

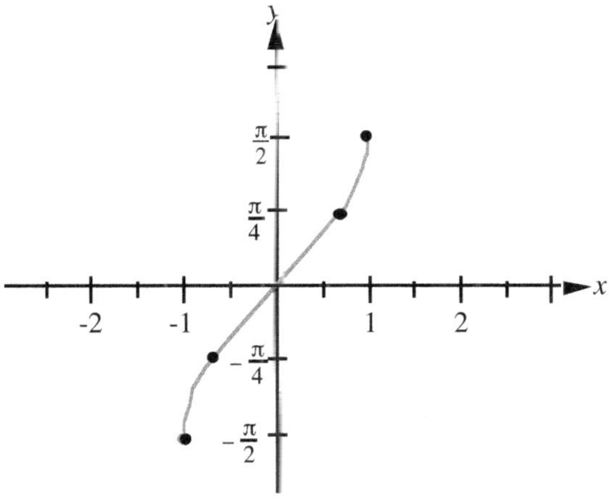

Figure: The graph of $y =$ Arcsin x or $\text{Sin}^{-1} x$

Practice sketching this graph and include it among your "trigonometric vocabulary".

OR Alternatively,

Step 2: Draw one sine curve between $-\pi \le y \le \pi$ **along the y-axis** using dotted (broken lines) lines, noting that the curve is in the 1st and 3rd quadrants.

Step 3: Darken the part of the curve for which $-\frac{\pi}{2} \le y \le \frac{\pi}{2}$

Example 2 Sketch the graph of $y = \text{Arcsin } 3x$
Method 1
Solution

Step 1: Solve for x by applying the definition, if $\text{Arcsin } x = y$ then $x = \sin y$

If $y = \text{Arcsin } 3x$ then

$\sin y = 3x$ ("Taking the sine of both sides of the equation")

$\frac{1}{3}\sin y = x$ where $-\frac{\pi}{2} \le y \le \frac{\pi}{2}$ Note that $\sin(\arcsin 3x) = 3x$.

Step 2: Choose convenient y-values from $-\frac{\pi}{2}$ to $\frac{\pi}{2}$, and calculate the corresponding values of x. using the equation $x = \frac{1}{3}\sin y$ to obtain ordered pairs. A table of the results is shown below.

$x = \frac{1}{3}\sin y$	-.33	-.24	0	.24	.33
y	$-\frac{\pi}{2}$	$-\frac{\pi}{4}$	0	$\frac{\pi}{4}$	$\frac{\pi}{2}$

Step 3: Plot the points $(-.33, -\frac{\pi}{2})$, $(-.24, -\frac{\pi}{4})$, $(0,0)$, $(.24, \frac{\pi}{4})$, $(.33, \frac{\pi}{2})$. (Points on $x = \frac{1}{3}\sin y$)

Step 4: Connect the points from Step 3 by a smooth curve to obtain the characteristic shape.

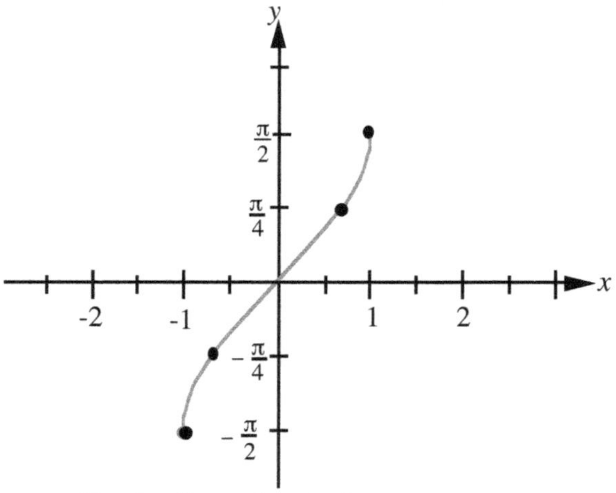

Graph of $y = \text{Arcsin } 3x$

Method 2 (Transformation Method)

Study the next example, Example 3, and imitate.

Example 3 Sketch the graph of $y = 4\,\text{Arcsin } 2x$

Method 1

Step 1 Solve $y = 4\,\text{Arcsin } 2x$ for $\text{Arcsin } 2x$, followed by solving for x by applying $x = \sin y$

$$\frac{y}{4} = \text{Arcsin } 2x$$

$$\sin\frac{y}{4} = 2x \qquad \text{("Taking the sine of both sides of the equation")}$$

$$x = \frac{1}{2}\sin\frac{y}{4} \qquad \text{where} \qquad -\frac{\pi}{2} \le \frac{y}{4} \le \frac{\pi}{2} \text{ < note the restricted range}$$

$$x = \frac{1}{2}\sin\frac{y}{4} \quad \text{where } -2\pi \le y \le 2\pi \text{ <---} -\frac{\pi}{2}\bullet\frac{4}{1}\frac{2}{1} \le \frac{y}{4}\bullet\frac{4}{1} \le \frac{\pi}{2}\bullet\frac{4}{1}\frac{2}{1} \text{ (solving for } y)$$

Step 2: Sketch the principal branch for $x = \frac{1}{2}\sin\frac{y}{4}$ for $-2\pi \le y \le 2\pi$.

Step 3: Determine the y-values to use to calculate x.

The distance from -2π to 2π is $2\pi - (-2\pi) = 4\pi$

Divide 4π by 4 to obtain the interval increment , π. Add π successfully to the y-values, beginning from -2π.

Step 4: Calculate the corresponding x-values, using $x = \frac{1}{2}\sin\frac{y}{4}$. (Results in the table below)

$x = \frac{1}{2}\sin\frac{y}{4}$	$-\frac{1}{2}$	-.35	0	.35	$\frac{1}{2}$
y	-2π	$-\pi$	0	π	2π

Step 5: Using this table plot and connect the points to obtain the characteristic shape.

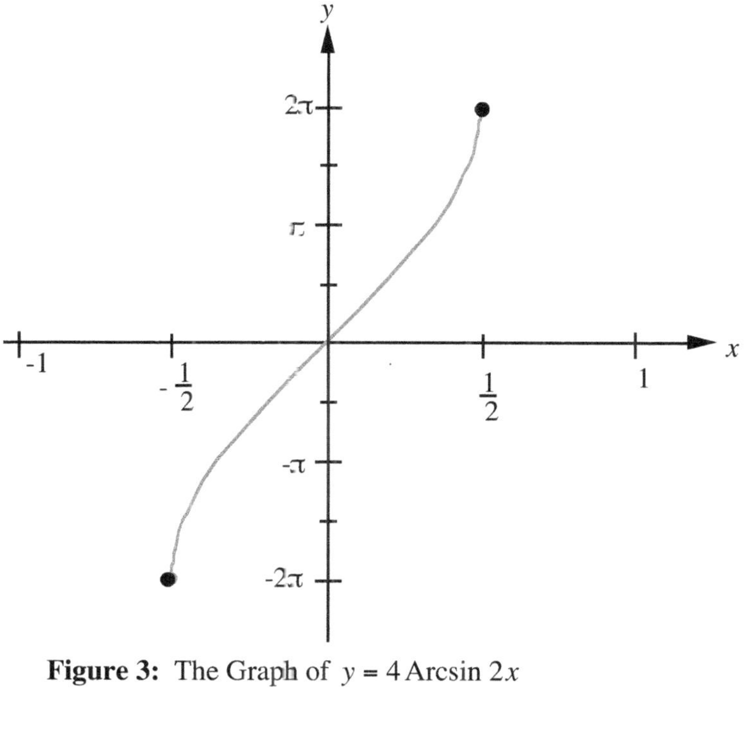

Figure 3: The Graph of $y = 4\,\text{Arcsin } 2x$

Method 2 (Transformation method)

Given $y = \text{Arcsin } x$, draw the graph of $y = 4\,\text{Arcsin } 2x$

Rewrite as $y = 4\,\text{Arcsin } 2(x - 0)$. Here, $k = 0$, $h = 0$, $a = 4$, $b = 2$,

$x = \sin y$		$y = 4\,\text{Arcsin } 2(x - 0)$		
x_0	y_0	$x_n = \frac{x_0}{b} + h$	$y_n = y_0(a) + k$	(x_n, y_n)
-1	$-\frac{\pi}{2}$	$x_n = \frac{-1}{2} + 0 = -\frac{1}{2}$	$-\frac{\pi}{2}(4) + 0 = -2\pi$	$(-\frac{1}{2}, -2\pi)$
−0.71	$-\frac{\pi}{4}$	$\frac{-0.71}{2} + 0 = -0.36$	$-\frac{\pi}{4}(4) + 0 = -\pi$	$(-0.36, -\pi)$
0	0	$\frac{0}{2} + 0 = 0$	$0(4) + 0 = 0$	$(0, 0)$
0.71	$\frac{\pi}{4}$	$\frac{0.71}{2} + 0 = 0.36$	$\frac{\pi}{4}(4) + 0 = \pi$	$(0.36, \pi)$
1	$\frac{\pi}{2}$	$\frac{1}{2} + 0 = \frac{1}{2}$	$\frac{\pi}{2}(4) + 0 = 2\pi$	$(\frac{1}{2}, 2\pi)$

Step 1: Quickly prepare a a table of values for x and y, using $x = \sin y$ and convenient y-values.

Step 2: Using the table of values from Step 1, determine the new coordinates x_n and y_n.

Step 3: Plot the new points and connect them by a solid curve (Figure 3 above)

Lesson 35 Exercises

Sketch the graphs of the following: **1**. $y = \text{Arcsin } x$; **2**. $y = 2\,\text{Arcsin } 4x$; .

Solution:

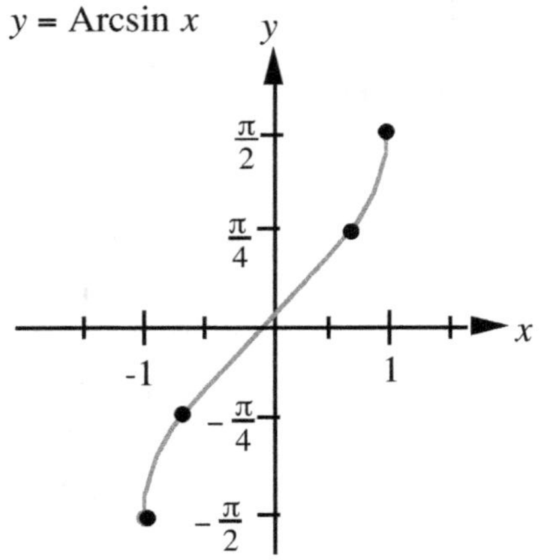

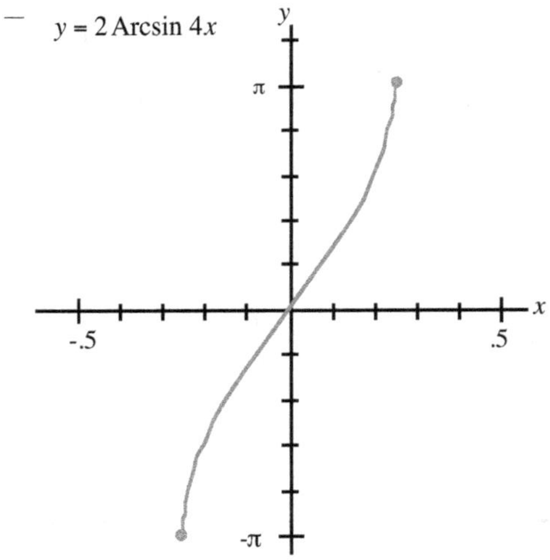

Lesson 36

How to Sketch the Graph of $y = \text{Arccos}\, x$

To sketch the graph of y = Arccos x we will sketch the graph of the equivalent equation
$x = \cos y$ for $0 \le y \le \pi$, by obtaining important points and connecting them.
The important points include 2 end points (a starting point and a stopping point), 1 mid-point and two quarter-points are needed. The points include the x- and y-intercepts.

Example 1 Sketch the graph of y = Arccos x

Solution We will sketch the graph of the equivalent equation
$x = \cos y$ for $0 \le y \le \pi$

Step 1: Choose convenient y -values from 0 to π, and calculate the corresponding values of x. using the equation $x = \cos y$ to obtain ordered pairs. It is more convenient to choose y and calculate x. We must be careful of how we list the ordered pairs obtained from the calculations, since we are choosing y and calculating x. A table containing these important points are shown below.

$x = \cos y$	1	.71	0	-.71	-1
y	0	$\frac{\pi}{4}$	$\frac{\pi}{2}$	$\frac{3\pi}{4}$	π

Step 2: Plot the points $(1, 0)$, $(.71, \frac{\pi}{4})$, $(0, \frac{\pi}{2})$, $(-.71, \frac{3\pi}{4})$. $(-1, \pi)$ (Points on $x = \cos y$)

Step 3: Connect the points from Step 2 by a smooth curve to obtain the characteristic shape.

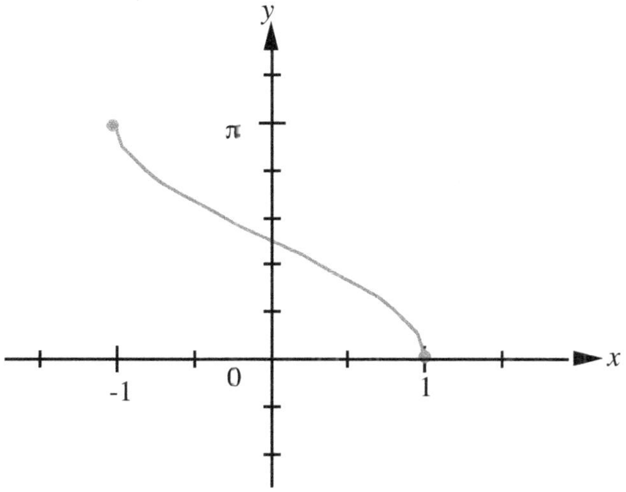

Graph of $x = \cos y$

Example 2 Sketch the graph of $y = \text{Arccos } 2x + \frac{\pi}{2}$

We cover two methods

Method 1

Step 1: Solve $y = \text{Arccos } 2x + \frac{\pi}{2}$ for Arccos $2x$,

Then $y - \frac{\pi}{2} = \text{Arccos } 2x$ (1)

Step 2: Solve (1) for $2x$ by applying if $y = \text{Arccos } x$ then $x = \cos y$, $0 \le y \le \pi$

Then, we obtain

$2x = \cos(y - \frac{\pi}{2})$, $0 \le y - \frac{\pi}{2} \le \pi$

Step 3 : Solve for x: $x = \frac{1}{2}\cos(y - \frac{\pi}{2})$ $0 \le y - \frac{\pi}{2} \le \pi$ <- - Note the form of the range

$x = \frac{1}{2}\cos(y - \frac{\pi}{2})$ $\frac{\pi}{2} \le y \le \frac{3\pi}{2}$ <- - - - (solving the inequality for y)

(The amplitude = $\frac{1}{2}$)

Step 4: Sketch the graph of the principal branch: $x = \frac{1}{2}\cos(y - \frac{\pi}{2})$ where $\frac{\pi}{2} \le y \le \frac{3\pi}{2}$

Choose convenient y -values from $\frac{\pi}{2}$ to $\frac{3\pi}{2}$, and calculate the corresponding x-values of x. using the equation $x = \frac{1}{2}\cos(y - \frac{\pi}{2})$ to obtain ordered pairs.

Starting y-value = $\frac{\pi}{2}$; end-point y-value = $\frac{3\pi}{2}$; incremental y-value = $\frac{1}{4}(\frac{3\pi}{2} - \frac{\pi}{2}) = \frac{\pi}{4}$.

Add $\frac{\pi}{4}$ successfully to the y-values, beginning from $y = \frac{\pi}{2}$.

Step 5: Plot and connect the points obtained in Step 4 by a smooth curve .

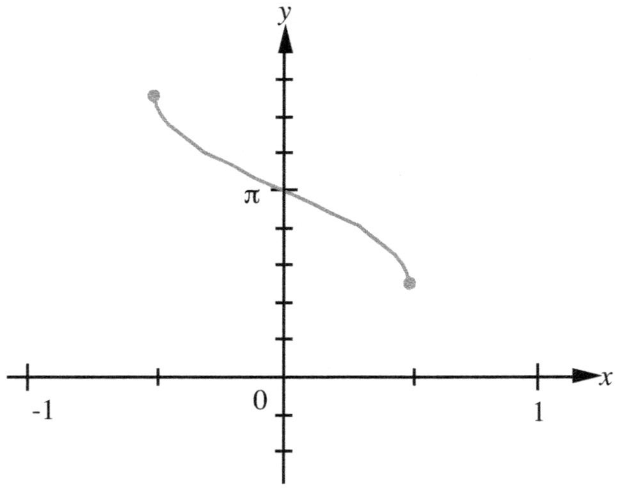

Figure 2: Graph of $y = \text{Arccos } 2x + \frac{\pi}{2}$

Method 2 (Transformation Method)

Given $y = \text{Arccos}\ x$, draw the graph of $y = \text{Arccos}\ 2x + \frac{\pi}{2}$ (see Example 1, for tables for $y = \text{Arccos}\ x$).

Rewrite as $y - \frac{\pi}{2} = \text{Arccos}\ 2(x - 0)$. Here, $k = \frac{\pi}{2}$, $h = 0$, $a = 1$, $b = 2$.

$x = \cos y$		$y - \frac{\pi}{2} = 4\,\text{Arccos}\ 2x$		
x_0	y_0	$x_n = \frac{x_0}{b} + h$	$y_n = y_0(a) + k$	(x_n, y_n)
1	0	$x_n = \frac{1}{2} + 0 = \frac{1}{2}$	$0(4) + \frac{\pi}{2} = \frac{\pi}{2}$	$(\frac{1}{2}, \frac{\pi}{2})$
0.71	$\frac{\pi}{4}$	$\frac{0.71}{2} + 0 = 0.36$	$\frac{\pi}{4}(1) + \frac{\pi}{2} = \frac{\pi}{4} + \frac{\pi}{2} = \frac{3\pi}{4}$	$(0.36, \frac{3\pi}{4})$
0	$\frac{\pi}{2}$	$\frac{0}{2} + 0 = 0$	$\frac{\pi}{2}(1) + \frac{\pi}{2} = \frac{\pi}{2} + \frac{\pi}{2} = \pi$	$(0, \pi)$
−0.71	$\frac{3\pi}{4}$	$\frac{-0.71}{2} + 0 = -0.36$	$\frac{3\pi}{4}(1) + \frac{\pi}{2} = \frac{3\pi}{4} + \frac{\pi}{2} = \frac{5\pi}{4}$	$(-0.36, \frac{5\pi}{4})$
-1	π	$\frac{-1}{2} + 0 = -\frac{1}{2}$	$\pi(1) + \frac{\pi}{2} = \pi + \frac{\pi}{2} = \frac{3\pi}{2}$	$(-\frac{1}{2}, \frac{3\pi}{2})$

Step 1: Quickly prepare a a table of values for x and y, using $x = \cos y$ and convenient y-values.

Step 2: Using the table of values from Step 1, determine the new coordinates x_n and y_n.

Step 3: Plot the new points and connect them by a solid curve (Figure 2 above)

Lesson 36 Exercises

Sketch the graphs of (a) $y = \text{Arccos}\ x$; (b) $y = \text{Arccos}\ 2x + \frac{\pi}{2}$ (try the transformation method)

Solution:

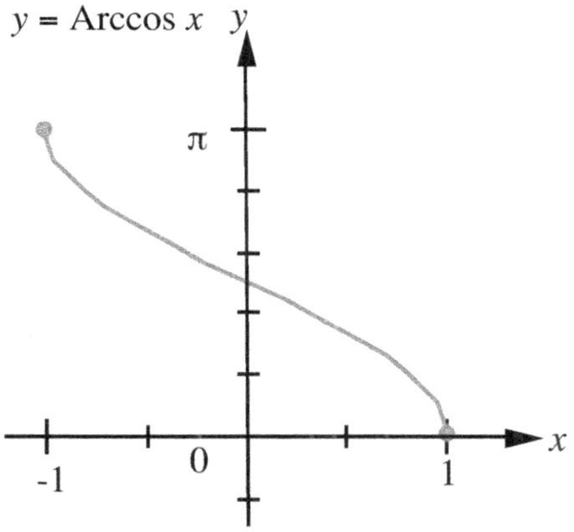

Lesson 37

How to Sketch the Graph of $y = \text{Arctan}\, x$

To sketch the graph of $y = \text{Arctan}\, x$ we will sketch the graph of the equivalent **equation**

$$x = \tan y \quad \text{for} \quad -\frac{\pi}{2} < y < \frac{\pi}{2}$$

The curve is not defined where $\cos x$ is zero: That is it is not defined at $y = -\frac{\pi}{2}$ and $y = \frac{\pi}{2}$. The curve is discontinuous at these points and the curve is asymptotic to the horizontal lines $y = -\frac{\pi}{2}$, and $\frac{\pi}{2}$. The tangent curve has points of inflection at the y-intercept.

Example Sketch the graph of $y = \text{Arctan}\, x$

Solution We will sketch the graph of the equivalent equation

$$x = \tan y \quad \text{for} \quad -\frac{\pi}{2} < y < \frac{\pi}{2}.$$

Step 1: Choose convenient y-values from $-\frac{\pi}{2}$ to $\frac{\pi}{2}$, and calculate the corresponding x-values using the equation $x = \tan y$ to obtain a table of x- and y-values. We must be careful of how we list x- and y-values obtained from the calculations, since we are choosing y and calculating x. A table containing these values are shown below.

$x = \tan y$	undefined	-1	0	1	undefined
y	$-\frac{\pi}{2}$	$-\frac{\pi}{4}$	0	$\frac{\pi}{4}$	$\frac{\pi}{2}$

Step 2: Plot the points $(-1, -\frac{\pi}{4})$, $(0,0)$, $(1, \frac{\pi}{4})$, and draw the horizontal asymptotes $y = -\frac{\pi}{2}$ and $y = \frac{\pi}{2}$.

Step 4: Connect the points from Step 2 by a smooth curve, noting the asymptotic behavior near $y = -\frac{\pi}{2}$ and $y = \frac{\pi}{2}$.

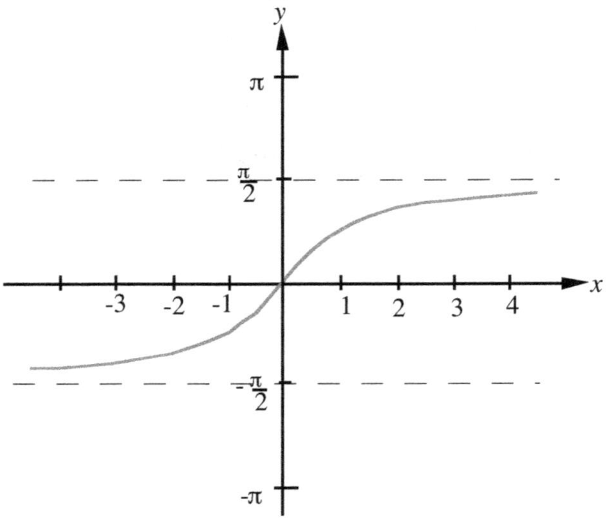

Figure $y = \text{Arctan}\, x$

Lesson 37 Exercises

Sketch the graphs of **1.** $y = 3\,\text{Arctan}\,x$.; . **2.** $y = 2\,\text{Arcsin}\,4x$; (Try the transformation method)

Solution:

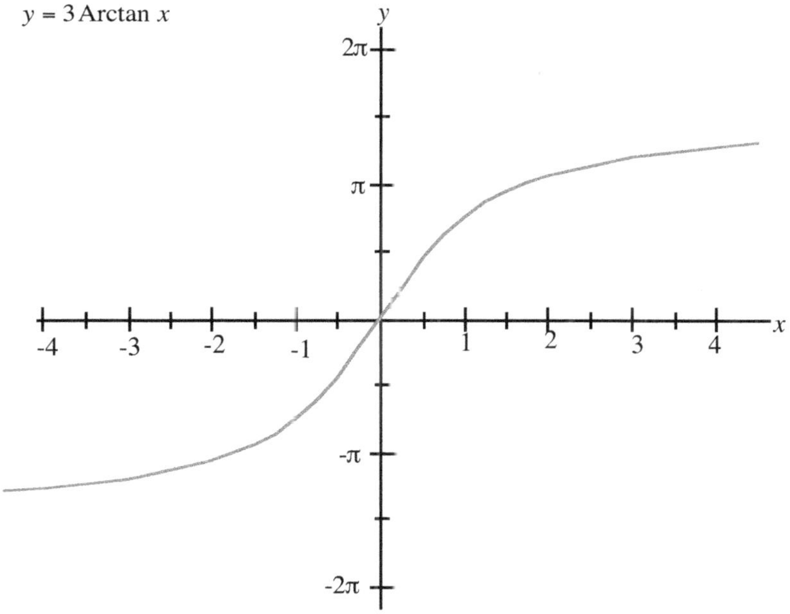

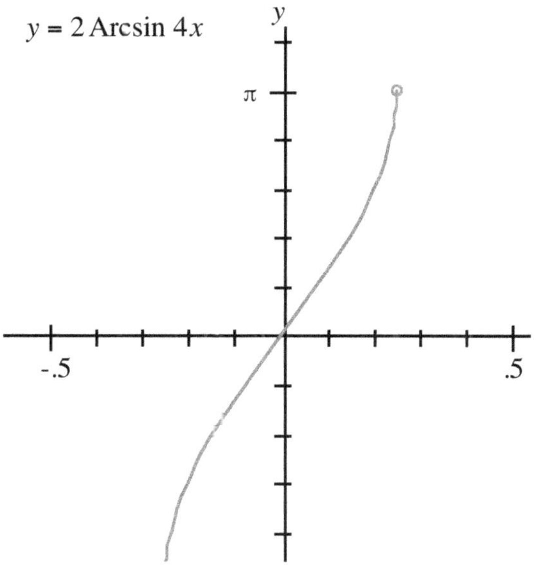

CHAPTER 13

Lesson 38: **Trigonometric Identities**

Lesson 39: **Proving Trigonometric Identities; Applications of Sum and Difference Identities**

Lesson 38
Trigonometric Identities

Introduction

1. Suppose we are asked to sketch the graph of $f(x) = \csc x \sec x$ (1).
At first glance, we might say that we do not know how to sketch its graph.
However, if we express this function in terms of a single basic function, we can readily sketch it.

$$f(x) = \csc x \sec x$$

$$= \frac{1}{\sin x \cos x}$$

$$= \frac{1}{\frac{1}{2}\sin 2x} \qquad (\sin x \cos x = \tfrac{1}{2}\sin 2x\text{, from the } \textbf{identity,}\ 2\sin x \cos x = \sin 2x).$$

$$= \frac{2}{\sin 2x}$$

$$= 2\csc 2x \qquad \left(\text{Note: } \frac{1}{\sin 2x} = \csc 2x\right)$$

$$f(x) = 2\csc 2x$$

Since we have shown that given function is identical with this familiar cosecant function, to sketch $f(x) = \sec x \csc x$, we shall sketch $f(x) = 2\csc 2x$.

2. We may also use trigonometric identities in solving trigonometric equations. Here, we use the trigonometric identities to express a given trigonometric equation of more than one function in terms of a single function or in terms of the product of factors, each factor containing only a single function: Example $\cos x\,(\sin x - 1) = 0$.

Trigonometric Identities

An **identity** is an equation that is true for all permissible replacements of the independent variable.

Trigonometric identities are based on eight fundamental identities which are grouped as reciprocal identities, ratio identities and the Pythagorean identities.

Reciprocal Identities

1. $\sec\theta = \dfrac{1}{\cos\theta}$ $(\cos\theta \neq 0)$

2. $\csc\theta = \dfrac{1}{\sin\theta}$ $(\sin\theta \neq 0)$

3. $\cot\theta = \dfrac{1}{\tan\theta}$ $(\tan\theta \neq 0)$

Ratio Identities

1. $\tan\theta = \dfrac{\sin\theta}{\cos\theta}$ $(\cos\theta \neq 0)$

2. $\cot\theta = \dfrac{\cos\theta}{\sin\theta}$ $(\sin\theta \neq 0)$

Pythagorean Identities

1. $\sin^2\theta + \cos^2\theta = 1$

2. $1 + \tan^2\theta = \sec^2\theta$

3. $1 + \cot^2\theta = \csc^2\theta$

For the Pythagorean identities, identities **2** and **3** can be derived from identity **1** by dividing identity **1** by $\cos^2\theta$ and by $\sin^2\theta$ respectively. Keep this in mind when trying to memorize these identities.

Sum and Difference Identities (Addition and Subtraction Formulas)

1. $\cos(A - B) = \cos A\cos B + \sin A\sin B$

2. $\cos(A + B) = \cos A\cos B - \sin A\sin B$

3. $\sin(A + B) = \sin A\cos B + \sin B\cos A$

4. $\sin(A - B) = \sin A\cos B - \sin B\cos A$

5. $\tan(A - B) = \dfrac{\tan A - \tan B}{1 + \tan A\tan B}$

6. $\tan(A + B) = \dfrac{\tan A + \tan B}{1 - \tan A\tan B}$

Double Angle Identities

1. $\sin 2\theta = 2\sin\theta\cos\theta$

2. $\cos 2\theta = \cos^2\theta - \sin^2\theta$

3. $\cos 2\theta = 2\cos^2\theta - 1$

4. $\cos 2\theta = 1 - 2\sin^2\theta$

5. $\tan 2\theta = \dfrac{2\tan\theta}{1 - \tan^2\theta}$

Half Angle Identities

1. $\cos\frac{1}{2}\theta = \pm\sqrt{\frac{1+\cos\theta}{2}}$ <---	+ sign for $\frac{\theta}{2}$ in quad I or IV but – sign for $\frac{\theta}{2}$ in quad II or III
2. $\sin\frac{1}{2}\theta = \pm\sqrt{\frac{1-\cos\theta}{2}}$ <---	+ sign for $\frac{\theta}{2}$ in quad I or II but – sign for $\frac{\theta}{2}$ in quad III or IV
3. $\tan\frac{1}{2}\theta = \frac{1-\cos\theta}{\sin\theta} = \frac{\sin\theta}{1+\cos\theta}$	+ sign for $\frac{\theta}{2}$ in quad I or III but – sign for $\frac{\theta}{2}$ in quad II or IV

The ± sign indicates the sign to use. The sign depends upon the quadrant in which $\frac{\theta}{2}$ is located.

For example, if $\frac{\theta}{2}$ is in the 1st or 4th quadrant, where the cosine function is positive, then

$$\cos\frac{1}{2}\theta = +\sqrt{\frac{1+\cos\theta}{2}}$$

However, if $\frac{\theta}{2}$ is in the 2nd or 3rd quadrant, where the cosine function is negative, then

$$\cos\frac{1}{2}\theta = -\sqrt{\frac{1+\cos\theta}{2}}$$

We will use either the plus sign or the minus sign but **never both** in any particular identity. Note also that the usage of the plus or minus sign is different from the ± signs involved in solving quadratic equations in algebra.

The following formulas will be useful in integral calculus.

Product or Product as a Sum Formulas

$$\left.\begin{array}{l} \mathbf{A}: 2\sin A\sin B = \cos(A-B) - \cos(A+B) \\ B: 2\cos A\cos B = \cos(A-B) + \cos(A+B) \end{array}\right\} \text{<-- cosine sums}$$

(Derived from sum and difference identities)

$$\left.\begin{array}{l} \mathbf{C}: 2\sin A\cos B = \sin(A+B) + \sin(A-B) \\ \mathbf{D}: 2\cos A\sin B = \sin(A+B) - \sin(A-B) \end{array}\right\} \text{<-- sine sums}\}$$

(Derived from sum and difference identities)

A **memory** device to help recall the above identities follows.

1. For **A** and **B** (sin sin or cos cos, the right hand side of each identity is in terms of **cos** function only.
Note in **B** that in some books, the order of the right side may be reversed

2. For **C**, and **D** (sin cos,) the right hand side is in terms of **sin** function only, Note above also that when the left side is in terms of the same function , the right side is in terms of **cos** only, while if the left side is in terms of cos and sin, the right side is in terms of **sin** only.

Sum Identities

$$\left.\begin{array}{ll} \text{1.} & \cos A + \cos B = 2\cos\frac{A+B}{2}\cos\frac{A-B}{2} \\ \text{2.} & \cos A - \cos B = -2\sin\frac{A+B}{2}\sin\frac{A-B}{2} \end{array}\right\} \text{For the cosine}$$

$$\left.\begin{array}{ll} \text{3.} & \sin A + \sin B = 2\sin\frac{A+B}{2}\cos\frac{A-B}{2} \\ \text{4.} & \sin A - \sin B = 2\sin\frac{A-B}{2}\cos\frac{A+B}{2} \end{array}\right\} \text{For the sine}$$

Cofunction Relationships

$\sin A = \cos(90^\circ - A)$
$\cos A = \sin(90^\circ - A)$ ← The sine and the cosine are cofunctions.

$\tan A = \cot(90^\circ - A)$
$\cot A = \tan(90^\circ - A)$ ← The tangent and the cotangent are cofuntions.

$\sec A = \csc(90^\circ - A)$
$\csc A = \sec(90^\circ - A)$ ← The secant and the cosecant are cofuntions.

Note above that: $(90^\circ - A) = (\frac{\pi}{2} - A)$

Other identities (Useful when finding functional values of negative angles)

For any θ .

1. $\cos(-\theta) = \cos\theta$
2. $\sin(-\theta) = -\sin\theta$
3. $\tan(-\theta) = -\tan\theta$
4. $\sec(-\theta) = \sec\theta$
5. $\csc(-\theta) = -\csc\theta$
6. $\cot(-\theta) = -\cot\theta$

In the last six identities, note that in finding the trigonometric functional value of a negative angle using these identities:

Step 1: Ignore the minus sign and find the trigonometric functional value as usual (this result may be positive or negative).

Step 2: If the trigonometric function is that of the cosine, or the secant, then the result in Step 1 is the answer; but if the trigonometric function is that of the sine, the tangent, the cosecant or the cotangent, then find the negative of the result from Step 1.

Examples 1. $\cos(-60^\circ) = \cos 60^\circ = \frac{1}{2}$ (applying $\cos(-\theta) = \cos\theta$)

2. $\sin(-30^\circ) = -\sin 30^\circ = -\left(\frac{1}{2}\right) = -\frac{1}{2}$ (applying $\sin(-\theta) = -\sin\theta$)

3. $\sin(-135^\circ) = -\sin 135^\circ = -(.7071) = -.7071$ (applying $\sin(-\theta) = -\sin\theta$)

4. $\sin(-210^\circ) = -\sin 210^\circ = -(-\frac{1}{2}) = +\frac{1}{2}$ Note: $\sin 210 = -\sin 30 = -\frac{1}{2}$

Notwithstanding the methods of the last examples, we must note that in finding functional values, we may still use the previous methods of drawing the negative angle, finding the reference angle and then finding the functional value according to the quadrant of the negative angle..

Lesson 38 Exercises

From memory, after practice, write the following down on paper:
Reciprocal identities; Pythagorean identities; Sum and difference identities; Double angle identities; half angle identities; and cofunction relationships.

Lesson 39

Proving Trigonometric Identities and Applications of the sum and difference identities

In proving trigonometric identities, the following guidelines will be helpful:

1. Express all functions in terms of sines and cosines and simplify.
   (However, sometimes, this may not be necessary.)
2. Work **separately** on the left-hand-side and **separately** on the right-hand side.
   Do **not** work on the equation as you would operate on both sides of a conditional equation This is not an "if-then statement" proof.
3. Apply the appropriate trigonometric identities or formulas (from above). For example, if there are double angles, use the double angle identities to change to single angles.
   (However, sometimes, this may not be necessary.)
4. If there are indicated operations such as addition or subtraction of fractions, combine these fractions and simplify. (You may have to go back and review how to add rational fractions)
   You may sometimes do the opposite of **4** as in **5** below.
5. If the numerator of a fraction consists of two or more terms and the denominator consists of only one term (especially, one function only), divide every term in the numerator by the denominator.
6. If there is a binomial in either the numerator or the denominator, multiply the binomial by its conjugate
7. Only practice will make the above guidelines meaningful. Do not be afraid to try and err. You will learn from your mistakes. After some mastery, you may even begin to enjoy proving trigonometric identities.

Example 1 Prove the following identity: $\frac{\sin x}{\tan x}+\frac{\cos x}{\cot x}=\cos x+\sin x.$

(**Note** that the right-hand side (RHS) is in terms sines and cosines. Therefore, we do not have to do any work on the RHS. We therefore work on the LHS and express the LHS in terms of sines and cosines and simplify.)

$$\frac{\sin x}{\tan x}+\frac{\cos x}{\cot x}\stackrel{?}{=}\cos x+\sin x$$ (We will remove the question mark" ? " when we have exactly the same expression on both sides of the equation)

$$\frac{\sin x}{\frac{\sin x}{\cos x}}+\frac{\cos x}{\frac{\cos x}{\sin x}}\stackrel{?}{=}\cos x+\sin x$$ $\left(\tan x=\frac{\sin x}{\cos x};\ \cot x=\frac{\cos x}{\sin x}\right)$

$$\frac{\sin x}{1}\cdot\frac{\cos x}{\sin x}+\frac{\cos x}{1}\cdot\frac{\sin x}{\cos x}\stackrel{?}{=}\cos x+\sin x$$ (Inverting the divisor and multiplying)

$$\frac{\cancel{\sin x}}{1}\cdot\frac{\cos x}{\cancel{\sin x}}+\frac{\cancel{\cos x}}{1}\cdot\frac{\sin x}{\cancel{\cos x}}\stackrel{?}{=}\cos x+\sin x$$

$$\cos x+\sin x=\cos x+\sin x.$$

QED

Example 2 Prove that $\dfrac{\sin 2x}{\sin x} = \sec x + \dfrac{\cos 2x}{\cos x}$

(In this problem, we will work on both sides.)

$$\frac{\sin 2x}{\sin x} \overset{?}{=} \sec x + \frac{\cos 2x}{\cos x}$$

$$\frac{2\sin x\cos x}{\sin x} \overset{?}{=} \frac{1}{\cos x} + \frac{\cos^2\theta\ -\sin^2\theta}{\cos x}$$ (sin $2x = 2\sin x\,\cos x$ for the LHS)

$$\frac{2\cancel{\sin x}\cos x}{\cancel{\sin x}} \overset{?}{=} \frac{1}{\cos x} + \frac{\cos^2 x - \sin^2 x}{\cos x}$$ ($\cos 2\theta = \cos^2\theta\ - \sin^2\theta$ for the RHS)

$$2\cos x \overset{?}{=} \frac{1 + \cos^2 x - \sin^2 x}{\cos x}$$ (Adding the fractions)

$$2\cos x \overset{?}{=} \frac{1 - \sin^2 x + \cos^2 x}{\cos x}$$ (Rewriting)

$$2\cos x \overset{?}{=} \frac{\cos^2 x + \cos^2 x}{\cos x}$$ ($1 - \sin^2\theta = \cos^2\theta$)

$$2\cos x \overset{?}{=} \frac{2\cos^2 x}{\cos x}$$

$$2\cos x \overset{?}{=} \frac{2\cos x\,\cancel{\cos x}}{\cancel{\cos x}}$$

$$2\cos x = 2\cos x$$ (Note the LHS = the RHS)

QED

Equivalents in Trigonometry

Quantities = 1	**Quantities = −1**	**Quantities = 0**	**Quantities = 2**
$\sin\frac{\pi}{2} = 1$;	$\sin\frac{3\pi}{2} = -1$	$\sin 0 = 0$	$\csc\frac{\pi}{6} = 2$
$\cos 0 = 1$	$\cos\pi = -1$	$\sin\pi = 0$	$\sec\frac{\pi}{3} = 2$
$\csc\frac{\pi}{2} = 1$;	$\sec\pi = -1$	$\cos\frac{\pi}{2} = \cos\frac{3\pi}{2} = 0$	----------
$\tan\frac{\pi}{4} = 1$	$\csc\frac{3\pi}{2} = -1$	$\cot\frac{\pi}{2} = \cot\frac{3\pi}{2} = 0$	**Quantities =** $\sqrt{3}$
$\cot\frac{\pi}{4} = 1$	----------	$\tan\pi = 0$	$\tan\frac{\pi}{3} = \sqrt{3}$
$\sin^2\theta + \cos^2\theta = 1$	**Quantities =** $\frac{1}{2}$	$\tan 0 = 0$	Also, If $A = B$, then
$\sec^2\theta - \tan^2\theta = 1$	$\sin\frac{\pi}{6} = \frac{1}{2}$; $\cos\frac{\pi}{3} = \frac{1}{2}$	----------	$A - B = 0$
$\csc^2\theta - \cot^2\theta = 1$			
$\cos 2\theta + 2\sin^2\theta = 1$			
$2\cos^2\theta - \cos 2\theta = 1$			

Proofs of some fundamental identities

The proofs of the fundamental identities can be constructed using the circle, below, and expressing the trigonometric functions in terms of x, y, and r. The techniques to be illustrated in the following example can also be applied to prove similar identities.

Example 1 Prove the reciprocal identity, $\sec\theta = \dfrac{1}{\cos\theta}$.

Step 1: Consider a point $P(x,y)$ at a distance r from the origin and on the terminal side of an angle θ, where θ is in standard position and also $r \neq 0$.

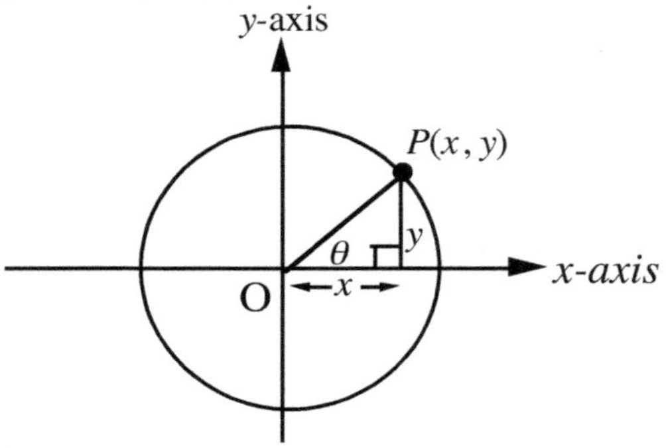

Proof: By definition, $\cos\theta = \dfrac{x}{r}$ (1)

$\sec\theta = \dfrac{r}{x}$ (2) (the reciprocal of $\cos\theta$)

$\dfrac{1}{\cos\theta} = \dfrac{1}{\frac{x}{r}}$ (from (1))

$\dfrac{1}{\cos\theta} = \dfrac{r}{x}$ (3)

The right-hand sides of (2) and (3) are identical.

Therefore $\sec\theta = \dfrac{1}{\cos\theta}$

(Quantities equal to the same quantity are equal to each other)

Example 2: Show that $\cos(A - B) = \cos A\cos B + \sin A\sin B$

Step 1: We first derive a formula for the length of a chord in a unit circle (Figure below)

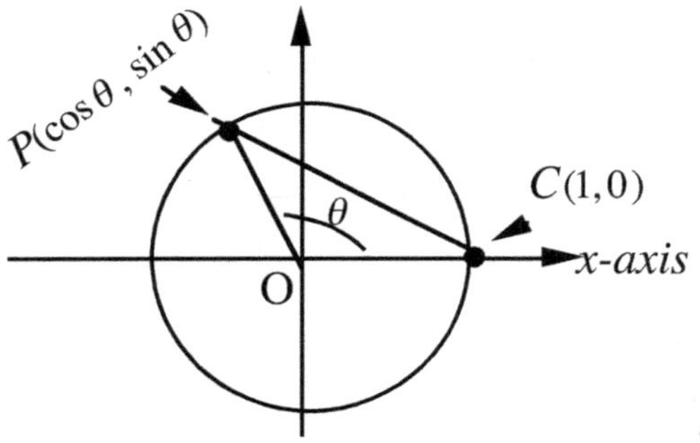

Consider a unit circle on the left. We find the length of chord PC whose corresponding arc subtends an angle θ at the center of the circle. Using the distance formula, the length of the PC is given by

$PC = \sqrt{(\cos\theta - 1)^2 + (\sin\theta - 0)^2}$

$= \sqrt{\cos^2\theta - 2\cos\theta + 1 + \sin^2\theta}$

$= \sqrt{\cos^2\theta + \sin^2\theta - 2\cos\theta + 1}$

$= \sqrt{1 - 2\cos\theta + 1} = \boxed{PC = \sqrt{2 - 2\cos\theta}}$ **(1)**

Step 2: Apply the formula derived in **Step 1** and also apply the distance formula in the figure below.

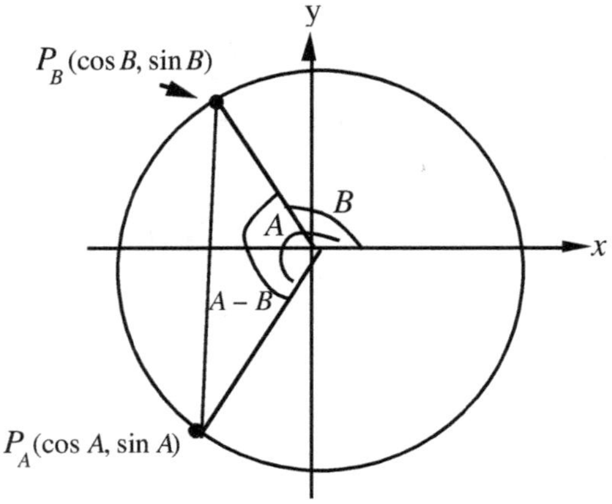

By applying (1) to chord P_AP_B,

$P_AP_B = \sqrt{2 - 2\cos(A - B)}$, where $\theta = A - B$ (2)

Also, applying the distance formula to the chord P_AP_B

The length of the chord is given by

$P_AP_B = \sqrt{(\cos B - \cos A)^2 + (\sin B - \sin A)^2}$

$= \sqrt{1 + 1 - 2\sin A\sin B - 2\cos A\cos B}$ (simplifying)

$= \sqrt{2 - 2\sin A\sin B - 2\cos A\cos B}$ (3)

Now, we equate the right hand-sides of (2) and (3)

$\sqrt{2 - 2\cos(A - B)} = \sqrt{2 - 2\sin A\sin B - 2\cos A\cos B}$

$\boxed{\cos(A - B) = \cos A\cos B + \sin A\sin B}$ (simplifying)

The other fundamental identities of the reciprocal identities, the ratio identities, and the Pythagorean identities can be similarly proved by imitating the above example.
Try to prove some of them.

Derivation of other identities

From the difference identity, $\cos(A-B)=\cos A\cos B+\sin A\sin B$ we can derive the other addition formulas as well as all the remaining identities (double-angle and half-angle identities; product as a sum identities, as well as the identities in the next section).

Applications of the sum and difference identities

From the sum and difference identities, we can prove the following identities

1. $\sin(-\theta)=-\sin\theta$
2. $\cos(-\theta)=\cos\theta$
3. $\tan(-\theta)=-\tan\theta$

We can also prove the following cofunctions which were derived previously (page 104, 235)

$\left.\begin{array}{l}\sin A=\cos(90^\circ-A)\\ \cos A=\sin(90^\circ-A)\end{array}\right\}\leftarrow$ The sine and the cosine are cofunctions.

$\left.\begin{array}{l}\tan A=\cot(90^\circ-A)\\ \cot A=\tan(90^\circ-A)\end{array}\right\}\leftarrow$ The tangent and the cotangent are cofuntions.

$\left.\begin{array}{l}\sec A=\csc(90^\circ-A)\\ \csc A=\sec(90^\circ-A)\end{array}\right\}\leftarrow$ The secant and the cosecant are cofuntions.

Note above that: $(90^\circ-A)=(\frac{\pi}{2}-A)$

Application of $\cos(A-B)=\cos A\cos B+\sin A\sin B$] to derive
$\cos(A+B)=\cos A\cos B-\sin A\sin B$

$\cos(A+B)=\cos[A-(-B)]$
$=\cos A\cos(-B)+\sin A\sin(-B)$ (applying $\cos(A-B)=\cos A\cos B+\sin A\sin B$)

$\boxed{\cos(A+B)=\cos A\cos B-\sin A\sin B}$ (Noting that $\cos(-B)=\cos B$; $\sin(-B)=-\sin B$)

Derivation of $\sin(A+B)=\sin A\cos B+\cos A\sin B$

Recall that $\sin\theta=\cos(\frac{\pi}{2}-\theta)$

let $\theta=(A+B)$

Then $\sin(A+B)=\cos[(\frac{\pi}{2}-(A+B)]$

$\sin(A+B)=\cos(\frac{\pi}{2}-A-B)$

$\sin(A+B)=\cos[(\frac{\pi}{2}-A)-B]$

$\sin(A+B)=\cos(\frac{\pi}{2}-A)\cos B+\sin(\frac{\pi}{2}-A)\sin B$ (applying $\cos(A-B)=\cos A\cos B+\sin A\sin B$)

$\boxed{\sin(A+B)=\sin A\cos B+\cos A\sin B}$ (Note: $\cos(\frac{\pi}{2}-A)=\sin A$; $\sin(\frac{\pi}{2}-A)=\cos A$)

Extra Example 1: Using $\tan\frac{\pi}{4} = 1$ and $\tan(A+B) = \frac{\tan A + \tan B}{1 - \tan A \tan B}$, show that

$$\ln\left|\frac{1+\tan\frac{x}{2}}{1-\tan\frac{x}{2}}\right| = \ln\left|\tan\left(\frac{\pi}{4}+\frac{x}{2}\right)\right|$$

Solution By applying $1 = \tan\frac{\pi}{4}$ to

$$\ln\left|\frac{1+\tan\frac{x}{2}}{1-\tan\frac{x}{2}}\right|,$$

$$\ln\left|\frac{1+\tan\frac{x}{2}}{1-\tan\frac{x}{2}}\right| = \ln\left|\frac{\tan\frac{\pi}{4}+\tan\frac{x}{2}}{1-\tan\frac{x}{2}\tan\frac{\pi}{4}}\right| \quad \text{(A)}$$

Similarly,, by applying

$\tan(A+B) = \frac{\tan A + \tan B}{1 - \tan A \tan B}$ to

$$\ln\left|\tan\left(\frac{\pi}{4}+\frac{x}{2}\right)\right|$$

$$\ln\left|\tan\left(\frac{\pi}{4}+\frac{x}{2}\right)\right| = \ln\left|\frac{\tan\frac{\pi}{4}+\tan\frac{x}{2}}{1-\tan\frac{x}{2}\tan\frac{\pi}{4}}\right| \quad \text{(B)}$$

Since the right sides of (A) and (B) are identical, the left sides are equal, and

$$\ln\left|\frac{1+\tan\frac{x}{2}}{1-\tan\frac{x}{2}}\right| = \ln\left|\tan\left(\frac{\pi}{4}+\frac{x}{2}\right)\right|$$

$$\therefore \ \ln\left|\frac{1+\tan\frac{x}{2}}{1-\tan\frac{x}{2}}\right| = \ln\left|\tan\left(\frac{\pi}{4}+\frac{x}{2}\right)\right|$$

Example 2

To exaggerate the substitution axiom:

If the cost of 2 books = \$100,
then, the cost of $\sin\frac{\pi}{2}$ book = \$ 50
(same as the cost of one book = \$50; since $\sin\frac{\pi}{2} = 1$)
We could also say that
the cost of $\tan\frac{\pi}{4}$ book = \$50, since
$\tan\frac{\pi}{4} = 1$.
Similarly, the cost of x^0 book = \$50.
Also $x^0 = \sin\frac{\pi}{2} = \tan\frac{\pi}{4} = 7 - 6 = \frac{5}{5} = 1$.
$(x \neq 0)$
Extra: Similarly,
$(\sin^2\theta + \cos^2\theta)$ book costs \$50.
since $\sin^2\theta + \cos^2\theta = 1$.

Lesson 39 Exercises

A Prove the following identities:

1. $\cot x + \tan x = \sec x \csc x$

2. $\tan x + \csc x = \frac{\sin x + \cot x}{\cos x}$

3. $\frac{\cos 2x}{\cos x} + \sec x = \frac{\sin 2x}{\sin x}$;

4. $\frac{\cos x}{\sec x} - \frac{\sin x}{\cot x} = \frac{\cos x \cot x - \tan x}{\csc x}$

5. $\csc^2 x \sec^2 x - \csc^2 x = \sec^2 x$;

6. $\sec x - \csc^2 x \sec x = -\cot x \csc x$.

7. $\frac{\tan x}{\sec x - 1} = \csc x + \cot x$.

B Given that $\sin A = \frac{1}{6}$ with the terminal side of $\angle A$ being in the second quadrant and $\cos B = \frac{1}{2}$ with the terminal side of $\angle B$ being in the fourth quadrant, Find
(a) $\cos(A - B)$; (b) $\sin(A + B)$; and (c) $\sin 2A$.

Answers: (a) $-\frac{\sqrt{35}+\sqrt{3}}{12}$; (b) $\frac{1+\sqrt{105}}{12}$; (c) $-\frac{\sqrt{35}}{18}$

CHAPTER 14

Lesson 40

Solutions of Trigonometric Equations

A trigonometric equation is an equation containing a trigonometric function of an unknown angle. A trigonometric equation is a conditional equation. By conditional we mean the equation is true for some but not all permissible replacements of the variable. We can solve a trigonometric equation either graphically or algebraically. We will cover only algebraic methods although on occasion we may complement the algebraic solutions by graphical methods. In solving trigonometric equations, we will make use of trigonometric identities as well as inverse trigonometric relations.
We will cover two types of trigonometric equations, namely,

1. The trigonometric equation which is linear in one trigonometric function of one variable.

2. The trigonometric equation which is quadratic in one variable of the function.
We will restrict our solutions for an unknown angle x such that $0 \le x < 360^o$ ($0 \le x < 2\pi$).

General Procedure for solving trigonometric equations:

Step 1: Collect all terms on one side of the equation, preferably on the left-hand side of the equation.

Step 2: Solve the equation for the trig function as exemplified below:
Examples in which the functions have been solved for: **1**. $\sin x = .5$
2. $\tan x = 1$
Examples in which the functions have **not** been solved for: **1**. $4\sin^2 x = 1$
2. $\cos^2 x = .5$

In trying to solve for the functions we may have to apply trigonometric identities if the equation contains multiple angles and / or more than one type of a trig function. If there are multiple angles such as double angles, we may apply the double angle identities or we may substitute a new variable for the multiple angle, solve for the function in terms of the new variable, and reconvert to the original variable after having found the measure of the angle.

Step 3: Solve for the angle: Use inverse trigonometric functions (from tables, memory or calculator) to find the measures of the angles involved.

Step 4: Check the solutions if the equation was squared or raised to a power, since we might have introduced extraneous solutions.

Example 1 Solve for x: $\sin x = \frac{1}{2}$ or .5 $0 \le x < 360^o$

Solution

Step 1: Find the angle whose sine is $\frac{1}{2}$ (from tables or calculator). The reference angle is 30°

Step 2: Determine the quadrants in which the sine is positive.
The sine is positive in quadrants I and II. ($\frac{1}{2}$ is positive)

Step 3 : Specify the measures of the angles (in standard position).
$x = 30^o$ or 150° (The angles are measured from the positive x-axis counterclockwise)

The solution set is $\{30^o, 150^o\}$ or $\left\{\frac{\pi}{6}, \frac{5\pi}{6}\right\}$ in radians.

Example 2 Solve for x: $2\sin x - 1 = 0 \quad 0 \le x < 360^\circ$

Step 1: Solve for $\sin x$

$2\sin x = 1$

$\sin x = \frac{1}{2}$ (1)

Step 2: Equation (1) of Step 1 is the same as Example 1 above. We therefore repeat the solution:

The solution set is $\{30^\circ, 150^\circ\}$ or $\left\{\frac{\pi}{6}, \frac{5\pi}{6}\right\}$ in radians

Example 3 Solve for x: $\cos 2x = \frac{1}{2}$.

Method 1

Step 1: Let $2x = \theta$

Then, $\cos 2x = \frac{1}{2}$ becomes

$\cos\theta = \frac{1}{2}$

Step 2: Determine the reference angle for θ.

The reference angle is 60° (From tables, memory or calculator)

Step 3: Determine the quadrant in which $\cos\theta$ is positive.

The cosine is positive in quadrant I and IV. ($\frac{1}{2}$ is positive)

Step 4: Specify the measures of the angles (in standard position).

$\theta = 60^\circ, \theta = 300^\circ$

Step 5: The general solutions are $\theta = 60^\circ + 360n$ or $\theta = 300^\circ + 360n$ ($n = 0, 1$ for this problem)

Step 6: Change back to x. That is, replace θ by $2x$

$2x = 60^\circ + n\cdot 360^\circ$ or $2x = 300^\circ + n\cdot 360^\circ$

$x = 30^\circ + n\cdot 180^\circ$ or $x = 150^\circ + n\cdot 180^\circ$ (Dividing by 2 and solving for x)

For $n = 0$, $\mathbf{x = 30^\circ}$ or $\mathbf{x = 150^\circ}$ (When $n = 0$, $n\cdot 180 = 0$)

For $n = 1$, $x = 30^\circ + 180^\circ$ or $x = 150^\circ + 180^\circ$

$\mathbf{x = 210^\circ}$ or $\mathbf{x = 330^\circ}$

The solution set is $\{30^\circ, 150^\circ, 210^\circ, 330^\circ\}$

Note that in Step 6, we considered two values of n ($n = 0, 1$) because of the "2" in $\cos 2x$. If we were given $\cos 3x$, we would have considered three values of n ($n = 0, 1, 2$) because of the "3" in $\cos 3x$.

Method 2: Using the double angle identity, $\cos 2x = 2\cos^2 x - 1$

The given equation $\cos 2x = \frac{1}{2}$ becomes

$$2\cos^2 x - 1 = \frac{1}{2}$$

$$2\cos^2 x = \frac{1}{2} + 1$$

$$2\cos^2 x = \frac{3}{2}$$

$$\cos^2 x = \frac{3}{4}$$

$$\cos x = \pm\sqrt{\frac{3}{4}}$$

$\cos x = \pm\frac{\sqrt{3}}{2}$

$\cos x = +\frac{\sqrt{3}}{2}$ or $\cos x = -\frac{\sqrt{3}}{2}$

For $\cos x = +\frac{\sqrt{3}}{2}$, the reference angle is 30°.
$\cos x$ is positive in quadrants I and IV.
Therefore, x = **30° or 330°** (angles in standard position).

Similarly, for $\cos x = -\frac{\sqrt{3}}{2}$, the reference angle is 30°.
$\cos x$ is negative in quadrants II and III.
Therefore, x = **150° or 210°** (angles in standard position).

The solution set is {30°, 150°, 210°, 330°}.

Again, we obtain the same solution set as by Method 1. You decide which method you prefer.

Example 4 Solve for x: $2\cos^2 x + 3\sin x = 3$, $0 \le x < 360°$

Solution $2\cos^2 x + 3\sin x - 3 = 0$

Step 1: $2(1 - \sin^2 x) + 3\sin x - 3 = 0$ $\quad (\cos^2 x = 1 - \sin^2 x)$

$2 - 2\sin^2 x + 3\sin x - 3 = 0$

$-2\sin^2 x + 3\sin x - 1 = 0$

$2\sin^2 x - 3\sin x + 1 = 0$ $\quad (1)$

*Step 2: Let $\sin x = t$, Then equation (1) becomes

$2t^2 - 3t + 1 = 0$ <--------quadratic equation.

We solve by factoring since the factors are easily recognizable.

$(2t - 1)(t - 1) = 0$

$2t - 1 = 0$ or $t - 1 = 0$

$t = \frac{1}{2}$ or $t = 1$

Now, replace t by $\sin x$.

Then $\sin x = \frac{1}{2}$ or $\sin x = 1$

Step 3: Solve for the angles:

When $\sin x = \frac{1}{2}$, the reference angle is 30°,

and x = **30°** or **150°**

When $\sin x = 1$, x = **90°**

The solution set is {30°, 90°, 150°}.

* Note that in Step 2, we could avoid substitution by doing the following: $(2\sin x - 1)(\sin x - 1) = 0$.
Then either $2\sin x - 1 = 0$ or $\sin x - 1 = 0$.

Solving, $\sin x = \frac{1}{2}$ or $\sin x = 1$

Example 5: Solve for x: $\tan x = 1$ $\qquad 0° \le x < 360°$

Solution

Step 1: Find the angle (or number) whose tangent is 1.

From tables or memory , the reference angle is $45°$ $\frac{\pi}{4}$.

Step 2: Determine the quadrant in which the tangent function is positive.

The tangent function is positive in the first and third quadrants

Step 3: Specify the measures of the angles (in standard position).

The solution set is $\{45°, 225°\}$ or $\left\{\frac{\pi}{4}, \frac{5\pi}{4}\right\}$

Extra

Solve the following system of equation simultaneously by graphing:

$$\left.\begin{aligned} y &= \tan x \\ y &= 1 \end{aligned}\right\}$$

Hint: Sketch the graphs of $y = \tan x$ and $y = 1$ on the same set of rectangular axes and read the coordinates of the point of intersection of the two curves.

A note about Solutions of Trigonometric Equations and Inverse Functions

Generally, when we find the inverse of a trigonometric function we shall restrict the domain of the function so that there will be only one value of x. Note that the value we read from tables is a reference angle and x must be expressed in standard position.

However, in solving the trigonometric equations, we usually restrict our solutions to $0 \le x < 2\pi$ or $0° \le x < 360°$, and here we usually have multiple solutions.

Comparison of equivalent functions, inverse functions and reciprocal functions

Equivalent functions

1. $y = 2x$ is equivalent to $\frac{y}{2} = x$
2. $\text{Arcsin } x = y$ is equivalent to $x = \sin y$
3. $\log_a x = y$ is equivalent to $a^y = x$
4. $y = x^2$, $x \ge 0$ and $x = \sqrt{y}$ are equivalent

Inverses

1. $y = 2x$ and $y = \frac{x}{2}$ are inverses.
2. $y = \sin x$ and $y = \arcsin x$ are inverses.
3. $y = a^x$ and $y = \log_a x$ are inverses.($y = a^x$ and $x = a^y$)
4. $y = x^2$, $x \ge 0$ and $y = \sqrt{x}$ are inverses.

Reciprocal functions

1. $y = x$ and $y = \frac{1}{x}$ are reciprocal functions.

2. $y = a^x$ and $y = \frac{1}{a^x} = a^{-x}$ are reciprocal functions.

3. $y = \sin x$ and $y = \frac{1}{\sin x} = \csc x$ are reciprocal functions.

4. $y = 2x$ and $y = \frac{1}{2x}$ are reciprocal functions.

5. $y = x^2$ and $y = \frac{1}{x^2}$ or x^{-2} are reciprocal functions

Lesson 40 Exercises

A Solve for x: 1. $4\cos^2 x - 8\cos x + 3 = 0$; **2.** $2\cos^2 x - \cos x = 1$

Answers: **1.** $\{60^\circ, 300^\circ\}$ or $\left\{\frac{\pi}{3}, \frac{5\pi}{3}\right\}$ **2.** $\{0^\circ, 120^\circ, 240^\circ\}$ or $\left\{0, \frac{2\pi}{3}, \frac{4\pi}{3}\right\}$

B **1.** Solve for x: $\cos^2 x - 1 = -\cos x - \cos^2 x$ for $0 \le x < 2\pi$

2. Solve for x: $2\cos^2 x = \sin x + 1$ for $0 \le x < 2\pi$.

Answers: **1.** $\left\{\frac{\pi}{3}, \frac{5\pi}{3}, \pi\right\}$; **2.** $\left\{\frac{\pi}{6}, \frac{5\pi}{6}, \frac{3\pi}{2}\right\}$

Extra

Example 6 Solve for x: $2\tan^{-1}(x) = \tan^{-1}\left(\frac{1}{4x}\right)$ (A)

Let $\tan^{-1}x = \theta$

Then (A) becomes

$2\theta = \tan^{-1}\left(\frac{1}{4x}\right)$

taking ("tan") of both sides to undo "$\tan^{-1}$"

Left side Right side

⇓ ⇓

$$\tan 2\theta = \tan\left(\tan^{-1}\frac{1}{4x}\right)$$

$$\frac{2\tan\theta}{1-\tan^2\theta} = \tan\left(\tan^{-1}\frac{1}{4x}\right)$$

$$\frac{2\tan[\tan^{-1}x]}{1-[\tan(\tan^{-1}x)]^2} = \tan\left(\tan^{-1}\frac{1}{4x}\right)$$

$$\frac{2x}{1-x^2} = \frac{1}{4x}$$

$$8x^2 = 1 - x^2$$

$$9x^2 = 1;$$

$$x^2 = \frac{1}{9};$$

$$x = \pm\sqrt{\frac{1}{9}};$$

$$x = \pm\frac{1}{3}$$

Trig Identity

$$\tan 2\theta = \frac{2\tan\theta}{1-\tan^2\theta}$$

$$2\tan^{-1}(x) = \tan^{-1}\left(\frac{1}{4x}\right)$$

$$2\tan^{-1}\left(\frac{1}{3}\right) \stackrel{?}{=} \tan^{-1}\left(\frac{1}{\frac{4}{3}}\right)$$

$$2\tan^{-1}\left(\frac{1}{3}\right) \stackrel{?}{=} \tan^{-1}\left(\frac{3}{4}\right)$$

Example 7 Solve for x: $2\tan^{-1}\sqrt{x-x^2} = \tan^{-1}x + \tan^{-1}(1-x)$

Solution

Step 1: $2\tan^{-1}\sqrt{x-x^2} = \tan^{-1}x + \tan^{-1}(1-x)$ (A)

$\tan(2\tan^{-1}\sqrt{x-x^2}) = \tan[\tan^{-1}x + \tan^{-1}(1-x)]$ (B)

(taking ("tan") of both sides to undo " "$\tan^{-1}$"

Step 2: Let $\tan^{-1}\sqrt{x-x^2} = \theta_1$

Let $\tan^{-1}x = \theta_2$; Let $\tan^{-1}(1-x) = \theta_3$

Then (B) becomes

Left side Right side

⇓ ⇓

$$\tan 2\theta_1 = \tan(\theta_2 + \theta_3)$$

$$\frac{2\tan\theta_1}{1-\tan^2\theta_1} = \frac{\tan\theta_2 + \tan\theta_3}{1-\tan\theta_2\tan\theta_3}$$

Trig. Identities

$$\tan 2A = \frac{2\tan A}{1-\tan^2 A}$$

$$\tan(A+B) = \frac{\tan A + \tan B}{1-\tan A\tan B}$$

Step 3:

$$\frac{2\tan[\tan^{-1}\sqrt{x-x^2}]}{1-[\tan(\tan^{-1}\sqrt{x-x^2})]^2} = \frac{\tan(\tan^{-1}x) + \tan[\tan^{-1}(1-x)]}{1-\tan(\tan^{-1}x)\bullet\tan[\tan^{-1}(1-x)]}$$

$$\frac{2\sqrt{x-x^2}}{1-[x-x^2]} = \frac{x+1-x}{1-[x(1-x)]}$$

$$\frac{2\sqrt{x-x^2}}{1-x+x^2} = \frac{1}{1-[x-x^2]}$$

$$\frac{2\sqrt{x-x^2}}{1-x+x^2} = \frac{1}{1-x+x^2}$$

$$2\sqrt{x-x^2} = 1$$

(Equating the numerators since the denominators are equal)

$$4(x-x^2) = 1$$

(squaring both sides of the equation

Step 4: Solve the resulting quadratic equation by factoring, since the factors are easily recognizable.

$$4x - 4x^2 - 1 = 0$$

$$4x^2 - 4x + 1 = 0$$

$$(2x-1)(2x-1) = 0$$

$$2x - 1 = 0$$

$$2x = 1$$

$$x = \frac{1}{2}$$

Solution check

Substitute $x = \frac{1}{2}$ in (A)

Left side Right side

⇓ ⇓

$$2\tan^{-1}\sqrt{\tfrac{1}{2} - (\tfrac{1}{2})^2} \stackrel{?}{=} \tan^{-1}\tfrac{1}{2} + \tan^{-1}(1-\tfrac{1}{2})$$

$$2\tan^{-1}\sqrt{\tfrac{1}{2} - \tfrac{1}{4}} \stackrel{?}{=} \tan^{-1}\tfrac{1}{2} + \tan^{-1}(\tfrac{1}{2})$$

$$2\tan^{-1}\sqrt{\tfrac{1}{4}} \stackrel{?}{=} 2\tan^{-1}\tfrac{1}{2}$$

$$2\tan^{-1}\tfrac{1}{2} \stackrel{?}{=} 2\tan^{-1}\tfrac{1}{2} \quad \text{Yes.}$$

Appendix A

How to change a decimal to a rational number

A. How to change a terminating decimal to a rational number

Procedure: Write each decimal as a decimal fraction and reduce to its lowest terms.

Example 1:

$$(a)\ .5 = \frac{5}{10} = \frac{1}{2}$$

$$(b)\ .25 = \frac{25}{100} = \frac{1}{4}$$

$$(c)\ .16 = \frac{16}{100} = \frac{4}{25}$$

$$(d)\ .125 = \frac{125}{1000} = \frac{1}{8}$$

B. How to Change a repeating decimal to a rational number

(In college algebra, we converted a repeating decimal to a rational fraction, using geometric series)

Example 2 Change the following to fractions;

(a) .333..., (b) .666..., (b) ..232323......, (c) .166...

Solution

(a) Step 1: Let $x = .333$ (1)

Step 2: Multiply equation (1) by 10 (Generally, the exponent on the 10 equals the number of digits in the repeating block.)

$10x = 3.333$ (2)

Step 3: Subtract equation (1) from equation (2) and solve for x.

$$10x - x = 3.333... - .333...$$

$$9x = 3$$

$$x = \frac{3}{9}$$

$$x = \frac{1}{3}$$

Therefore $.333... = \frac{1}{3}$

(b) Step 1: Let $x = .666...$ (1)

Step 2: Multiply equation (1) by 10

$10x = 6.666...$ (2)

Step 3: Subtract equation (1) from equation (2) and solve for x:

$$10x - x = 6.666... - .666...$$
$$9x = 6$$
$$x = \frac{6}{9}$$
$$x = \frac{2}{3}$$

Therefore $.666... = \frac{2}{3}$

(c) Step 1: Let $x = .232323...$ (1)

Step 2: Multiply equation (1) by 10^2 or 100 (The exponent on the 10 is 2 since there are two digits in the repeating block.)

$100x = 23.232323...$ (2)

Step 3: Subtract equation (1) from equation (2) and solve for x:

$$100x - x = 23.232323... - .232323...$$
$$99x = 23$$
$$x = \frac{23}{99}$$

d) **Method 1**

Let $x = .166...$ (1)

Step 2: Multiply equation (1) by 10.

$10x = 1.666$ (2) (Note: $1.66... = 1.666...$))

Step 3: Subtract equation (1) from equation (2) and solve for x:

$$10x - x = 1.666 - .166$$
$$9x = 1.5$$

$90x = 15$ (Multiplying by 10 to eliminate the decimal point)

$$x = \frac{15}{90}$$
$$x = \frac{1}{6}$$

Note: $.166... = .1666...$

Method 2 Let $x = .166...$ (1)

Step 2: Multiply equation (1) by 10.

$$10x = 1.66 \quad (2)$$

Also multiplying (1) by 100,

$$100x = 16.66... \quad (3)$$

Step 3: Subtract (2) from (3) and solve for x:

$$100x - 10x = 16.666... - 1.6666...$$
$$90x = 15$$
$$x = \frac{15}{90}$$
$$x = \frac{1}{6}$$

Appendix B

Definitions of Inverse Trigonometric Functions

We summarize the definitions of the **inverse trigonometric functions** below
Note that the restrictions made in obtaining the above inverse functions are arbitrary, and that other branches may be used. However, the branches presented here are the ones that are usually used.

Restricted Function	Domain	Range	Inverse	Notation	Domain	Range
$y = \sin x$,	$-\frac{\pi}{2} \le x \le \frac{\pi}{2}$,	$-1 \le y \le 1$	$x = \sin y$,	$y = \text{Arc}\sin x$ or $\text{Sin}^{-1}x$;	$-1 \le x \le 1$;	$-\frac{\pi}{2} \le y \le \frac{\pi}{2}$
$y = \cos x$,	$0 \le x \le \pi$,	$-1 \le y \le 1$	$x = \cos y$,	$y = \text{Arc}\cos x$ or $\text{Cos}^{-1}x$;	$-1 \le x \le 1$;	$0 \le y \le \pi$
$y = \tan x$,	$-\frac{\pi}{2} < x < \frac{\pi}{2}$,	$-\infty < y < \infty$	$x = \tan y$,	$y = \text{Arc}\tan x$ or $\text{Tan}^{-1}x$;	$-\infty < x < \infty$;	$-\frac{\pi}{2} < y < \frac{\pi}{2}$
$y = \cot x$,	$0 < x < \pi$,	$-\infty < y < \infty$	$x = \cot y$,	$y = \text{Arc}\cot x$ or $\text{Cot}^{-1}x$;	$-\infty < x < \infty$;	$0 < y < \pi$
(Also: $\text{Tan}^{-1}x + \text{Cot}^{-1}x = \frac{\pi}{2}$)				or $y = \text{Cot}^{-1}x = \text{Tan}^{-1}\left(\frac{1}{x}\right) = \frac{\pi}{2} - \text{Tan}^{-1}x$;		$0 < y < \pi$
$y = \sec x$,	$0 \le x \le \pi$, $x \ne \frac{\pi}{2}$,	$\lvert y \rvert \ge 1$	$x = \sec y$,	$y = \text{Arc}\sec x$ or $\text{Sec}^{-1}x$;	$\lvert x \rvert \ge 1$,	$0 \le y \le \pi$, $y \ne \frac{\pi}{2}$
				Also, $y = \text{Sec}^{-1}x = \text{Cos}^{-1}\left(\frac{1}{x}\right)$;	$\lvert x \rvert \ge 1$	$0 \le y \le \pi$, $y \ne \frac{\pi}{2}$
$y = \csc x$,	$-\frac{\pi}{2} \le x \le \frac{\pi}{2}$, $x \ne 0$,	$\lvert y \rvert \ge 1$	$x = \csc y$,	$y = \text{Arc}\csc x$ or Csc^{-1};	$\lvert x \rvert \ge 1$,	$-\frac{\pi}{2} \le y \le \frac{\pi}{2}$, $y \ne 0$
(Also: $\text{Sec}^{-1}x + \text{Csc}^{-1}x = \frac{\pi}{2}$)				Also. $y = \text{Csc}^{-1}x = \text{Sin}^{-1}\left(\frac{1}{x}\right)$;	$\lvert x \rvert \ge 1$,	$-\frac{\pi}{2} \le y \le \frac{\pi}{2}$, $y \ne 0$

More on quadrants for the definitions of the inverse functions

$\text{Sin}^{-1}x$, $\text{Csc}^{-1}x$ and $\text{Tan}^{-1}x$ are defined in the **first** and **fourth** quadrants or appropriately quadrantal.. If $\text{Sin}^{-1}x$,$\text{Csc}^{-1}x$,or $\text{Tan}^{-1}x$ is in the fourth quadrant, then it is a **negative** angle or appropriately quadrantal. $\text{Cos}^{-1}x$**,** $\text{Sec}^{-1}x$,and $\text{Cot}^{-1}x$ are defined in the **first** and **second** quadrants or appropriately quadrantal. $\text{Cos}^{-1}x$, $\text{Sec}^{-1}x$, and $\text{Cot}^{-1}x$ are **never** negative.

Note: Saying that $-\frac{\pi}{2} \le y \le \frac{\pi}{2}$ implies that y is either in the first quadrant or in the fourth quadrant or appropriately quadrantal. Also, saying that $0 \le y \le \pi$ implies that y is either in the first quadrant or in the second quadrant or appropriately quadrantal.

Note the restricted domains and ranges of these inverse functions. In all operations involving inverse functions as well as in sketching their graphs, we will be guided by these domains and ranges. For example, $\text{Sin}^{-1}x$ (read: inverse sine x) means the angle (in radians) or the number between $-\pi/2$ and $\pi/2$, whose sine is x. Similarly, $\text{Cos}^{-1}x$ (read: inverse cosine x) means the angle (in radians) or the number between 0 and π, whose cosine is x.. In each definition, two conditions must be satisfied simultaneously. For example, the conditions for $\text{Sin}^{-1}x$ are:

1. The sine of the angle (or number) $= x$, **2**. The angle must be between $-\frac{\pi}{2}$ and $\frac{\pi}{2}$.

Inverse Cofunction Identities

1. $\text{Sin}^{-1}x + \text{Cos}^{-1}x = \frac{\pi}{2}$, $[-1,1]$; **2.** $\text{Tan}^{-1}x + \text{Cot}^{-1}x = \frac{\pi}{2}$, $(-\infty,\infty)$; **3** $\text{Sec}^{-1}x + \text{Csc}^{-1}x = \frac{\pi}{2}, |x| \geq 1$

Each identity is subject to the domain of definition

We prove these identities assuming $0 \leq x \leq 1$.

Proof:

Let θ and β be two angles of a right triangle.

$\sin\theta = \frac{x}{1} = x$

and $\sin^{-1}x = \theta$. (A)

Similarly, $\cos\beta = \frac{x}{1} = x$ and

$\cos^{-1}x = \beta$ (B)

$\theta + \beta = \frac{\pi}{2}$ (C) (Sum of the other two angles of a right triangle)

$\sin^{-1}x + \cos^{-1}x = \theta + \beta$ (A) + (B)

$\sin^{-1}x + \cos^{-1}x = \frac{\pi}{2}$ (From (C) $\theta + \beta = \frac{\pi}{2}$)

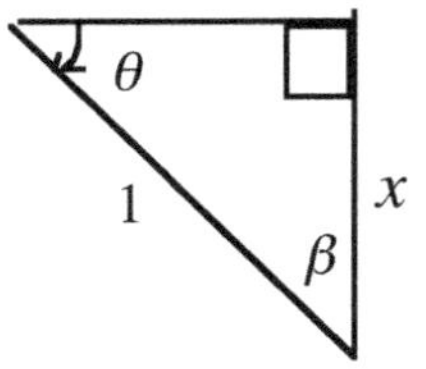

Proof:

Let θ and β be two angles of a right triangle.

$\tan\theta = \frac{x}{1} = x$

and $\tan^{-1}x = \theta$. (A)

Similarly, $\cot\beta = \frac{x}{1} = x$ and

$\cot^{-1}x = \beta$ (B)

$\theta + \beta = \frac{\pi}{2}$ (C) (Sum of the other two angles of a right triangle)

$\tan^{-1}x + \cot^{-1}x = \theta + \beta$ (A) + (B)

$\text{Tan}^{-1}x + \text{Cot}^{-1}x = \frac{\pi}{2}$ (From (C) $\theta + \beta = \frac{\pi}{2}$)

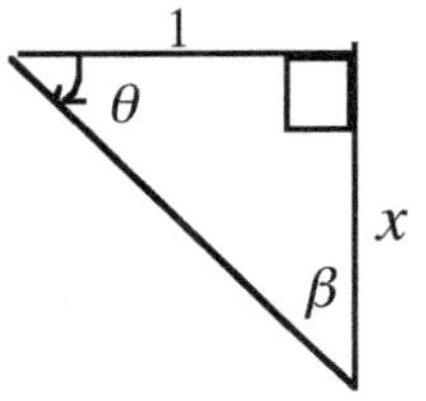

Proof:

Let θ and β be two angles of a right triangle.

$\sec\theta = \frac{x}{1} = x$

and $\sec^{-1}x = \theta$. (A)

Similarly, $\csc\beta = \frac{x}{1} = x$ and

$\csc^{-1}x = \beta$ (B)

$\theta + \beta = \frac{\pi}{2}$ (C) (Sum of the other two angles of a right triangle)

$\sec^{-1}x + \csc^{-1}x = \theta + \beta$ (A) + (B)

$\text{Sec}^{-1}x + \text{Csc}^{-1}x = \frac{\pi}{2}$ (From (C) $\theta + \beta = \frac{\pi}{2}$)

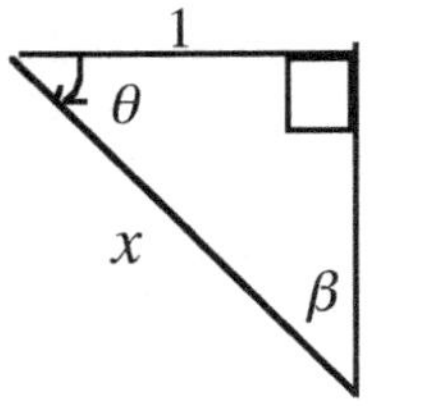

Appendix C

Area and Perimeter of a Circle; Arc Length and Sector of a Circle; Area and Perimeter of Composite Figures

Area and Perimeter of a Circle

Area of a circle

The area, A , of a circle of radius , r, is given by $A = \pi r^2$.
π is approximately equal to 3.1416; in calculations, we will leave our answers in terms of π, unless instructed otherwise.

Note also that the radius of a circle $= \frac{1}{2}$ of the diameter of the circle.

Area of a semicircle (half-circle) $= \frac{1}{2}$ of the area of the given circle.

Area of a quarter-circle $= \frac{1}{4}$ of the area of the given circle.

Perimeter (or Circumference) of a Circle

The perimeter of a circle is the distance around the circle (i.e., the length of the circle).
The perimeter, P, of a circle of radius, r, is given by $P = 2\pi r$.

The perimeter of a semicircle $= \frac{1}{2}$ of the perimeter of the given circle. However, if there is a diameter joining the end points of the curved part, then add the diameter. Therefore, the shape of the figure determines whether or not to add the diameter. The circle formula is for the curved part only.

The perimeter of a quarter circle is $\frac{1}{4}$ of the perimeter of the given circle plus $2r$ if two radii connect the end points of the curved part.

Example Find the area and the perimeter of a circle of radius 6 units.

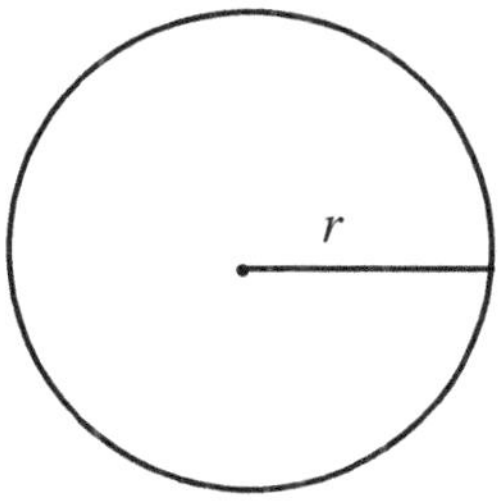

Finding the area

$r = 6$, Area of a circle $= \pi r^2$.
Area of the given circle $= \pi(6)^2 = \pi(36) = 36\pi$ sq. units

Finding the perimeter

$r = 6$ units
Perimeter of a circle $= 2\pi r$
Perimeter of the given circle $= 2\pi(6) = 12\pi$ units

Note: In the above problem, the area of a **semicircle** of radius, $r = 6$ is $\frac{1}{2}(36\pi) = 18\pi$ sq. units; and the perimeter of the semicircle (curved part only) of radius, $r = 6$, is $\frac{1}{2}(12\pi) = 6\pi$ units. If the diameter connects the end points of the curved part, then add the diameter, 2(6) to obtain $(6\pi + 12)$ units.

Arc Length and Sector of a Circle

Arc length: Arc length is the length of a part of a circle between any two points on the circle. In a circle of radius r, the arc length s cut off by a central angle θ (θ in degrees) is given by

$$s = \frac{\theta}{360} \text{ of the circumference of the circle.}$$

$$\boxed{s = \frac{\theta}{360}(2\pi r) \text{ or } \frac{\theta\pi r}{180}}.$$ (From the proportion, s is to θ as 2πr is to 360)

However, if the central angle is in radians, replace 360 by 2π and simplify to obtain ($360^\circ = 2\pi$ radians, or $180^\circ = \pi$ radians)

$$\boxed{s = r\theta}$$

Sector of a circle: A sector of a circle is the part of the interior of the circle bounded by two radii and the intercepted arc.

In a circle of radius r, the area A of a sector with central angle θ (θ in degrees) is given by

$$A = \frac{\theta}{360} \text{ of the area of the circle.}$$

$$\boxed{A = \frac{\theta}{360}(\pi r^2)}$$ (From the proportion, A is to θ as πr^2 is to 360)

However, if θ is in radians, replace 360 by 2π and simplify to obtain

$$\boxed{A = \frac{1}{2}r^2\theta}$$

Example In the circle shown below, if the diameter is 6 and the central angle is 30°,

(a) Find the circumference of the circle.
(b) Find the arc length of CB
(c) Find the area of sector COB.

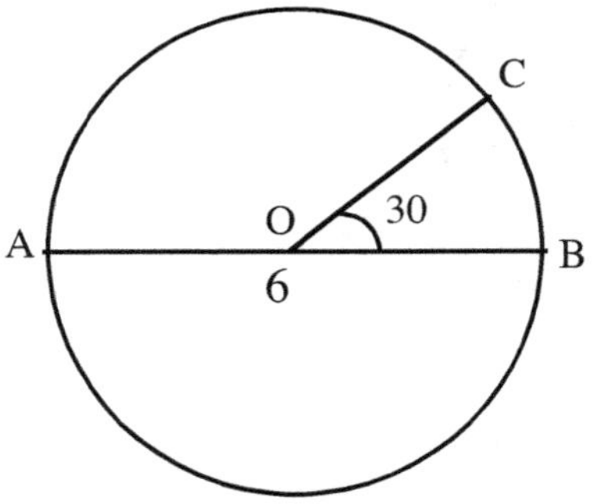

Figure 1

Solution

(a) Circumference of a circle of radius r is given by $2\pi r$.

Radius of a circle $= \frac{1}{2}$ of the diameter of the circle.

Therefore, $r = \frac{1}{2}(6) = 3$

Circumference of the given circle is $2\pi r. = 2\pi(3) = 6\pi$ units.

(b) arc length, s, of CB = $\frac{\theta}{360}$ of the circumference of the circle

$$s = \frac{30}{360} \cdot 2\pi(3) = \frac{\pi}{2} \text{ units}$$

OR if θ is in radians, $s = r\theta = 3 \cdot \frac{\pi}{6}$ (**Note**: $\theta = 30° = \frac{\pi}{6}$; $r = 3$.)

$$s = \frac{\pi}{2} \text{ units.}$$

(c) Area, A, of the shaded sector = $\frac{30°}{360°}$ of the area of the circle.

$$A = \frac{30°}{360°}(\pi r^2)$$
$$= \frac{30°}{360°}(\pi \cdot 3^2)$$
$$= \frac{30}{360}(9\pi)$$
$$= \frac{3\pi}{4} \text{ sq. units}$$

OR if θ is in radians, $A = \frac{1}{2}r^2\theta$ (**Note**: $\theta = 30° = \frac{\pi}{6}$; $r = 3$.)

$$= \frac{1}{2} \cdot 3^2 \cdot \frac{\pi}{6}$$
$$= \frac{1}{2} \cdot \frac{9}{1} \cdot \frac{\pi}{6}$$
$$= \frac{3\pi}{4} \text{ sq. units.}$$

Areas and Perimeters of Composite Figures

Example 1 In the figure below:
(a) Find the area of the figure.
(b) Find the perimeter of figure.

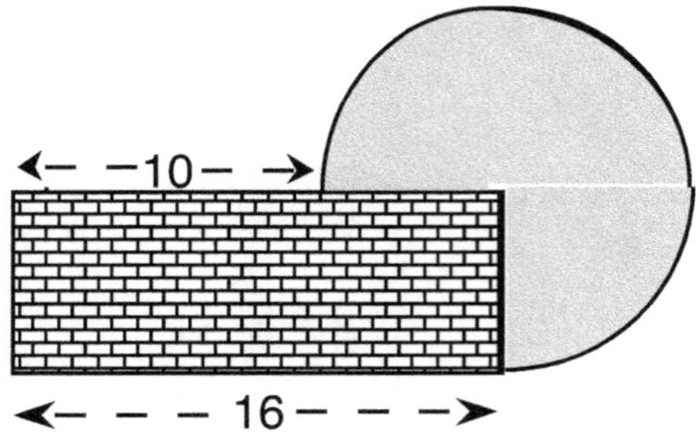

Solution

(a) Finding the area

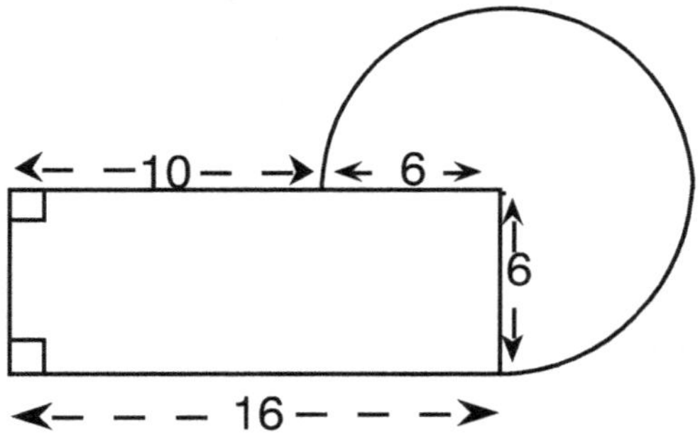

Radius, r, of the three-quarter circle = 6

Area of the three-quarter circle $= \frac{3}{4}\pi r^2$ ($\frac{3}{4}$ of the area of a circle of radius 6 units)

$= \frac{3}{4}\pi(6)^2$

$= \frac{3}{\cancel{4}_1}\pi(\cancel{36})^9$

$= 27\pi$ sq. units

Area of the 16 by 6 rectangle = 16×6 sq. units
= 96 sq. units

Area of the **whole figure** = area of rectangle + area of three-quarter circle
= $(96 + 27\pi)$ sq. units

(b) Finding the perimeter

Perimeter of the three-quarter circle $= \frac{3}{4} 2\pi r$

$= \frac{3}{4}(2\pi)(6)$

$= 9\pi$ units

Perimeter of the whole figure $= 6 + 16 + 9\pi + 10$

$= (32 + 9\pi)$ units

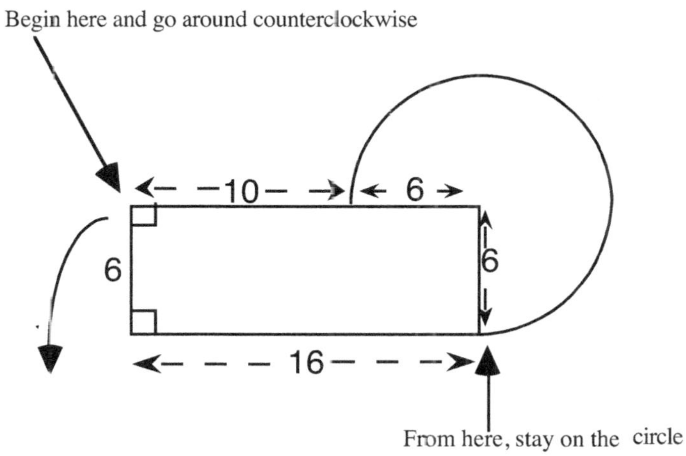

Note that in finding the perimeter we did **not** add the parts (6 and 6) of the rectangle which are also radii of the three-quarter circle, because the perimeter is the distance around the figure.

Appendix C Exercises

A

In the circle below, if the diameter of the circle is 18 units and the central angle has a measure 60°,

(a) Find the circumference of the circle.

(b) Find the area of sector COB.

(c) Find the arc length of CB.

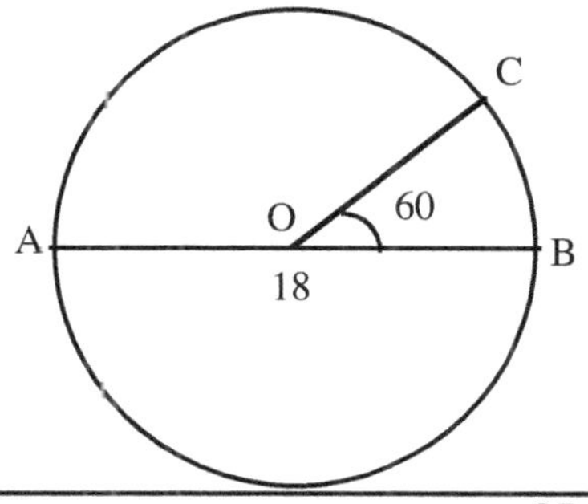

Answers (a) 18π units; (b) $\frac{27\pi}{2}$ sq. units; (c) 3π units

B

(a) Find the area of the figure.
(b) Find the perimeter of figure.

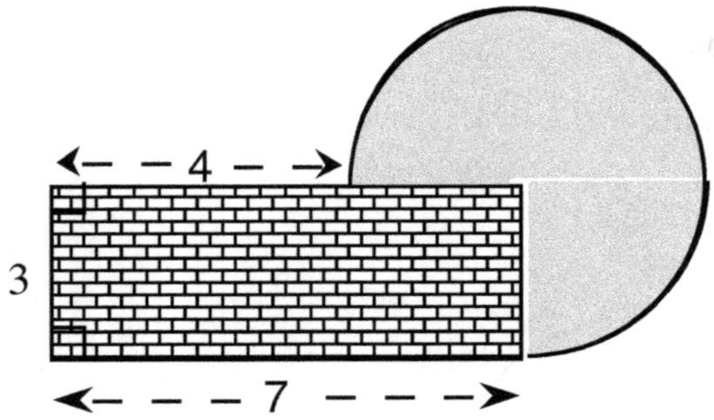

Answers: **(a)** $(21 + \frac{27\pi}{4})$ sq. units; **(b)**. $(14 + \frac{9\pi}{2})$ units

Appendix D
About Measurements

Standard Unit, Error, Rounding-off Numbers, Significant Digits, Scientific Notation

To determine the size of a physical quantity, we compare its size with a standard quantity called a unit. Example: To determine the length of the cover of a book in inches or in meters, we can use a ruler with its scale in inches or in meters. A measurement is the ratio of the magnitude of a physical quantity to that of a standard unit

Standard unit
A standard unit is a measure with which other quantities are compared. A standard unit of measure is defined by a legal authority (such as the US Bureau of Weights and Measures) or by a conference of scientists.

Some universally accepted standards:
1. For mass, the standard (primary standard) is the kilogram (kg).
2. For length, the standard is the meter (m).
3. For time, the standard is the second (s)

Some devices for taking measurements
Examples: Rulers (for length), chemical balances (for mass), stop watches (for time), ammeters, (for electric current) voltmeters (for electric voltage) , thermometers (for temperature) and barometers (for pressure).

Experimental Errors (or Uncertainties)
There are two main types of errors, namely, systematic errors, and random errors.

Systematic Errors (constant errors):
These errors are due to faulty measuring devices. Systematic errors make the measurements either too small or too large:

Examples of faulty devices:
1. Instruments with needles off the zero mark; 2. Faulty clocks (stop clock)
3, Corroded weights; 4. Faulty thermometers. Heat leaking equipment

Random errors (accidental errors or indeterminate errors)
Random errors may be due to chance variations of the physical quantity being measured, or chance variations in the measuring device. Random errors may also be due to failure to take into account variables such as temperature fluctuations; and environmental effects. We can reduce random errors by making a large number of measurements and taking the average of the measurements.

Error (or uncertainty)

Error = Experimental value − accepted (true value)
Example: In an experiment to determine the acceleration due to gravity, g, the experimental value of g, was 986 cm/s^2. The accepted value of g is 980 cm/s^2. Find the error.

Solution Experimental value = 986 cm/s^2

Accepted value = 980 cm/s^2
Error = experimental value - accepted value
= (986 - 980) cm/s^2
= 6 cm/s^2

The positive value indicates that the experimental value is greater than the accepted value.

Note: If the experimental value = 964 cm/s^2
The error = (964 - 980) cm/s^2
= -16 cm/s^2

The negative value means that the experimental value is less than the accepted value.

Relative error (or relative uncertainty)

Relative error = $\frac{\text{error}}{\text{accepted value}}$

Example 1: If the error = 6 cm/s^2, and

the accepted value = 980 cm/s^2,

Relative error = $\frac{\text{error}}{\text{accepted value}}$

Then relative error = $\frac{6\ cm\ /\ s^2}{980\ cm\ /\ s^2}$

= 0.00612

Example 2: If the error = -16 cm//s^2 and

the accepted value = 980 cm/s^2

Then relative error = $\frac{-16\ cm\ /\ s^2}{980\ cm\ /\ s^2}$

= - 0.00612
= - 0. 0163

Note: If an accepted value is not known, and we have two or more experimental values, then the average of the experimental values would be used as the "accepted value" in calculations.

Rounding-Off Numbers

The rules for rounding-off a number may differ slightly depending upon the field. For instance, in accounting the rule may be slightly different from the rule in chemistry.

1. If the digit or group of digits to be dropped is more than 500...,hen drop that digit or group and add 1 to the last digit retained.

2. If the digit or group of digits to be dropped is less than 500...,then drop that digit or group and leave the last digit retained unchanged.

3. If the digit or group of digits to be dropped is exactly 500..., then drop that portion and add 1 to the last digit retained if this digit is odd but if this digit is even, then this digit remains unchanged.

Rounding off Whole numbers

Procedure:

Step 1: Locate the digit in the round-off place.
(The round-off place is the place to which we want to round-off the number)

Step 2: Drop all digits to the right of the round-off place, and if the digit immediately to the right of the round-off place is more than 5 or is 5 followed by non-zero digits, , add 1 to the round-off place digit (i.e. we round-up); but if the digit immediately to the right of the round-off place is less than 5, the round-off place digit remains unchanged. However, if the digit immediately to the right of the round-off place digit is 5 or 5 followed by zeros, we add 1 to the round-off place digit if it is odd, but if it is even, it remains unchanged.(i.e. we round-down). Also, replace each digit dropped by a zero.

Rounding-off Decimals

The procedure is the same as that for rounding-off whole numbers, except that after the decimal point, we do not replace any digits dropped by zeros.

Procedure:

Step 1: Locate the digit in the round-off place.
(The round-off place is the place to which we want to round-off the number)

Step 2: Drop all digits to the right of the round-off place, and if the digit immediately to the right of the round-off place is more than 5 or is 5 followed by non-zero digits, add 1 to the round-off place digit (i.e., we round-up); but if the digit immediately to the right of the round-off place is less than 5, the round-off place digit remains unchanged. However, if the digit immediately to the right of the round-off place digit is 5 or 5 followed by zeros, we add 1 to the round-off place digit if it is odd, but if it is even, it remains unchanged.(i.e., we round-down).

Rounding-off (Alternatively)

When we round-off a number, we drop some of the digits explicitly or implicitly specified. We must distinguish between rounding-off to a specified number of decimal places (or significant digits) and the implicit rounding-off which we must determine from the numbers involved in the calculation.

The rules for rounding-off a number may differ slightly depending upon the field. For instance, in accounting the rule may be slightly different from the rule in chemistry.

1. If the digit or group of digits to be dropped is more than 500...then drop that digit or group and add 1 to the last digit retained.

2. If the digit or group of digits to be dropped is less than 500...then drop that digit or group and leave the last digit retained unchanged.

3. If the digit or group of digits to be dropped is exactly 500..., then drop that portion and add 1 to the last digit retained if this digit is odd but if this digit is even, then this digit remains unchanged.

Example: The following have been rounded-off to three decimal places.

(1) .4398. ≈ **.440**

(2) .43652 ≈ **.437**

(3) .43637 ≈ **.436**

(4) .43750 ≈ **.438**

(5) .43650 ≈ **.436**

(6) .43946 ≈ **.439**

(7) .43650001 ≈ **.437**

(8) .4365001 ≈ **.437**

Example We round off **85376.7463** to the following places, using the simple "5 or greater or less than 5 rule"

1. 85376.7463 to the nearest **thousandth** becomes **85376.746** (We do **not** replace the 3 dropped by a zero)
2. 85376.7463 to nearest **hundredth** becomes **85376.75** (We added 1 to the digit in the round-off place)
3. 85376.7463 to the nearest **tenth** becomes **85376.7** (The 7 is unchanged since the 4 dropped is less than 5)
4. 85376.7463 the nearest **unit** becomes **85377**. (Adding 1 to the 6)
5. 85376.7463 to the nearest **ten** becomes **85380**. (Replacing the 6 dropped by a zero)
6. 85376.7463 to the nearest **hundred** becomes **85400.** (Replacing the digits (6 and 7) dropped by zeros)
7. 85376.7463 to the nearest **whole number** becomes **85377**. (same as to the nearest unit)

Estimation
In estimation, we round-off the numbers before carrying out the operations of addition, subtraction, multiplication , division etc. For convenience, we will round-off each number to the first non-zero digit, unless specified.

Approximate Numbers, Significant Digits, Scientific Notation,

A measurement consists of a numerical value and a unit of that measurement. Example: 4 kilograms, where the 4 is the numerical value and the kilograms is the unit of the measurement.

Numbers obtained from a measurement are never exact (i.e., are approximate) due to the limitations of the measuring instrument as well as the skill of the person making the measurement. As such, when one records a measurement, one should indicate the reliability of the measurement. All measurements may be assumed to have an uncertainty in at least one unit in the last digit of the measurement, since in making a measurement, we usually estimate the last digit.

Results obtained from calculations using measurements are also as uncertain as the measurements themselves.
In summary, the numbers that we deal with in calculations are obtained from observations. Some of the numbers are exact and some are approximate, The approximate numbers are those numbers obtained from making measurements.

Exact Numbers: An exact number is a number that contains no uncertainties. It is assumed to be infinitely accurate. We can obtain exact numbers from definitions and from direct count.
For example, the number of students in a math class by count is 25. In this case, there is no uncertainty, since we know that there are exactly 25 students. Similarly, when one counts 200 dollars, one knows that one has exactly 200 dollars, and there is therefore, no uncertainty.
Also by definition, 60 minutes = 1 hour; 2.54 centimeters = 1 inch. Since these numbers are defined, this 60 and 2,54 are exact and contain no uncertainties. We can also add that this 60 has an infinite number of significant digits, (we can write 60 as 60.000...) and therefore the zero in the 60 is significant. However if you make your own measurement and by coincidence obtain 2.54 centimeters, then this 2.54 would not be exact. We can generalize that all the conversion factors (from tables) are exact.

Significant Digits (Significant Figures), Digits obtained in a measurement

A significant digit (or figure) is one which is known to be reasonably reliable (or correct).
When we make measurements, the digits we read and estimate on a scale are also called significant digits (or significant figures). These digits include digits that we are certain of, and one additional digit that we are uncertain of. This uncertain digit is obtained by the estimation of the fractional part of the smallest subdivision on the scale being used. As such, the rightmost digit is assumed to be uncertain.

Significant figure notation is an approximate method of indicating the uncertainty of a measurement. when recording a measurement.
We agree to the following:

1. The digits 1, 2, 3, 4, 5, 6, 7, 8, 9 are always significant.

2. The digit zero, 0. may or may not be significant according to its position in the number as follows:

(a) Zeros before the first non-zero digits are **not** significant.

For example: (i) .0450 has **three** significant digits: The first zero is not significant; but the last zero is significant since if it were not we would not write it.

(ii) .0012 has **two** significant digits. The first two zeros are not significant

. (b) Zeros between non-zero digits **are** significant.

Example: 3.045 has **four** significant digits. The zero between 3 and 4 is significant.
40.240 has **five** significant digits. The last zero is significant because if it were not we would not write it.

More examples: The numbers referred to below are assumed to have been obtained from measurements.
23.00 has four significant digits: the zeros in this case are significant since we do not have to write the zeros if they were not significant. If the zeros were not significant we would have written 23.
2300. has two significant digits, and 600 has one significant digit; however, in each of these two examples, the number of significant digits is sometimes ambiguous.
It is suggested that when the number of significant digits is in doubt , the maximum number of significant digits is to be assumed. Also, in recording data, if we know the number of significant digits, we will use the scientific notation.
Note above that if 600 and 2300 had been obtained by counting, or by definition, all the zeros would have been significant. As a reminder, significant notation generally pertains to numbers obtained from measurements.
Using **scientific notation** avoids all ambiguities with respect to the number of significant digits.. For instance, if we know that the above number, 600 were measured to two significant digits, then we would write 6.0×10^2. If the measurement were to three significant digits, we would write 6.00×10^2 When we deal with very large or very small numbers we prefer to write the numbers in scientific notation form. In this form, the significant digits (digits) including the zeros can be unambiguously indicated. In scientific notation:

(1) 125000 would be written 1.25×10^5

(2) 1467 would be written 1.467×10^3

(3) .032500 would be written 3.2500×10^{-2}

(4) .0325 would be written 3.25×10^{-2}

(5) If 125000 were known to four significant digits it would be written 1.250×10^5.

Note: Some authors indicate which zeros are significant by underlining the last significant zero. For example, in 230$\underline{0}$0 the first two zeros are significant but the last zero is not significant.
Note also that in some books, a decimal point placed after the last zero makes all the zeros significant. For example, 23000. has five significant digits, but 23000 has two significant digits. However, but there may still be ambiguity if the number is at the end of a sentence.

Accuracy and Precision in Measurements

Two contributions to uncertainty in measurement are limitations of precision, and limitations of accuracy. **Accuracy** indicates how close a measured value is to the true value but **precision** indicates how close two measurements of the same quantity are close to each other. Generally, more precision implies more accuracy. However, there are instances in which numbers may be more precise, but may not be more accurate.. For example, if a measuring device is incorrectly calibrated (having incorrect scale).

Accuracy and Precision in Calculations

With respect to significant digits, **accuracy** refers to the number of significant digits but **precision** refers to the number of decimal places. The larger the number of significant digits in a number, the more accurate the number. The larger the number of decimal places, the more precise the number.

Note: In calculations, the approximate numbers determine how the rounding-off is done

Rounding-off to Significant Digits or Figures in Arithmetic Operations

(Implicit Specification)

1. In **multiplication and division** involving significant digits, the product or quotient (answer) should be rounded-off so that the number of significant digits in the answer is equal to the number of significant digits of the number with the least number of significant digits. (In other words, the answer should not be more accurate than the number with the least accuracy)

Example 1: Multiply 2.34 cm by 5.6 cm

Step 1: $2.34 \times 5.6 = 13.104$

Step 2: The number with the fewest number of significant digits is 5.6 and it has two significant digits. Therefore, the product (answer) should contain only two significant digits.

Thus, 13.104 becomes 13.

Answer: 13 cm^2

In the above case the "4" in 2.34 is considered to be reasonably reliable. The " 6" in 5.6 is considered to be reasonably reliable.

Example 2: Maria determined the length of a piece of wood to be 6.47 yards. What is the length of this wood in feet?

Solution

By definition, 3 feet = 1 yard

6.47 yards is used to determine the number of significant digits in the answer,

(The 3 feet is exact and has an infinite number of significant digits)

$$\frac{6.47 \text{ yards}}{1} \times \frac{3 \text{ feet}}{1 \text{ yard}}$$

$= 19.41$ feet

$= 19.4$ feet (6.47 has three significant figures)

2. In addition and subtraction, we shall round-off so that the number of decimal place in t he answer equals the number of decimal places in the number with the least number of decimal places.(That is,, the answer should not be more precise than the number with the least precision)

Example 1: Add: 143.54, 172.3, and 64.62

Solution

143.54 <---- has two decimal places
172.3 <-----has one decimal place (determines the number of decimal places in answer)
64.62 <----- has two decimal places
380.46
Answer: 380.5 (has one decimal place)

Example 2 Subtract 12.4 from 143.63

Solution

143.63 <---- has two decimal places
12.4 <-----has one decimal place (determines the number of decimal places in answer)
131.23
Answer: 131.2 (has one decimal place)

3. In finding powers and roots, the root or power should be rounded-off so that the number of significant digits in the answer is equal to the number of significant digits in the number.

Example 1: Find $\sqrt{26.9}$

Radicand has three significant digits , and therefore the root should have three significant digits.
From a calculator, $\sqrt{26.9} = 5.1865$

$\sqrt{26.9}$ = 5.19 (has three significant digits. Same as in the radicand)

Note: When two or more different operations are involved, the final operation determines how the final result is rounded-off.

Example Simplify: $38.3 + 12.9(3.58)$

Solution According to order of operations, we multiply 12.9 by 3.58 first and then add 38.3

$38.3 + 12.9(3.58)$

$= 38.3 + 46.182$

$= 84.485$

$= 84.5$ (has one decimal place as in 38.3)

Since addition was the last step, we use precision (the number of decimal places) to round-off.

Addition and Subtraction Involving Scientific Notation

Before adding or subtracting the numbers must have the **same powers** of 10. We will rewrite the expression so that the power of 10 is that of the highest power in the expression

Example: **1**. $4\times10^2+3\times10^2=(4+3)\times10^2$
$=7\times10^2$

Example: **2**. $4\times10^2+2\times10^3=0.4\times10^3+2\times10^3$
$=(0.4+2)\times10^3$
$=2.4\times10^3$

or

$$\begin{aligned}4\times10^2+2\times10^3 &= 4\times10^2+20\times10^2\\ &= (4+20)\times10^2\\ &= 24\times10^2\\ &= 2.4\times10^3\end{aligned}$$ (Again, we obtain the same result)

Order of Magnitude (for comparing relative sizes using powers of 10).
The order of magnitude is the power of 10 closest to the given number.
(It is an approximation to the number. Note the sequence, ..., $10^{-2}, 10^{-1}, 10^0, 10^1, 10^2$,...)

If a given quantity is 1000 times another quantity, the given quantity is larger by three orders of magnitude.

Examples:

1. The order of magnitude of 123 is 10^2, since 123 is closer to 100 than to 1000.

2. Find the order of magnitude of 0.00352.

Solution

Step 1: Write the number in scientific notation.

$0.00352 = 3.52\times10^{-3}$.

Step 2: Since the integer before the decimal point, 3, is less than 5, we replace 3.52 by 1 (since this is closer to 1 or 10^0, than it is to 10 or 10^1)

Step 3: 3.52×10^{-3}
$=10^0\times10^{-3}$
$=1\times10^{-3}$
$=10^{-3}$

The order of magnitude of 0.00352 is 10^{-3}. (since by definition, the order of magnitude is the power of 10 closest to the given number.).
We can use the order of magnitude in estimation by rounding-off to the orders of magnitude.

More examples: Round-off to the nearest order of magnitude.

1. 1.32×10^2
2. 8.02×10^4
3. 0.0009
4. 0.0302

Solution:

1. Since 1.32 is closer to 1 than to 10,

1.32×10^2
$= 10^0 \times 10^2$
$= 1 \times 10^2$
$= 10^2$

The order of magnitude of 1.32×10^2 is 10^2

2. 8.02 is closer to 10 than to 1

8.02×10^4
$= 10^1 \times 10^4$
$= 10^5$

The order of magnitude is 10^5

3. $0.0009 = 9 \times 10^{-4}$
$= 10^1 \times 10^{-4}$
$= 10^{-3}$

The order of magnitude is 10^{-3}.

4. $0.0201 = 2.01 \times 10^{-2}$
$= 10^0 \times 10^{-2}$
$= 1 \times 10^{-2}$
$= 10^{-2}$

The order of magnitude is 10^{-2}

Summary for rounding-off a number to the order of magnitude.

Step !: Write the number in scientific notation

Step 2: Ignoring the power of 10, if the integer before the decimal point is 5 or greater, replace the non-power of 10 part by 10 ((i.e. 10^1); but if the integer is less than 5, replace the non-power of 10 part by 1 (10^0)

Step 3: Simplify

International System of Units

The International System of Units (SI) has adopted a set of seven base (or primary) units.

Quantity	Unit	Symbol
Length	meter	m
Mass	kilogram	kg
Time	seconds	s
Electric current	ampere	A
Temperature	Kelvin	K
Amount of substance	mole	mol
Luminous Intensity	candela	cd

Derived units

In addition to the seven base units, there are derived units which are combinations of the base units

Example:

From the SI base unit, m (meter). for length, the unit for area is $m \times m = m^2$,
Since area = length $\times$ width, and the unit of length is *m* and the unit of with is *m*.

For more practical or convenient units, we use prefixes and multiplication factors (in powers of 10) to express other units.

Example: 1 kilometer = $10^3 m$, where kilo = 10^3

1 km = 10^3 m or I km = 1000m.

APPENDIX E

Complex Numbers

Lesson 41: **Basic Definitions; Basic Operations with Complex Numbers**

Lesson 42: **Equality of Complex Numbers; Roots of Complex Numbers; Equations Involving Complex Numbers**

Lesson 43: **Graphical Representation and Addition of Complex Numbers**

Lesson 44: **Polar (Trigonometric) Form of Complex Numbers**

Lesson 45 **Powers of Complex Numbers; De Moivre's Theorem; Roots of Complex Numbers**

Lesson 41

Definitions; Basic Operations with Complex Numbers

We use complex numbers in the study of electricity and magnetism, in the analysis of feed-back systems especially by the root-locus method, Bode analysis and Nyquist analysis.

Definition, Powers of i, Square Root of Negative Numbers

Sometimes, in attempting to solve certain polynomial equations, we arrive at situations in which we have to find the square roots of negative numbers.

Example: If $x^2 = -1$, then $x = \pm\sqrt{-1}$

Since, for the set of real numbers, there is no provision for the square root of a negative number, we introduce a number "i" which we call the imaginary unit, with the following definition:

Definition: $i = \sqrt{-1}$ or $i^2 = -1$ (We will use both forms of the definition; memorize them)

For example, $(\sqrt{-1})(\sqrt{-1}) = (i)(i) = i^2 = -1$

Powers of i (Cyclical property of i or i^2) **:** All powers of i are either equal to ±1 or ± i.

Examples

(a) $i^2 = -1$

(b) $i^3 = (i^2)i = -1(i) = -i$

(c) $i^4 = (i^2)(i^2) = (-1)(-1) = 1$

(d) $i^{10} = (i^4)(i^4)(i^2) = (1)(1)(-1) = -1$

Note above: Even powers of i are equal to ±1. Odd powers of i are equal to ± i.

Note that the imaginary unit "i" is only a tool in mathematics, and that it is not more imaginary (in the literal sense) than the real number 3. The introduction of the imaginary unit allows us to find the square roots of negative numbers.

Example Find the square root: (a) -4 ; (b) -25 ; (c) Simplify: $\sqrt{-15}$.

Solution

(a) $\sqrt{-4} = (\sqrt{-1})(\sqrt{4})$; (b) $\sqrt{-25} = \sqrt{-1}(\sqrt{25})$; (c) $\sqrt{-15} = (\sqrt{-1})(\sqrt{15})$

$= i(2)$ $\quad$ $= i(5)$ $\quad$ $= i\sqrt{15}$ or $\sqrt{15}i$

$\sqrt{-4} = 2i$ $\quad$ $\sqrt{-25} = 5i$.

Note: In (c) we prefer the first form of the answer; because, sometimes, if we are not careful in writing "i", the "i" may look as if it is under the radical sign . However, if no radical is involved we shall leave the answers as in (a) and (b) above. In some old textbooks, you may find "i" written after the radical.

Generally, $\sqrt{-b} = (\sqrt{-1})(\sqrt{b})$ $\quad$ $(b \geq 0)$

$= i\sqrt{b}$

The introduction of the imaginary unit helps us to expand the real number system to a more general system called the complex number system.

If we denote a complex number by z, then $z = a + bi$, where a and b are real numbers; a is called the real part and b is called the imaginary part. If $b = 0$, we have a pure real number (e.g., $z = 3 + 0i = 3$). If $a = 0$, we have a pure imaginary number ($z = 0 + 2i = 2i$). Thus, the product bi is called a pure imaginary number.

Distinction Between Roots of Negative Numbers and Roots of Equations

Roots of numbers (Principal Roots)**:** $\sqrt{-b} = i(\sqrt{b})$ $\quad$ $(b \geq 0)$

Example (a) $\sqrt{-4} = (\sqrt{-1})(\sqrt{4})$

$= 2i$.

(b) $\sqrt{-25} = \sqrt{-1}(\sqrt{25})$

$= 5i$.

Roots of equations: Here, we must note that we have more than one root. (two roots.)

Example 1: Consider the solution to the equation $x^2 = -25$.

Solution $\quad$ $x^2 = -25$

$x = \pm\sqrt{-25}$

$= \pm 5i$, which means $x = +5i$ or $x = -5i$.

Example 2 Solve for x:: $x^2 = -4$

Solution

$x = \pm\sqrt{-4}$

$x = \pm 2i$ (or $x = +2i$ or $x = -2i$)

Addition and Subtraction of Complex Numbers

In adding complex numbers, we shall add the real parts, and then add the imaginary parts (in much the same way as we add like terms in polynomial addition).

Perform the indicated operations, leaving the answers in the form $a + bi$.

Example 1 Simplify: $(-3 + 5i) + (-2 + 7i)$

Step 1: Remove the parentheses.

$$(-3 + 5i) + (-2 + 7i)$$

$$= -3 + 5i - 2 + 7i$$

Step 2: Add the real parts, and add the imaginary parts.

$$-3 + 5i - 2 + 7i$$
$$= -5 + 12i.$$

Scrapwork:
For the real parts: $-3 - 2 = -5$
For the imaginary parts: $5 + 7 = 12$

Example 2 Simplify: $(6 - 5i) - (-3 + 2i)$

Solution Remove the parentheses and add.

$$(6 - 5i) - (-3 + 2i)$$
$$= 6 - 5i + 3 - 2i$$

$$= 9 - 7i$$

Example 3 Simplify: $(5 + 2i) + (-3 - 6i)$
Solution
Remove the parentheses and add.
$(5 + 2i) + (-3 - 6i)$
$= 5 + 2i - 3 - 6i$
$= 5 - 3 + 2i - 6i$ <-------you may skip this step.
$= 2 - 4i$

or adding vertically:

$$\begin{array}{r} 5 + 2i \\ \underline{-3 - 6i} \\ 2 - 4i \end{array} \quad \text{(Adding).}$$

Example 4 Simplify : $(3 + 4i\} - (2 - 5i)$ (subtraction)
Solution
Remove the parentheses and add.
$(3 + 4i) - (2 - 5i)$
$= 3 + 4i - 2 + 5i$
$= 3 - 2 + 4i + 5i$ <-------you may skip this step.
$= 1 + 9i$

Example 5 Simplify : $(6 - 3i) + (2 + 3i)$

Solution

$(6 - 3i) + (2 + 3i)$
$= 6 - 3i + 2 + 3i$
$= 8 + 0i$
$= 8$

Multiplication of Complex Numbers

The approach here is multiply, replace i^2 by -1 (or higher powers of i by ± 1 or $\pm i$), add the real parts and add the imaginary parts.

Example 1 Multiply $-4 - 2i$ and $-5 + i$

Solution

Step 1: Multiply as you multiply binomials

$(-4 - 2i)(-5 + i)$
$= -4(-5 + i) + (-2i)(-5 + i)$ <--- You may skip this step.
$= 20 - 4i + 10i - 2i^2$

Step 2: (Replace i^2 by -1, since $i^2 = -1$ by definition)

$= 20 - 4i + 10i - 2(-1)$
$= 20 + 6i + 2$

Step 3: (Add the like terms: add the real parts; and add the imaginary parts)

$= 22 + 6i$

Example 2 Multiply $3 + 2i$ and $4 + 5i$

Procedure: Multiply, replace i^2 by -1, and simplify.

$(3 + 2i)(4 + 5i)$
$= 12 + 15i + 8i + 10i^2$
$= 12 + 23i + 10(-1)$
$= 12 + 23i - 10$
$= 2 + 23i$ (Adding)

Note: All powers of i are either equal to ± 1 or $\pm i$. For example $i^3 = -i$, $i^4 = 1$, $i^{10} = -1$

Example 3 Simplify : $4(6 - 3i)$

Solution

$4(6 - 3i)$
$= 24 - 12i$

Example 4 Simplify: $(4 + 2i)(3 - 6i)$

Solution The above implies multiplication.

Multiply, replace i^2 by -1, and add the like terms.

$(4 + 2i)(3 - 6i)$
$= 12 - 24i + 6i - 12i^2$
$= 12 - 24i + 6i - 12(-1)$
$= 12 - 24i + 6i + 12$
$= 24 - 18i$ (Note: $12 + 12 = 24$ and $-24i + 6i = -18i$)

Example 5 Simplify: $(7 + 2i)(7 - 2i)$

Solution

$(7 + 2i)(7 - 2i)$
$= 49 - 14i + 14i - 4i^2$
$= 49 - 4(-1)$
$= 49 + 4$
$= 53$

Multiplication of the square roots of negative numbers

Example 1 Find the product of $(\sqrt{-8})$ and $(\sqrt{-2})$

Solution

Step 1: Change to complex number forms.

$\sqrt{-8} = \sqrt{-1}(\sqrt{8})$
$= i\sqrt{8}$

Similarly , $\sqrt{-2} = i\sqrt{2}$

Step 2: Multiply the complex forms now.

$(\sqrt{-8})(\sqrt{-2}) = i\sqrt{8}\, i\sqrt{2}$
$= i^2\sqrt{16}$ $(\sqrt{16} = 4)$
$= i^2\ (4)$
$= (-1)(4)$ $(i^2 = -1)$
$= -4$

In the above problem (Example 1), you may skip Step 1 and show only Step 2.

Example 2 Find the product $(\sqrt{-49})(\sqrt{-25})$

Solution

Step 1: Change to complex number forms.

Then, $\sqrt{-49} = i\sqrt{49} = i(7) = 7i$

<---You may do Step 1 mentally and show only Step 2.

$\sqrt{-25} = i\sqrt{25} = i(5) = 5i$

Step 2: Multiply the complex forms now.

Then, $(\sqrt{-49})(\sqrt{-25}) = (7i)(5i)$
$= (7)(5)i^2$
$= 35(-1)$ $(i^2 = -1)$
$= -35$

We must note in the last two examples that it was necessary first to express the square roots of the negative numbers in terms of *i* before proceeding to multiply. Failure to do this may result in error such as the following:

Wrong procedure---> $(\sqrt{-8}(\sqrt{-2}) = \sqrt{(-2)(-8)}$
$= \sqrt{16}$
$= 4$, **which is a wrong answer**.

Generally, it is true that $(\sqrt{a})(\sqrt{b}) = \sqrt{ab}$ if *a* and *b* are positive but it is not true if *a* and *b* are negative.

Complex Conjugates

Definition : The **conjugate** of a given binomial is another binomial that differs from the given binomial only in the sign of one of the terms. The conjugate of $a + b$ is $a - b$; and the conjugate of $a - b$ is $a + b$.

The **conjugate** of $a + bi$ is $a - bi$. The conjugate of $a - bi$ is $a + bi$. A complex number and its conjugate differ only in the sign of the imaginary part. To find the conjugate of a complex number, change the sign of the imaginary part and keep the sign of the real part unchanged.

Examples: The conjugate of $2 + 3i$ is $2 - 3i$. The conjugate of $4 - 2i$ is $4 + 2i$.

The conjugate of $7 - 5i$ is $7 + 5i$; The conjugate of $-3 + 6i$ is $-3 - 6i$.

The conjugate of $-2 - 3i$ is $-2 + 3i$.; The conjugate of $4i$ is $- 4i$.

The conjugate of 7 is 7

The product of a complex number and its conjugate is a real number.

The product of $-2 - 3i$ and $-2 + 3i$. $= 4 - 6i + 6i - 9i^2 = 4 - 9(-1) = 4 + 9 =$ **13**, a real number.

The product of $a + bi$ and $a - bi = a^2 + b^2$

Some uses of complex conjugates: We may use the complex conjugate of a complex number to rationalize the denominator of a fraction or to divide by a complex number.

Division of Complex Numbers

Example 1 Simplify : $\dfrac{2 + 3i}{4 - 5i}$ or divide $2 + 3i$ by $4 - 5i$

To simplify the above complex number is meant we are to write it in the form $z = a + bi$.
The operation here is similar to that of the rationalization of denominators of radical expressions.
To simplify the above expression, we **multiply both the denominator and the numerator** by the **conjugate of the denominator.**

Solution The conjugate of $4 - 5i$ is $\mathbf{4 + 5i}$. (See also the note below)
Multiply both the denominator and the numerator by $4 + 5i$.

Then, we obtain:

$$\frac{2 + 3i}{4 - 5i}$$
$$= \frac{(2 + 3i)}{(4 - 5i)} \cdot \frac{\mathbf{(4 + 5i)}}{\mathbf{(4 + 5i)}}$$
$$= \frac{8 + 10i + 12i + 15i^2}{16 + 20i - 20i - 25i^2}$$
$$= \frac{8 + 22i + 15(-1)}{16 + 0 - 25(-1)}$$
$$= \frac{8 + 22i - 15}{16 + 25}$$
$$= \frac{8 + 22i - 15}{41}$$
$$= \frac{-7 + 22i}{41}$$
$$= -\frac{7}{41} + \frac{22}{41}i$$

We may observe above that the **denominator** does **not** contain the imaginary unit "i", even though the numerator contains the imaginary unit.

Note above that we could have multiplied by -4 - 5i. (Try it.) It is therefore not critical in this problem which terms should differ in sign, (so far as the rationalization of the denominator is concerned) provided that either the real parts differ in sign or the imaginary parts differ in sign, but **not** both. We may produce a minus sign in the denominator which we can take care of as usual.

Example 2 Simplify: $\frac{5 + 4i}{-3 + 2i}$

Solution Multiply both the denominator and the numerator by -3 - 2i (The conjugate of -3 + 2i).

$$\frac{5 + 4i}{-3 + 2i}$$

$$= \frac{(5 + 4i)}{(-3 + 2i)} \cdot \frac{\mathbf{(-3 - 2}\boldsymbol{i}\mathbf{)}}{\mathbf{(-3 - 2}\boldsymbol{i}\mathbf{)}}$$

$$= \frac{-15 - 10i - 12i - 8i^2}{9 + 6i - 6i - 4i^2}$$

$$= \frac{-15 - 10i - 12i - 8(-1)}{9 + 0 - 4(-1)}$$

$$= \frac{-15 - 22i + 8}{9 + 4}$$

$$= \frac{-7 - 22i}{13}$$

$$= -\frac{7}{13} - \frac{22}{13}i \qquad (-15 + 8 = -7)$$

Example 3 Simplify: $\frac{4 - 2i}{i}$

Solution

$$\frac{4 - 2i}{i}$$

$$= \frac{(4 - 2i)}{i} \frac{\mathbf{(-}\boldsymbol{i}\mathbf{)}}{\mathbf{(-}\boldsymbol{i}\mathbf{)}} \qquad \text{(The conjugate of } i \text{ is } -i \text{ ; but you could also use } i\text{)}$$

$$= \frac{(4 - 2i)(-i)}{(i)(-i)}$$

$$= \frac{-4i + 2i^2}{-i^2}$$

$$= \frac{-4i + 2(-1)}{-(-1)} \qquad (i^2 = -1)$$

$$= \frac{-4i - 2}{+1}$$

$$= -2 - 4i$$

Lesson 41 Exercises

A Find the square root or simplify:

1. $\sqrt{-9}$; **2** $\sqrt{-49}$; **3.** $\sqrt{-36}$; **4.** $\sqrt{-18}$; **5.** $\sqrt{-\frac{4}{9}}$; **6.** i^{40}; **7.** i^{93} ; **8.** i^{100}
9. $\sqrt{-28}$; **10.** $\sqrt{-32}$

Answers: **1.** $3i$; **2.** $7i$; **3.** $6i$; **4.** $3i\sqrt{2}$; **5.** $\frac{2}{3}i$; **6.** 1 ; **7.** i ; **8.** 1 ; **9.** $2i\sqrt{7}$; **10.** $4i\sqrt{2}$

B Simplify: **1.** $4 + 3i - 6 + 5i$; **2.** $(7 - 8i) + (2 + 9i)$ **3.** $(2 - 3i) - (7 - 4i)$;
4. Subtract $(1 - i)$ from $4 - 2i$; **5.** $4 + 2i - 3(4 - 6i) + i$; **6.** $-i^5 + i^2$; **7.** $4 - i + i^2$

Answers: **1.** $-2 + 8i$; **2.** $9 + i$; **3.** $-5 + i$; **4.** $3 - i$; **5.** $-8 + 21i$; **6.** $-1 - i$; **7.** $3 - i$

C Evaluate the following:
1. $(-2i)^2$; **2.** $6i^2$; **3.** $(-2i)(3i)$; **4.** $(-5i)(-6i)$; **5.** $(-i)(-2i)$; **6.** $i^3(-4i^3 + 2i^2)$

Answers: **1.** -4; **2.** -6; **3.** 6; **4.** -30; **5.** -2; **6.** $4 + 2i$..

D **Multiply**: $4 + 3i$ and $2 + 5i$; **2.** Multiply $2 - 5i$ and $-2 + 6i$; **3. Simplify**: $(6 - 7i)(2 + 3i)$;
4. Simplify $(1 - i)(1 + i)$; **5.** $(2 + 3i)^2$ **6.** $(4i^2)(-8i)(i^2)$; **7.** $(5 + 3i)(5 - 3i)$

Answers: **1.** $-7 + 26i$; **2.** $26 + 22i$; **3.** $33 + 4i$; **4.** 2 ; **5.** $-5 + 12i$; **6.** $-32i$; **7.** 34.

E Find the product of each of the following:
1. $(\sqrt{-4})(\sqrt{-1})$; **2.** $(\sqrt{-16})(\sqrt{-4})$; **3.** $(\sqrt{-25})(\sqrt{49})$; **4.** $(\sqrt{-8})(\sqrt{-2})$

Answers: **1.** -2 **2.** -8; **3.** $35i$; **4.** -4.

F Find the complex conjugate of each of the following:
1. $3 + 2i$; **2.** $4 - 5i$; **3.** $6i$; **4.** $-9i$; **5.** $-5 - 3i$; **6.** $-5 + 4i$.

Answers: **1.** $3 - 2i$; **2.** $4 + 5i$; **3.** $-6i$; **4.** $9i$; **5.** $-5 + 3i$; **6.** $-5 - 4i$..

G Divide: **1.** $\frac{8 - 6i}{2}$; **2.** $\frac{6 + 8i}{2i}$; **3.** $\frac{\sqrt{16}}{\sqrt{-16}}$; **4.** $\frac{3}{4 - \sqrt{-9}}$; **5.** $\frac{4 - 2i}{3 + 5i}$

Answers: **1.** $4 - 3i$; **2.** $4 - 3i$; **3.** $-i$; **4.** $\frac{12}{25} + \frac{9}{25}i$; **5.** $\frac{1}{17} - \frac{13}{17}i$

H Simplify: **1.** $\frac{2 + 3i}{5 - 2i}$; **2.** $\frac{4}{3 - 4i}$; **3.** $\frac{5 + 2i}{-i}$ **4.** Divide $4 + 3i$ by $-2 + 3i$;
Simplify: **5.** $\frac{2 - 6i}{4 + 3i}$; **6.** $\frac{i - 2}{2 - i}$; **7.** $(3 - 2i)(4 + 2i)(i^2)$; **8.** $(6 - 4i)(6 + 4i)$

Answers: **1.** $\frac{4}{29} + \frac{19}{29}i$; **2.** $\frac{12}{25} + \frac{16}{25}i$; **3.** $-2 + 5i$; **4.** $\frac{1}{13} - \frac{18}{13}i$; **5.** $-\frac{2}{5} - \frac{6}{5}i$; **6.** -1 ; **7.** $-16 + 2i$; **8.** 52

I Find the quotient and simplify: $\frac{(2 + 5i)(1 - i)}{(4 - 2i)(3 + i)}$

Answer: $\frac{23 + 14i}{50}$ or $\frac{23}{50} + \frac{7}{25}i$

Lesson 42

Equality of Complex Numbers; Roots of Complex Numbers; Equations Involving Complex Numbers

Equality of two complex numbers

Two complex numbers $a + bi$ and $c + di$ are equal if and only if $a = c$ (that is the real parts are equal) and $b = d$ (that is the imaginary parts are equal).

Example 4 Find x and y if $2x + 5yi = 8 + 15i$

Solution

Step 1: Set the real parts equal to each other and solve for x.

Then $2x = 8$
$x = 4$

Step 2: Set the imaginary parts equal to each other and solve for y.

Then $5y = 15$ and $y = 3$

Therefore $x = 4, y = 3$.

Roots of Complex Numbers

Example 5 Find the square root of 5 - 12i

Step 1 Let $x + yi$ be the square root of 5 - 12i (Where x and y are real numbers; Try a nd b instead)

$$\text{Then } (x + yi)^2 = 5 - 12i$$

$$x^2 + 2xyi + y^2i^2 = 5 - 12i$$

$$x^2 - y^2 + 2xyi = 5 - 12i \qquad (i^2 = -1 \text{ and } y^2i^2 = y^2(-1) = -y^2)$$

We apply the equality property of two complex numbers:

For the real parts: $x^2 - y^2 = 5$ (2)

For the imaginary parts:: $2xy = -12$ (3)

We shall now solve equations (2) and (3) simultaneously.
From equation (3),

$$y = -\frac{12}{2x}$$

$$y = -\frac{6}{x}$$

Substituting for y in equation (2)

$$x^2 - \left(-\frac{6}{x}\right)^2 = 5$$

$$x^2 - \frac{36}{x^2} = 5$$

$$x^4 - 36 = 5x^2$$

$$x^4 - 5x^2 - 36 = 0$$

We shall use the quadratic formula to solve for x. .

Step 2:: By the substitution method (Also, see p.38)

Let $x^2 = u$

Then $x^4 - 5x^2 - 36 = 0$ becomes

$u^2 - 5u - 36 = 0 \qquad (u = x^2)$

Solving, $u = \dfrac{5 \pm \sqrt{169}}{2}$

$u = \dfrac{5 \pm 13}{2}$

Now, converting back to x by letting $u = x^2$

$$x^2 = \frac{5 \pm 13}{2}$$

$$x^2 = \frac{5 + 13}{2} \quad \text{or} \quad x^2 = \frac{5 - 13}{2}$$

$$x^2 = \frac{18}{2} \quad \text{or} \quad x^2 = \frac{-8}{2}$$

$$x^2 = 9 \quad \text{or} \quad x^2 = -4$$

$$x = \pm\sqrt{9} \quad \text{or} \quad x = \sqrt{-4}$$

$$x = \pm 3 \quad \text{or} \quad x = \pm 2i$$

We reject the imaginary solution since x and y must be real.

Therefore, $x = \pm 3$.

Step 3: Substitute $x = \pm 3$ in $y = -\dfrac{6}{x}$ to obtain the corresponding y-values.

When $x = +3$, $y = -\dfrac{6}{3} = -2$, and

when $x = -3$, $y = -\dfrac{6}{-3} = 2$.

Thus, when $x = 3$,, $y = -2$, and
when $x = -3$, $y = 2$.

Substituting these values in $x + yi$, the two square roots are $(3 - 2i)$ and $(-3 + 2i)$.

These roots can be verified by squaring each root.

Note above that **in Step 2,** we could have solved the quadratic equation by factoring as follows:

$$x^4 - 5x^2 - 36 = 0$$

$$(x^2 - 9)(x^2 + 4) = 0$$

$$x^2 - 9 = 0; \quad \text{or} \quad x^2 + 4 = 0$$

$$x^2 = 9; \quad \text{or} \quad x^2 = -4$$

$$x = \pm 3 \quad \text{or} \quad x = \pm 2i;$$

This approach would be faster than by substitution.

Example 6 If $z^2 = 8i$, find z satisfying this equation.

Solution: $z = \pm\sqrt{8i}$

Thus, we are to find the square root of $8i$. We can use the same method as in Example 5 , above.

Let $a + bi$ be the square root of $8i$. (where a and b are real)

Then $(a + bi)^2 = 8i$ (By definition, $\sqrt{8i} = a + bi$ if $(a + bi)^2 = 8i$)

$a^2 + 2abi - b^2 = 8i + 0$ (Note: $b^2i^2 = b^2(-1) = -b^2$)

$a^2 - b^2 + 2abi = 8i + 0$ (Also, note that $(a^2 - b^2,\ 2abi) = (0,\ 8i)$)

Equating the real parts, $a^2 - b^2 = 0$ (1)

Equating the imaginary parts, $2ab = 8$ (2)

We shall now solve equations (1) and (2) simultaneously for a and b.

From equation (2), $b = \dfrac{4}{a}$ (3)

Substituting for b from (3) in equation (1),

$$a^2 - \left(\frac{4}{a}\right)^2 = 0$$

$$a^2 - \frac{16}{a^2} = 0$$

$$a^2 \bullet a^2 - \frac{16 \bullet a^2}{a^2} = 0 \bullet a^2$$ (multiplying each term of the equation by a^2 to undo the denominator)

$$a^4 - 16 = 0$$

$$a^4 = 16$$

$$a^2 = \pm 4$$

$$a^2 = +4 \quad \text{or } a^2 = -4$$

$$a = \pm 2 \quad \text{or} \quad a = \pm 2i.$$

Since a must be a real number, we reject the imaginary a-values, $\pm 2i$, and accept the real a-values, ± 2.
Therefore a = +2 or -2

When a= +2, b= $\dfrac{4}{2} = 2$

When a= -2, $b = \dfrac{4}{-2}$ = -2

Substituting the values of a and b in $a + bi$, the two square roots of $8i$ are $(2 + 2i)$ and $(-2 - 2i)$.
Verify these roots by squaring each root.

Note: In the above problem, do not confuse $\pm\sqrt{8i}$ with $\pm\sqrt{-8}$ which is $\pm i\sqrt{8}$ or $\pm 2i\sqrt{2}$.

Equations involving complex numbers

Example 7 Solve for x, given that

$$3ix + 12 - 8i + 2x = 2(6 - i) + 2x$$

Solution We shall get x alone on one side of the equation and the other side should not contain x:

$$3ix + 12 - 8i + 2x = 2(6 - i) + 2x$$
$$3ix - 8i + 12 + 2x = 12 - 2i + 2x$$
$$3ix - 8i = -2i$$
$$(3x - 8)i = -2i$$
$$3x - 8 = -2$$
$$x = 2$$

Example 8 Solve for x, given that $2ix + 3 - 4i = (3 + 2i)x + 5i$

Solution We shall get x alone on one side of the equation and the other side should not contain x:

$$2ix + 3 - 4i = (3 + 2i)x + 5i$$
$$2ix + 3 - 4i = 3x + 2ix + 5i$$
$$2ix - 3x - 2ix = -3 + 9i$$
$$2ix - 3x - 2ix = -3 + 9i$$
$$-3x = -3 + 9i$$
$$x = 1 - 3i$$

Example 9 Solve for x, if x is a real number, given that

Solution

$$2ix + 3 - 4i = (3 + 2i)x + 5i$$
$$2ix + 3 - 4i = 3x + 2ix + 5i$$
$$2ix - 3x - 2ix = -3 + 9i$$
$$2ix - 3x - 2ix = -3 + 9i$$
$$-3x = -3 + 9i$$
$$x = 1 - 3i$$

Since on solving for x, the solution contains the imaginary unit, x is not real.
Therefore, there is **no** solution. (From the original problem, for a solution, x must be real)
.

Lesson 42 Exercises

A Find the square roots of the following **1.** $16 - 30i$; **2.** $-5 - 12i$

Answers: **1.** $5 - 3i$ and $-5 + 3i$; **2.** $2 - 3i$ and $-2 + 3i$

B Find z satisfying the given equation **1.** $z^2 = 16 - 30i$; **2.** $z^2 = -5 - 12i$

Answers: **1.** $5 - 3i$ and $-5 + 3i$; **2.** $2 - 3i$ and $-2 + 3i$

C Solve for x: **1.** $3ix - 30 + 3i = (12 + 3i)x + 3i$; **2.** $x^2 = 16i$

3. Determine the real values of x and y if $4x + 7yi = 12 - 35i$

Answers: **1.** $x = -\frac{5}{2}$; **2.** $x = 2\sqrt{2} + 2i\sqrt{2}$, $x = -2\sqrt{2} - 2i\sqrt{2}$; **3.** $x = 3$, $y = -5$.

Lesson 43

Graphical Representation and Addition of Complex Numbers

A complex number may be considered as an ordered pair of real numbers (a, b). Graphically, we represent a complex number z by a point P whose x-coordinate $= a$, and y-coordinate $= b$.

We graph a complex number in a rectangular coordinate system of axes. We call this system the complex plane. The horizontal axis (x-axis) represents the real axis and the vertical axis (y-axis) represents the imaginary axis.

Example: Graph the following in the complex plane:

$(a)\ \ 2-3i;\ \ (b)\ 3+2i;\ \ (c)\ -4-3i;\ \ (d)-i;\ \ (e)\ 5i;\ \ (f)\ 6.$

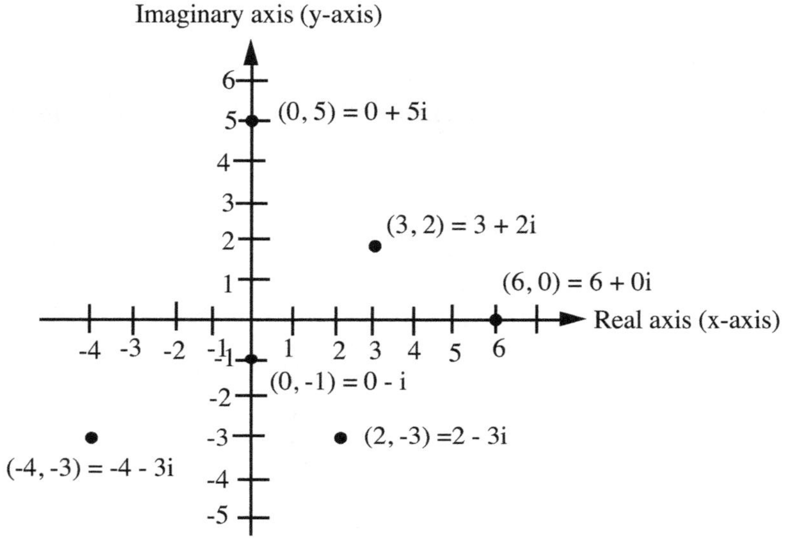

Figure: The graphs of the points, $2-3i,\ 3+2i,\ -4-3i,\ -i,\ 5i,\ 6$, in the complex plane.

We may also think of a complex number as a vector. We represent a vector quantity by an arrow drawn to scale. The length of the arrow represents the magnitude of the vector and the direction of the arrow head represents the direction of the vector. In Figure we represent the complex number

$a + bi$ as a vector drawn from the origin to the point $(a, b) = a + bi$..

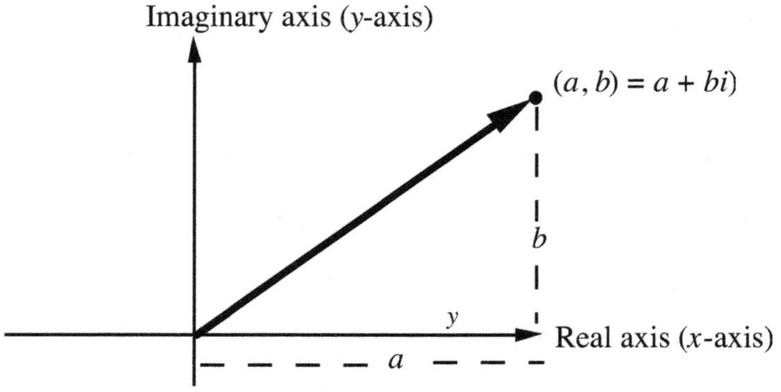

Figure: Graphical representation of a complex number as a vector.

Graphical Addition of Complex Numbers

We may use the parallelogram law of **vector addition** to add complex numbers.

In the complex plane, we draw each of the two vectors $a + bi$ and $c + di$ to scale. The **sum** of the complex numbers is represented by the **diagonal** of the parallelogram (Figure).

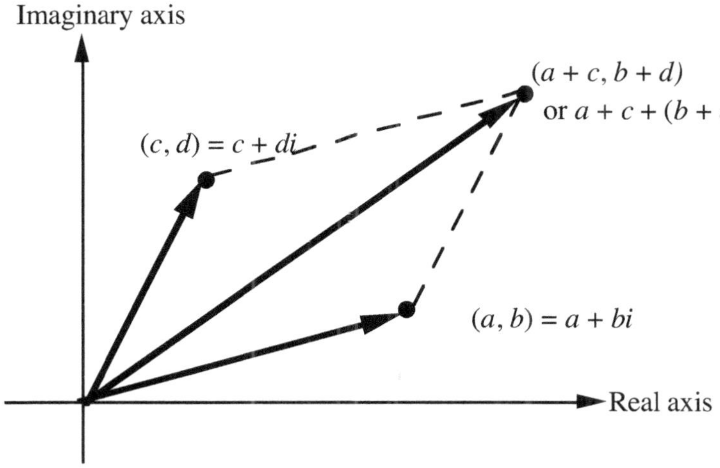

Figure: Graphical addition of complex numbers: $(a + bi) + (c + di)$

Subtraction of Complex Numbers: $(a + bi) - (c + di) = (a + bi) + (-c - di)$

By definition, to subtract $(c + di)$ from $(a + bi)$ means add the negative of $(c + di)$ to $(a + bi)$. Graphically, $(-c - di)$ has the same magnitude as $(c + di)$ except that its direction is opposite to that of $(c + di)$ as shown in Figure

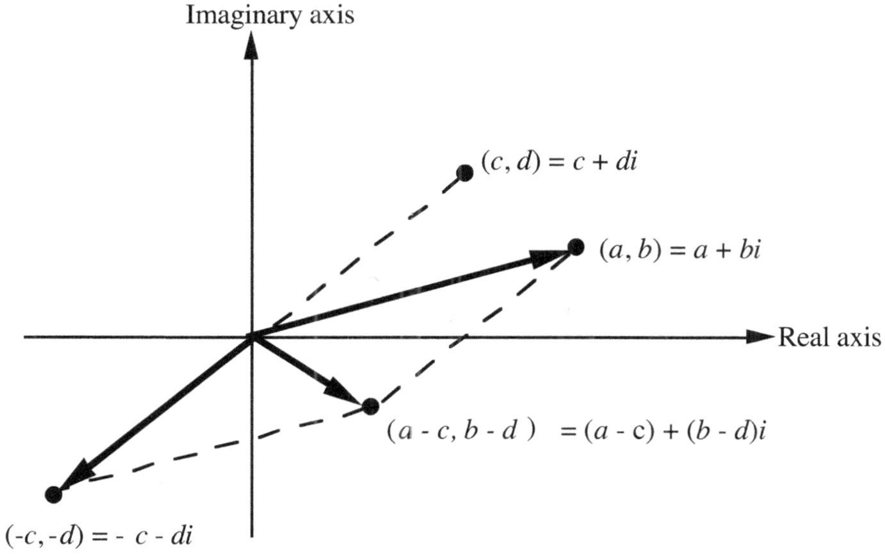

Figure: Graphical subtraction of complex numbers: $(a + bi) - (c + di)$

Example 1 Add algebraically and graphically: $(2 + 3i) + (7 + 5i)$.

Algebraic method

$$(2 + 3i) + (7 + 5i)$$
$$= 2 + 3i + 7 + 5i$$
$$= 9 + 8i$$

Graphical method

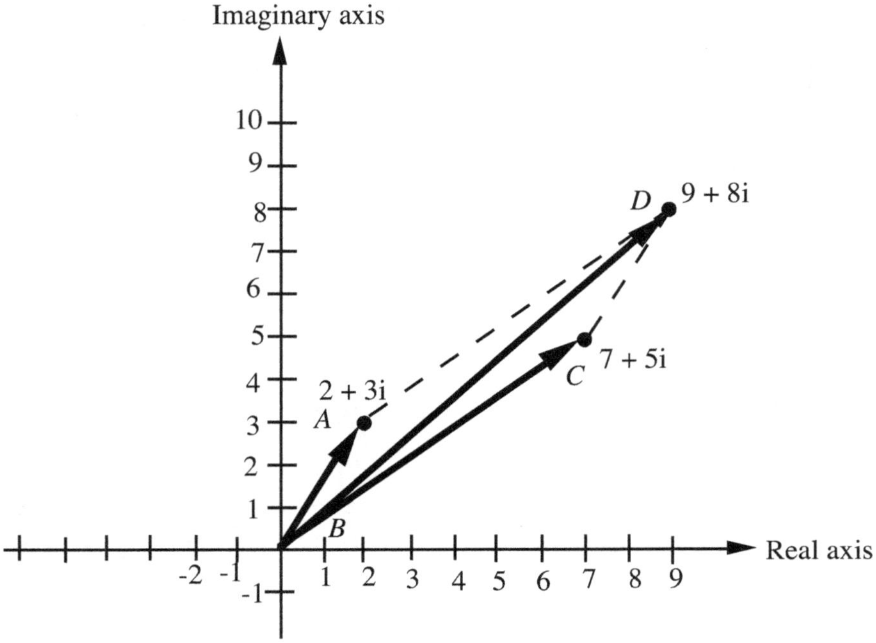

:

Figure : $(2 + 3i) + (7 + 5i)$ is represented by BD

Lesson 43 Exercises

Graph each of the following in the complex plane:

1. $3 - 2i$; **2.** $3 + 2i$; **3.** $-3 - 3i$; **4.** $-4i$; **5.** $6i$; **6.** 5

Solution

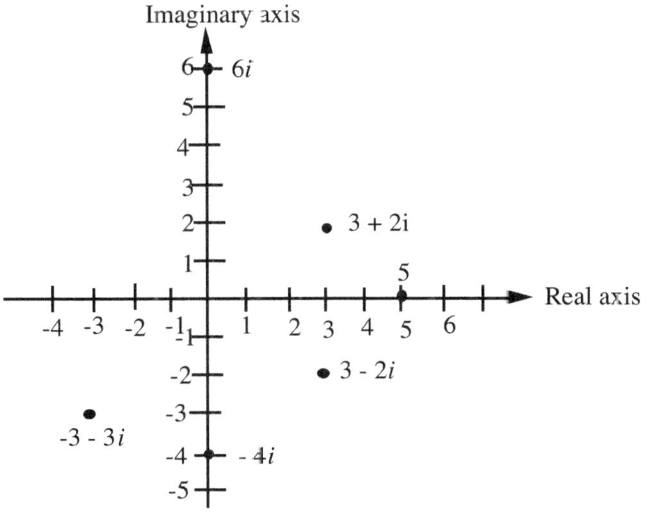

Lesson 44

Polar (Trigonometric) Representation of Complex Numbers.

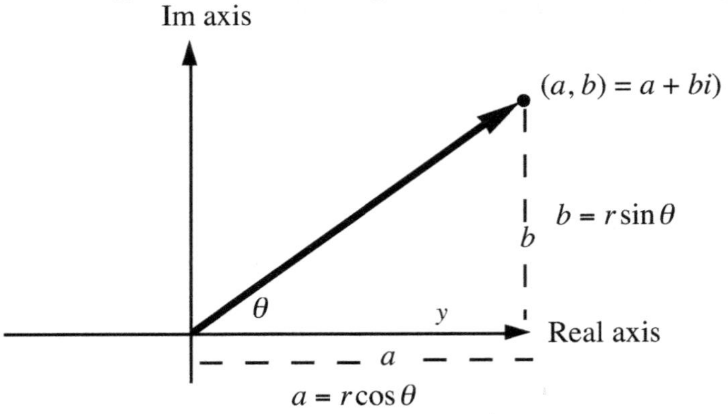

Figure : The graph of polar or trigonometric form of a complex number.

A complex number $z = a + bi$ is represented by a vector drawn from the origin to the point (a, b) and in direction θ with the x-axis. The standard position angle, θ, is called the argument of the complex number.

The **length** r of the vector is given by $r = \sqrt{a^2 + b^2}$

From trigonometric definitions,

$$a = r\cos\theta \qquad (1)$$

$$b = r\sin\theta \qquad (2)$$

$$\theta = \arctan\left(\frac{b}{a}\right) \qquad (3)$$

By substituting for a and b from equations (1) and (2) respectively, in the rectangular form, $z = a + bi$, we obtain

$z = r\cos\theta + r\sin\theta i$ or

$z = r\cos\theta + ir\sin\theta$; and if we factor out the r, we obtain

$$\boxed{z = r(\cos\theta + i\sin\theta)} \qquad (4)$$

Sometimes, the quantity $(\cos\theta + i\sin\theta)$ is abbreviated by cis θ and then equation (4) becomes

$$z = rcis\theta \qquad (5)$$

We can view this abbreviation this way:

$$\begin{array}{l} z = r\,(\,\cos\theta + i\,\sin\theta\,) \\ \;\;\;\;\;\;\uparrow\;\uparrow\;\;\uparrow\;\;\;\uparrow\uparrow\;\uparrow \\ z = r\;\;c\;\;\;\;\;\underline{\;\;i\;s\;\;} \end{array}$$

Although it might be cumbersome to write equation (4) repeatedly instead of equation (5), it is recommended that we use the form of equation (4) a number of times before switching to the abbreviated form, $z = rcis\theta$.

Changing from rectangular form to polar form

Rectangular form: $z = a + bi$
Polar form: $z = r(\cos\theta + i\sin\theta)$ or $z = rcis\theta$

$$r = \sqrt{a^2 + b^2}$$

$$a = r\cos\theta \qquad (1)$$

$$b = r\sin\theta \qquad (2)$$

$$\theta_{ref} = \arctan\left(\frac{b}{a}\right) \qquad (3)$$

Example 1 Change the complex number $z = 3 - 4i$ to polar form.

Solution In polar form, we want the form $z = r(\cos\theta + i\sin\theta)$. So we must find r and θ from the rectangular form.

Step 1: Plot $z = 3 - 4i$ as a vector (Figure)

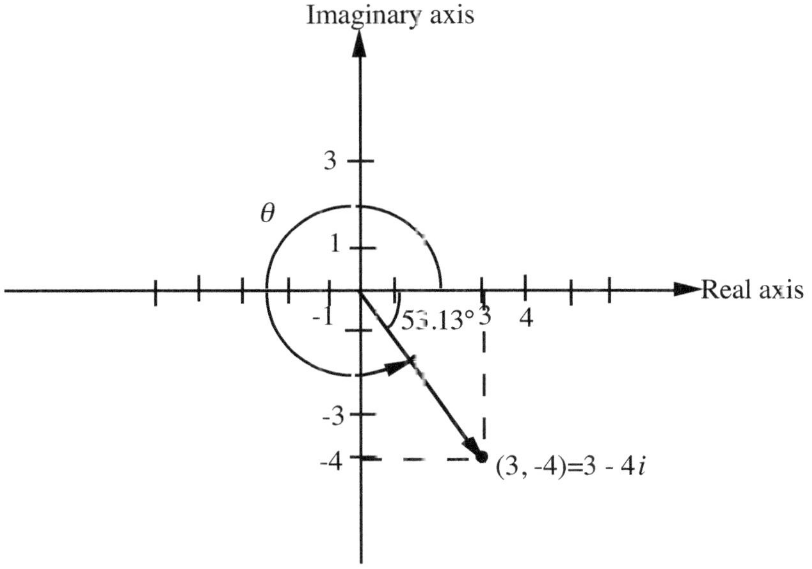

Figure: The graph of $z = 3 - 4i$

Step 2: Find r.

$a = 3, b = -4$

$$r = \sqrt{3^2 + (-4)^2}$$

$$r = \sqrt{9 + 16}$$

$$r = \sqrt{25}$$

$$r = 5.$$

Step 3: Find θ.

$\tan\theta_{ref} = \frac{b}{a}$ (θ_{ref} is the reference angle)

$= -\frac{4}{3}$

$= -1.3$

Now, from tables or calculator, find the angle whose tangent is 1.3

$\theta_{ref} = 53.13°$ (Terminal side of θ is in the 4th quadrant)

$\theta = 360° - 53.13°$

$\theta = 306.87°$

(**Note** above that there is a difference between θ and θ_{ref})

Step 4: Substitute $r = 5, \theta = 305.87°$ in $z = r(\cos\theta + i\sin\theta)$.

Then $z = 5(\cos 306.87° + i\sin 306.87°)$

or $z = 5cis306.87°$.

We must note above that the polar form was **not** specified in terms of $53.13°$ but rather in terms of $306.87°$, this angle being in standard position, and as such, it is measured counter-clockwise from the positive x-axis. ($53.13°$ is the reference angle from tables or a calculator.)

Example 2 Change to polar (trigonometric) form: $z = -3 - 5i$

Step 1: Plot $z = -3 - 5i$ as a vector (Figure)

Step 2: Find r.

$a = -3, b = -5$ (from $z = -3 - 5i$) (**Note**: General form is $z = a + bi$)

$r = \sqrt{(-3)^2 + (-5)^2}$ ($r = \sqrt{a^2 + b^2}$)

$r = \sqrt{9 + 25}$

$r = \sqrt{34}$

Step 3: Find θ

$\tan\theta_{ref} = \frac{-5}{-3}$ $\left(\frac{b}{a}\right)$

$= 1.67$

$\theta_{ref} = 59.03°$ (Now, from tables or calculator, find the angle whose tangent is 1.67)

$\theta = \theta_{ref} + 180°$ (The terminal side of θ is in the 3rd quadrant.)

$\theta = 59.03° + 180°$

$\theta = 239.03°$

Step 4: Substitute $r = \sqrt{34}$, $\theta = 239.03°$ in

$z = r(\cos\theta + i\sin\theta)$ to obtain

$z = \sqrt{34}(\cos 239.03° + i\sin 239.03°)$. <------polar (or trigonometric) form

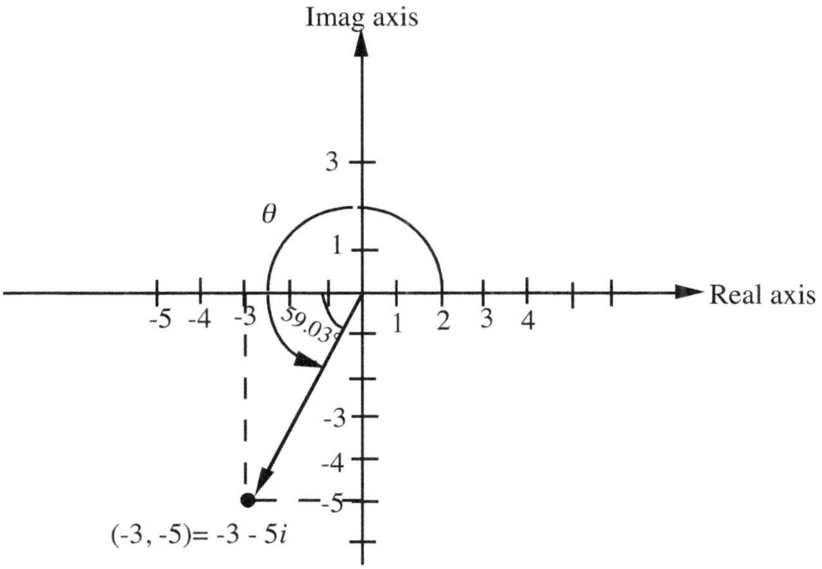

Example 3 Change to polar form.
(a) 9 ; (b) -9 ; (c) $-3i$.

Solution

(a) Step 1: Sketch $z = 9 + 0i$ (Figure)

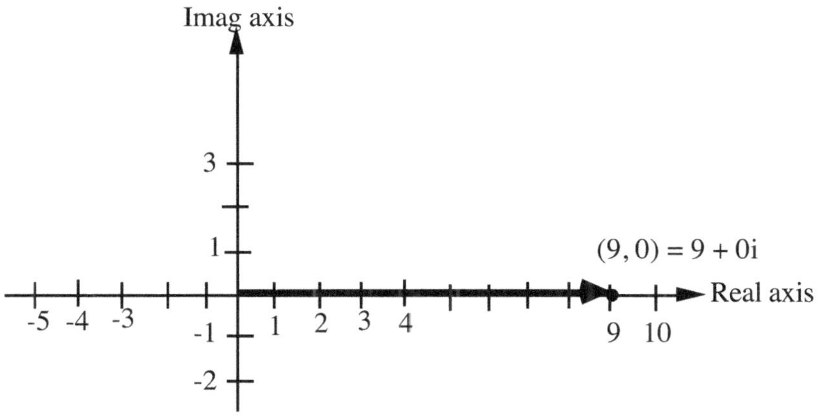

Step 2: Find r

$r = 9$ (since θ is quadrantal; terminal side is along the positive x-axis)
$a = 9$
$b = 0$

Step 3: Find θ

$$\tan\,\theta_{ref} = \frac{0}{9} = 0$$

$\theta_{ref} = 0°$

In this problem, $\theta = \theta_{ref}$.

$\theta = 0°$.

Step 4: Substitute $r = 9$, $\theta = 0°$ in the general equation ($z = r(\cos\theta + i\sin\theta)$) to obtain

$z = 9(\cos 0° + i\sin 0°)$

$z = 9\cos 0°$ or $9\,cis 0°$ $\qquad$ ($\sin 0° = 0$)

(b) Step 1: Sketch $z = -9 + 0i$ (Figure)

Step 2: Find r and θ.

$r = 9$ $\qquad$ (Note that r is positive.)

By inspection, since the terminal side falls along the negative x-axis, $\theta_{ref} = 0°$

Therefore, $\theta = 180°$

Substituting $r = 9$, $\theta = 180°$ in the general equation, we obtain

$z = 9(\cos 180° + i\sin 180°)$

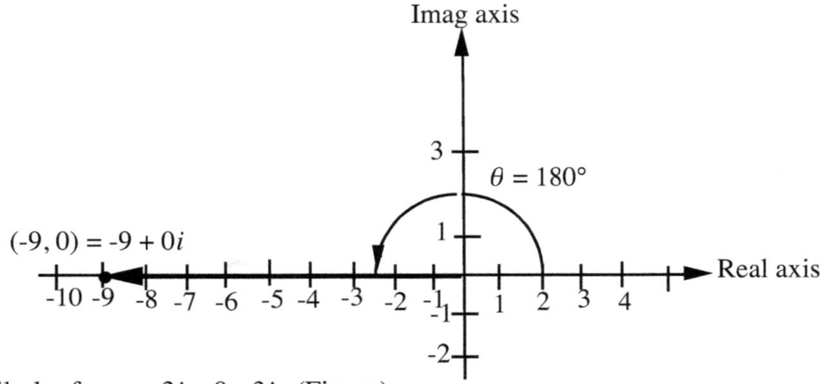

(c) Similarly, for $z = -3i = 0 - 3i$ (Figure)

$r = 3$, $\theta = 270°$ and

$z = 3(\cos 270° + i\sin 270°)$ or $z = 3\,cis 270°$

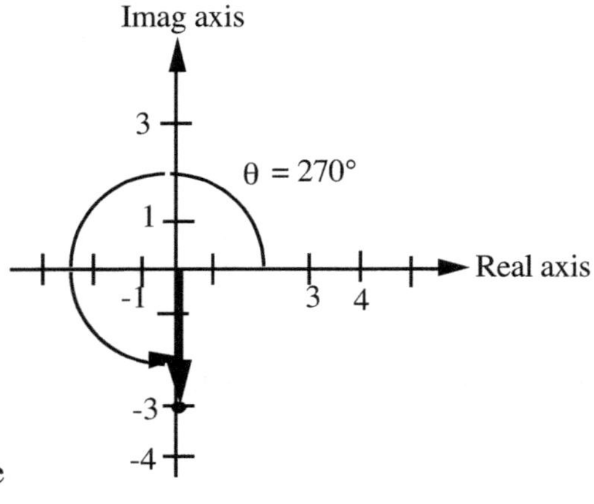

Figure

We must note above that by agreement, we restricted the angular measurement to $0° \le \theta \le 360°$.

Changing from polar form to rectangular (algebraic) form

Example 1 Change $z = 10(\cos 220° + i\sin 220°)$ to rectangular form.

The general rectangular form is given by

$$z = a + bi.$$

We must therefore find a and b from the polar form.

By inspection and comparison of $z = r(\cos\theta + i\sin\theta)$, with

$$z = 10(\cos 220° + i\sin 220°)$$

$$r = 10, \quad \theta = 220°$$

$$a = 10\cos 220° \qquad (a = r\cos\theta)$$

$$a = -7.7 \qquad (\cos 220° = -.77)$$

$$b = 10\sin 220°$$

$$b = -6.4 \quad (\sin 220° = -.64)$$

Substituting $a = -7.7$, $b = -6.4$ in $z = a + bi$, we obtain

$$z = -7.7 - 6.4i$$

$$10\,cis\,220° = -7.7 - 6.4i$$

↑ (Polar form) ↑ (Rectangular form)

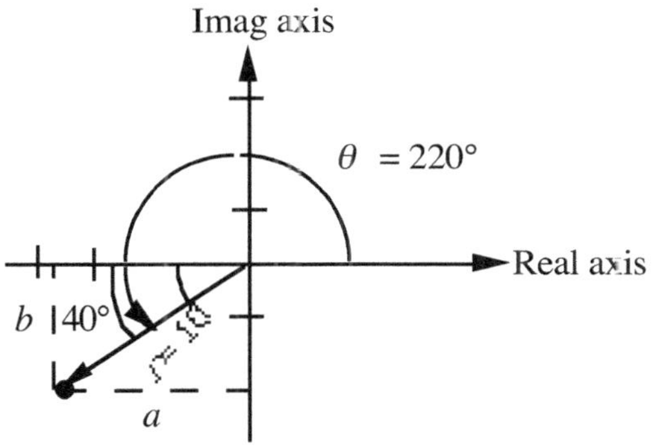

Figure

It may be remarked that it easier to change from the polar form to the rectangular form than vice versa

Exponential form of a complex Number

The exponential form of a complex number, z, is given by $z = re^{i\theta}$ where θ is in radians.

Multiplication of Complex Numbers in Polar Form

Consider the polar (trigonometric) forms of two complex numbers,

$z_1 = r_1(\cos\theta_1 + i\sin\theta_1)$ and

$z_2 = r_2(\cos_2 + i\sin\theta_2)$.

The product z_1z_2 of the two complex numbers in polar form is given by

$$z_1z_2 = r_1r_2[\cos(\theta_1 + \theta_2) + i\sin(\theta_1 + \theta_2)] \qquad (1)$$

Therefore, to multiply two complex numbers in polar forms, multiply the radii (moduli) and add the angles (arguments).

Example Given that $z_1 = 3(\cos 30° + i\sin 30°$
$z_2 = 4(\cos 40° + i\sin 40°)$

(a) Find by analytic means (using equation (1) above) the product of z_1 and z_2.

(b) Show the graphical form of the above method.

Solution

(a) By equation (1) above,

Apply $z_1 z_2 = r_1 r_2[\cos(\theta_1 + \theta_2) + i\sin(\theta_1 + \theta)]$ with

$r_1 = 3,\ r_2 = 4,\ \theta_1 = 30°,\ \theta_2 = 40°$ to obtain

$$z_1z_2 = (3)(4)[\cos(30° + 40°) + i\sin(30° + 40°)$$
$$= 12[\cos 70° + i\sin 70°] \quad (2)$$
$$z_1z_2 = 12cis70°.$$

(From tables or calculator we can simplify equation (2) for z_1z_2)

(b) Graphical form:

Step 1: Find the sum of the angles $30°$ and $40°$

Then $\theta = 70°$

Step 2: Find the product $r = r_1 r_2$

Then $r = (3)(4) = 12$

Step 3: Draw $\theta = 70°$ in standard position and $r = 12$ (Figure)

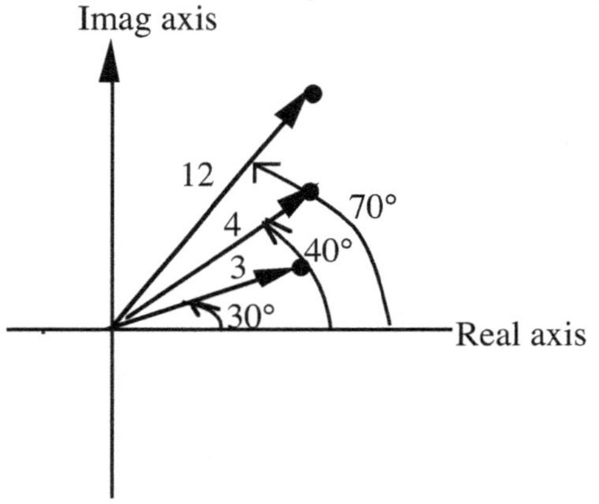

Division of Complex Numbers in Polar Form

Consider the polar (trigonometric) form of two complex numbers, say

$z_1 = r_1(\cos\theta_1 + i\sin\theta_1)$ and

$z_2 = r_2(\cos_2 + i\sin\theta_2)$.

The quotient $\frac{z_1}{z_2} = \frac{r_1}{r_2}[\cos(\theta_1 - \theta_2) + i\sin(\theta_1 - \theta_2)]$

or $\frac{z_1}{z_2} = \frac{r_1}{r_2}\ \text{cis}(\theta_1 - \theta_2)$.

Therefore, to divide two complex numbers in polar forms, divide the radii (moduli) and subtract the angles (arguments).

Example 1 Given that $z_1 = 15(\cos 50° + i\sin 50°)$ and $z_2 = 5(\cos 30° + i\sin 30°)$, find the quotient $\frac{z_1}{z_2}$

Solution
$$\frac{z_1}{z_2} = \frac{15}{5}[\cos(50°-30°) + i\sin(50°-30°)]$$
$$= 3[\cos 20° + i\sin 20°] \text{ or } 3\text{cis}\,20°$$

Example 2 If $z_1 = 6(\cos 15° + i\sin 15°)$ and $z_2 = 12(\cos 50° + i\sin 50°)$ find the quotient $\frac{z_1}{z_2}$.

Solution
$$\frac{z_1}{z_2} = \frac{6}{12}[\cos(15° - 50°) + i\sin(15° - 50°)]$$
$$= \frac{1}{2}[\cos(-35) + i\sin(-35)] \text{ or } \frac{1}{2}\text{cis}(-35°)$$

But since we want to specify θ so that $0° \le \theta \le 360°$
$$\frac{z_1}{z_2} = \frac{1}{2}\text{cis}(325°)$$

Lesson 44 Exercises

A Plot each complex number and then change to polar form.

1. $3 + 4i$; **2.** $1 + i$; **3.** $1 - 3i$; **4.** $3 - i$; **5.** $1 - i\sqrt{3}$; **6.** 7;
7. $-i$; **8.** $-3 - 4i$; **9.** $-5i$; **10.** -3; **11.** $-.69 + 3.94i$

Answers:

1. $z = 5cis53.13°$; 2. $z = \sqrt{2}\,cis45°$; **3.** $z = \sqrt{10}\,cis288.43°$; **4.** $z = \sqrt{10}\,cis341.57°$; **5.** $z = 2cis300°$; **6.** $z = 7cis0°$; **7.** $z = cis270°$; **8.** $z = 5cis233.13°$; **9.** $z = 5cis270°$; **10.** $z = 3cis180°$; **11.** $4cis100$

B Plot each complex number and then change to rectangular form.

1. $4(\cos 30° + i\sin 30°)$; **2.** $2(\cos 210° + i\sin 210°)$; **3.** $5cis30°$;
4. $3(\cos 0° + i\sin 0°)$; **5.** $2cis150°$; **6.** $9cis\frac{2\pi}{3}$.

Answers:

1. $z = 2\sqrt{3} + 2i$; 2. $z = -\sqrt{3} - i$; **3.** $z = \frac{5\sqrt{3}}{2} + \frac{5}{2}i$; **4.** $z = 3$; **5.** $z = -\sqrt{3} + i$; **6.** $z = -\frac{9}{2} + \frac{9i\sqrt{3}}{2}$

C Perform the indicated operation and leave the answer in polar form.

1. $(\cos 20° + i\sin 20°)(\cos 10° + i\sin 10°$; **2.** $3(\cos 25° + i\sin 25) \bullet 4(\cos 125° + i\sin 125°)$;
3. $[2(\cos 60° + i\sin 60°)]^3$; **4.** $(3cis45°)(4cis360°)$

Answers: **1.** $z = cis30°$; 2. $z = 12cis150°$; **3.** $z = 8cis180°$; **4.** $z = 12cis405°$.

D Perform the indicated operations and leave the answers in polar form.

1. $\frac{14cis170°}{2cis60°}$; **2.** $\frac{10cis60°}{5cis90°}$ **3.** $\frac{15(\cos 100° + i\sin 100°)}{3(\cos 25° + i\sin 25°)}$

Answers: **1.** $z = 7cis110°$; 2. $z = 2cis330°$; **3.** $z = 5cis75°$.

Lesson 45

Powers of Complex Numbers; De Moivre's Theorem; and Roots of Complex Numbers

Powers of Complex Numbers

Let us consider the squaring $z = r\text{cis}\,\theta$

By the multiplication rule, we multiply the moduli (radii) and add the arguments (angles).

Thus $z^2 = (r\text{cis}\,\theta)^2$

$= (r\text{cis}\,\theta)(r\text{cis}\,\theta)$

$= (r)(r)\text{cis}(\theta + \theta)$

$= r^2\text{cis}\,2\theta$

Similarly,

$z^3 = r^3\text{cis}\,3\theta$

$z^4 = r^4\text{cis}\,4\theta$

From the above examples, we arrive at a generalization known as **De Moivre's** theorem or formula.

De Moivre's Theorem: If n is a positive integer, then

$$z^n = (r\,\text{cis}\theta)^n = r^n\text{cis}\,n\theta$$

or $[r(\cos\theta + i\sin\theta)]^n = r^n(\cos n\theta + i\sin n\theta)$

Example 1 Use De Moivre's theorem to find $(3 - 4i)^6$.

Step 1: Change the complex number to polar (trigonometric) form.
Thus, we want to change 3 - 4i to the form $z = r(\cos\theta + i\sin\theta)$

Step 2: Previously, $(3 - 4i) = 5(\cos 306.87° + i\sin 306.87°)$

Applying $z^n = r^n(\cos n\theta + i\sin n\theta)$, we obtain

$(3 - 4i)^6 = [5(\cos 306.87° + i\sin 306.87°)^6$

$= 5^6[\cos(6)(306.87°) + i\sin(6)(306.87°)]$

$= 5^6[\cos 1841.22° + i\sin 1841.22°]$

$= 15625[\cos 41.22° + i\sin 41.22°]$

$= 15625\,\text{cis}\,41.22°$ <--- polar form

In exponential form:: $z = re^{i\theta}$, where θ is in radians

$$41.22° = \frac{41.22°}{180°} \times \frac{\pi \text{ rad}}{1} = 0.72 \text{ rad}$$

$(3 - 4i)^6. = 15625e^{0.72i}$<----exponential form ($r = 15625$,$\theta = 0.72$ rad)

$(3 - 4i)^6. = 11753 + 10296i$ **<---** rectangular form

Comparison

Rectangular form	**Polar form**	**Exponential form**
$z = a + bi$ or $z = x + iy$	$z = r(\cos\theta + i\sin\theta = r\text{cis}\theta$	$z = re^{i\theta}$

Roots of Complex Numbers

We may apply De Moivre's theorem to find the roots of complex numbers.

if $x^n = A$, (where n is a positive integer)

then $x = A^{\frac{1}{n}}$

Example: If $x^3 = 8$, then $x = \sqrt[3]{8} = 8^{\frac{1}{3}}$

Similarly, for a complex number A if $x^n = A$, then $x = A^{\frac{1}{n}}$ (where n is a positive integer).
There are n distinct roots, where x is an nth root of A.

According to De Moivre's Theorem (which deals with powers of complex numbers in polar form),

$$[r(\cos\theta + i\sin\theta)]^n = r^n(\cos n\theta + i\sin n\theta) .$$

If we replace n by $\frac{1}{n}$ in De Moivre's Theorem, and take into account the periodic nature of the cosine and sine functions, we obtain

$$[r(\cos\theta + i\sin\theta)]^{\frac{1}{n}} = r^{\frac{1}{n}}[\cos\frac{(\theta - 360°k)}{n} + i\sin\frac{(\theta + 360°k)}{n}] \qquad \text{where } k = 0, 1, 2, ...$$

$$= \sqrt[n]{r}[\cos\frac{(\theta + 360°k)}{n} + i\sin\frac{(\theta + 360°k)}{n}]$$

$$(r\operatorname{cis}\theta)^{\frac{1}{n}} = r^{\frac{1}{n}}\operatorname{cis}\frac{(\theta + 360°k)}{n}$$

$$= \sqrt[n]{r}\operatorname{cis}\frac{(\theta + 360°k)}{n}$$

If $n = 2$, there are two distinct roots and we shall use two k-values, namely, $k = 0, k = 1$.
For example, if $k = 0$,

$$\frac{\theta + 360°k}{n} = \frac{\theta + 360°(0)}{n} = \frac{\theta + 0}{n} = \frac{\theta}{n}.$$

If $k = 1$,

$$\frac{\theta + 360°k}{n} = \frac{\theta + 360°(1)}{n} = \frac{\theta + 360°}{n}.$$

Similarly, if $n = 3$, there shall be three distinct roots and we use three k-values, namely , $k = 0$, 1, and 2.

We should note that the k-values, $k = 0, 1, 2, ..., n - 1$ are used only in finding the roots of the complex numbers but are **not** used in computing the powers (De Moivre's Theorem).
For $k = n - 1$, the results repeat one of the first n values (previous values).

Applying De Moivre's Theorem to Find the Roots of Numbers

Step 1: Change the given expression to polar (trigonometric) form (if it is already not in polar form).

Step 2: Apply the formula

$$[r(\cos\theta + i\sin\theta)]^{\frac{1}{n}} = \sqrt[n]{r}[\cos\frac{(\theta + 360^\circ k)}{n} + i\sin\frac{(\theta + 360^\circ k)}{n}] \qquad \text{where } k = 0,1,2,...$$

with the appropriate k-values.

Example 1 Find the 4-th roots of $16(\cos 120^\circ + i\sin 120^\circ)$.

Solution

The wording of this problem is equivalent to:

if $z^4 = 16(\cos 120^\circ + i\sin 120^\circ)$, solve for z.

Since the given expression is already in polar form, we go ahead to Step 2 and apply the formula

$$[r(\cos\theta + i\sin\theta)]^{\frac{1}{n}} = \sqrt[n]{r}[\cos\frac{(\theta + 360^\circ k)}{n} + i\sin\frac{(\theta + 360^\circ k)}{n}] \qquad \text{with } k = 0,\ 1,\ 2, \text{ and } 3.$$

$n = 4,\ r = 16,\ \theta = 120^\circ$

Substituting these values

$$z = \sqrt[4]{16}[\cos\frac{(120^\circ + 360^\circ k)}{4} + i\sin\frac{(120^\circ + 360^\circ k)}{4}]$$

$$z = 2[\cos\frac{(120^\circ + 360^\circ k)}{4} + i\sin\frac{(120^\circ + 360^\circ k)}{4}] \qquad \text{(A)}$$

We can leave the answer in either the trigonometric (polar) form or the rectangular form , $z = a + bi$

Since $n = 4$, we shall use k-values of $k = 0, 1, 2, 3$.

When $k = 0$ in (A), we obtain the root,

$$z = 2[\cos\frac{(120^\circ + 0)}{4} + i\sin\frac{(120^\circ + 0)}{4}]$$

$= 2[\cos 30^\circ + i\sin 30^\circ]$ <------------------Polar form

$$= 2[\frac{\sqrt{3}}{2} + \frac{1}{2}i]$$

$= \sqrt{3} + i.$ <-----------------------------Rectangular form

When $k = 1$ in (A), we obtain the root

$$z = 2[\cos\frac{(120^\circ + 360^\circ(1))}{4} + i\sin\frac{(120^\circ + 360^\circ(1))}{4}]$$

$$= 2[\cos\frac{(120^\circ + 360^\circ}{4} + i\sin\frac{(120^\circ + 360^\circ)}{4}]$$

$$= 2[\cos\frac{480^\circ}{4} + i\sin\frac{480}{4}]$$

$= 2[\cos 120^\circ + i\sin 120^\circ]$ <----------------polar form

$= 2[-\cos 60° + i\sin 60°]$

$= 2[-\left(\frac{1}{2}\right) + i\left(\frac{\sqrt{3}}{2}\right)]$

$= -1 + \sqrt{3}i$ or $-1 + i\sqrt{3}$ <-------------------rectangular form

For $k = 2$ in (A), we obtain the root.

$z = 2[\cos\frac{(120° + 360°(2))}{4} + i\sin\frac{(120° + 360°(2))}{4}]$

$= 2[\cos\frac{(120° + 720°}{4} + i\sin\frac{(120° + 720°)}{4}]$

$= 2[\cos\frac{840°}{4} + i\sin\frac{840°}{4}]$

$= 2[\cos 210° + i\sin 210°]$ <----------------polar form

$= 2[-\cos 30° + i\sin 30°]$

$= 2[-\left(\frac{\sqrt{3}}{2}\right) + i\left(-\frac{1}{2}\right)]$

$= -\sqrt{3} - i$ <-------------------rectangular form

Similarly, for $k = 3$ in (A) we obtain the root

$2[\cos 300° + i\sin 300°]$ <-----------------polar form

$= 2[\cos 60° + i(-\sin 60°)]$

$1 - \sqrt{3}i$ or $1 - i\sqrt{3}$ <-------------------rectangular form

The 4th roots of $16(\cos 120° + i\sin 120°)$ are $\sqrt{3} + i$; $-1 + i\sqrt{3}$; $-\sqrt{3} - i$; and $1 - i\sqrt{3}$

Example 2 If $z^4 = -8 + 8\sqrt{3}i$, solve for z.

Solution

Step 1: Change the right-hand side of the equation to polar form.

$r = 16, \quad \theta = 120°$

$-8 + 8\sqrt{3}i = 16(\cos 120 + i\sin 120)$

$\therefore\ z^4 = 16(\cos 120 + i\sin 120)$.

The rest of the steps are the same as those in Example 1 , above.

Example 3 (Method 1) Find the square root of $5-12i$ (This problem was done previously.

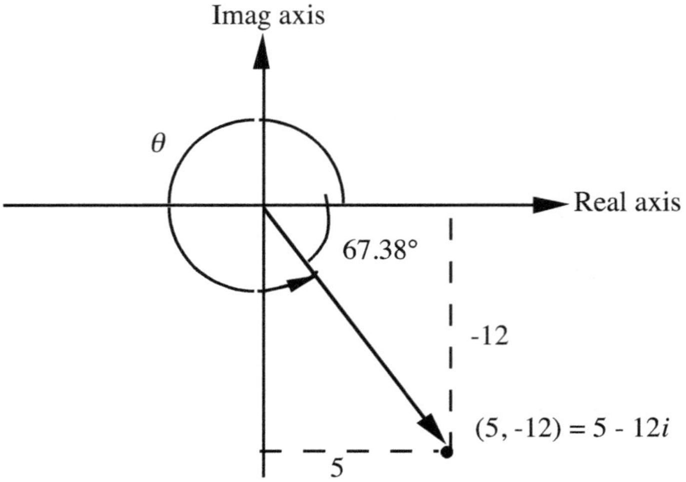

Step 1: Change to polar form.

$a=5, b=-12$

$r=\sqrt{5^2+(-12)^2}=13$

$\tan\theta_{\text{ref}} = -\frac{12}{5}$

$\theta_{\text{ref}} = 67.38°$

$\theta = 360° - 67.38°$

$\theta = 292.62°$

Now, $r=13$, $\theta = 292.62°$ and

$5-12i = = 13(\cos 292.62° + i\sin 292.62°)$

Step 2: Let $z^2 = 13(\cos 292.62° + i\sin 292.62°)$.

Then $z=\sqrt{13}\left(\cos\frac{292.62°+360°k}{2} + i\sin\frac{292.62°+360°k}{2}\right)$, $k=0,1$. (Two k-values for $n=2$)

When $k=0$, $z=\sqrt{13}\left(\cos\frac{292.62°}{2} + i\sin\frac{292.62°}{2}\right)$

$= \sqrt{13}\left(\cos 146.31° + i\sin 146.31°\right)$

$= \sqrt{13}(-.83+.55i)$

$z = -2.99 + 1.98i$ (1)

When $k=1$, $z=\sqrt{13}\left(\cos\frac{292.62°+360°(1)}{2} + i\sin\frac{292.62°+360°(1)}{2}\right)$

$z=\sqrt{13}\left(\cos\frac{292.62°+360°}{2} + i\sin\frac{292.62°+360°}{2}\right)$

$= \sqrt{13}\left(\cos 326.31° + i\sin 326.31°\right)$

$= \sqrt{13}(.83 + i(-.55))$ ($\cos 326.31° = .83$; $\sin 326.31 = -.55$)

$= \sqrt{13}(.83 - .55i)$

$= 2.99 - 1.98i$ (2)

The square roots are $-2.99+1.98i$ and $2.99-1.98i$ (Combining the roots for $k=0$ and $k=1$)

Another Method of Finding Square Root (Short-Cut Method)

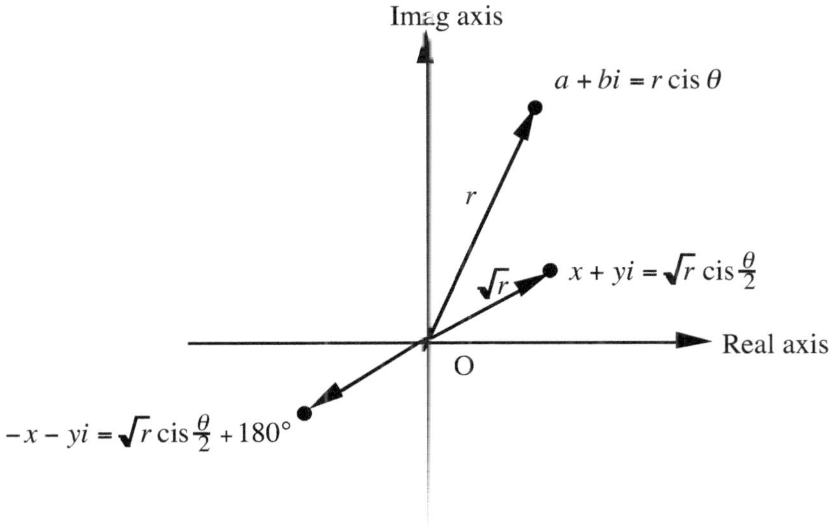

Let $x + yi$ be the square root of $a + bi$

Then $(x + yi)^2 = a + bi$ (1)

$x^2 - y^2 + 2xyi = a + bi$ (2)

Equating the real parts: $x^2 - y^2 = a$ (3)

Equating the imaginary parts: $2xy = b$ (4)

$(\sqrt{r})^2 = x^2 + y^2$ (From figure)

$\therefore \quad r = x^2 + y^2$ (5)

Also, $r^2 = a^2 + b^2$ (From figure)

$r = \sqrt{a^2 + b^2}$ (6)

Equating (5) to (6),

$x^2 + y^2 = \sqrt{a^2 + b^2}$ (7)

We shall apply equations (3), (4), (6), and (7) to find the square root of a complex number, by redoing Example 3, above

Example 3 (Method 2) Find the square root of $5 - 12i$.

Step 1: Let $x + yi$ be the square root of $5 - 12i$.

From $r^2 = 5^2 + (-12)^2$,

$r = 13$ (1)

Applying $x^2 + y^2 = \sqrt{a^2 + b^2}$

$x^2 + y^2 = 13$ (2)

Step 2: $x^2 - y^2 + 2xy = 5 - 12i$

Equating the real parts: $x^2 - y^2 = 5$ (3)

Equating the imaginary parts: $2xy = -12$ (4)

We now solve for x and y using equations (2) , (3) and (4)

Step 3: By adding (2) and (3); we eliminate y^2.

$2x^2 = 18$

$x^2 = 9$ and $x = +3$ *or* -3

Step 4: We find y by substituting the x-values in equation (4)

When $x = 3$, $2(3)y = -12$ and from which $y = -2$.

When $x = -3$, $2(-3)y = -12$ and from which $y = 2$.

Thus, when $x = 3, y = -2$, and when $x = -3, y = 2$

Substituting these pairs (of values) in $x + yi$, the square roots are $3 - 2i$ and $3 + 2i$.

Plotting Roots in the Complex Plane

If we graph the roots, all the roots will be found to be equally spaced on a circle of radius, $\sqrt[n]{r}$ (or $r^{\frac{1}{n}}$), and the roots are $\frac{360}{n}$ degrees from one another. The first root is known as the principal nth root. The roots form the vertices of a regular n-sided polygon with its center at the origin.

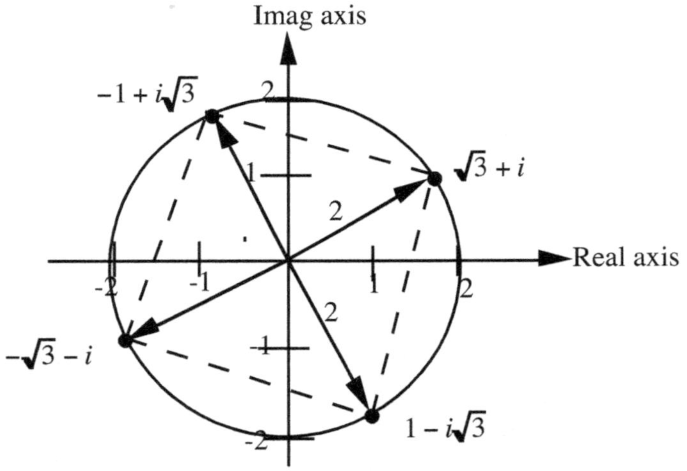

Figure: Graph of the roots for Example 1: $\sqrt{3} + i$; $-1 + i\sqrt{3}$; $-\sqrt{3} - i$; $1 - i\sqrt{3}$.

Lesson 45 Exercises

A Use De Moivre's theorem to evaluate the following:

1. $(2-3i)^4$; **2.** $(4+4i)^5$; **3.** $(-3+i)^6$; **4.** $(4\,cis\,30°)^5$; **5.** $(-i)^{11}$

Answers: **1.** $-119+120i$; **2.** $-4096-4096i$; **3.** $-352-934i$; **4.** $-886.81+512i$; **5.** i.

B Find the indicated roots.

1. The cube roots of 8; **2.** The fifth roots of 1; **3.** The fifth roots of 3;

4. The square roots of $2-3i$; **5.** The cube roots of $-1+i$

Find all the roots of the following: **6** $16x^4=-81$; **7.** $x^3+8=0$; **8.** $x^3+8i=0$

Answers:

1. $2,\ -1+i\sqrt{3},\ -1-i\sqrt{3}$; **2.** $1,\ 0.31+0.95i,\ -0.81+0.59i,\ -0.81-0.59i,\ .31-0.95i$;

3. $\sqrt[5]{3},\ \sqrt[5]{3}(0.31+0.95i),\ \sqrt[5]{3}(-0.81+0.59i),\ \sqrt[5]{3}(-0.81-59i),\sqrt[5]{3}(.31-0.95i)$;

4. $\sqrt[4]{13}(-0.88+0.47i),\sqrt[4]{13}(0.88-0.47i)$;

5. $\sqrt[6]{2}(0.71+0.71i),\ \sqrt[6]{2}(-0.96+0.26i),\ \sqrt[6]{2}(0.26-0.96i)$;

6. $1.061+1.061i,\ -1.061+1.061i,-1.061-1.061i,\ 1.061-1.061i$;

7. $-2,\ 1-i\sqrt{3},\ 1+i\sqrt{3}$; **8.** $2i,\ -\sqrt{3}-i,\ \sqrt{3}-i$.

APPENDIX F

Lesson 46: Graphing Polar Coordinates
Lesson 47: Graphing Polar Equations

Introduction to Polar Coordinates

We use polar coordinates in the design of control systems in engineering. Some equations, say, in x and y which are fairly complex in the rectangular forms become relatively simple when changed to the equivalent polar forms. For example, let us consider the equation of the circle

given by $x^2 + y^2 = 9$ <------------rectangular form.

In polar form, $r^2 = 9$ or simply $r = \pm 3$ <------------polar form (a circle of radius r)

Similarly, in polar form, the equation of the line $y = x$ is given simply by $\theta = 45°$ (for all r)

Lesson 46
Graphing Polar Coordinates

The Polar Coordinate System

As shown in Figures 1 and 2, the **rectangular coordinate system** has two axes, namely the x-axis and the y-axis. These two axes intersect at right angles at the origin which we label (0, 0). A point P in the plane of this system has just one pair of coordinates (x, y) which is an ordered pair. Also an ordered pair of coordinates (x, y) represents only one point.

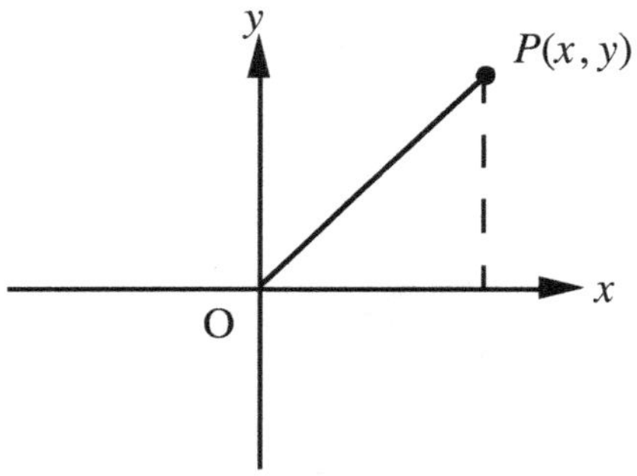

Figure 1: The rectangular coordinate system

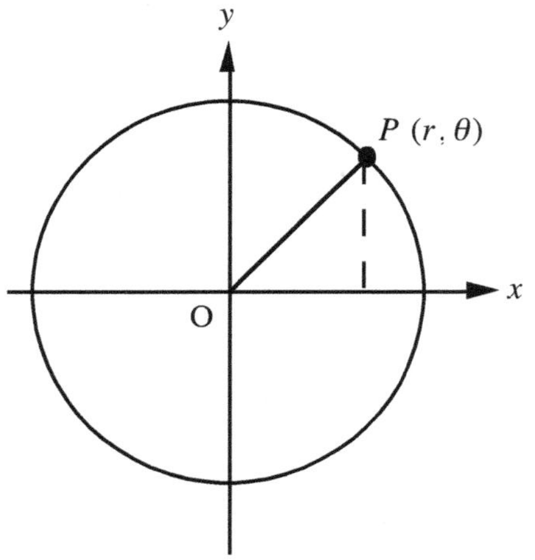

Figure 2: The polar coordinate System

Similarly, the **polar coordinate system** (Figure) has two axes, the x-axis (the **polar-axis** or 0°-axis) and the y-axis ($\frac{\pi}{2}$ or 90° axis) . These two axes intersect at right angles at the origin (or **the pole**) which we label $(0, \theta)$, where θ is any angle. A **given point** P in the polar coordinate system does **not** have just one pair of coordinates (in contrast to the coordinates in the rectangular coordinate system), but has an unlimited number of polar coordinate pairs. Each polar coordinate pair (r, θ) is ordered. However, we must note that a given **ordered pair** represents only one point.

Thus a given point has many names, but any of the possible names represents only one point, the given point.

A given point in the polar coordinate system has an unlimited number of coordinate pairs because different angular rotations to the same point are possible; and also, the radius can be generated in more than one assumed direction.

For the ordered pair (r, θ), r is the **directed distance** (may be positive or negative) from the origin (the pole) and θ, the **angle of rotation**, may be in degrees or in radians, and may be positive or negative. The angle is positive if measured counterclockwise but negative if measured clockwise. The angles are measured in the same way as done in trigonometry.

Although a point has an unlimited polar coordinates, we shall restrict θ so that $0 \leq \theta \leq 2\pi$. We then obtain what is called the **two primary representations** of the polar coordinates. The primary representations are (r, θ)) and $(-r, \theta + \pi)$.

How to Plot Polar Coordinates

We shall now learn how to **locate a point** in the polar plane given the coordinates (r, θ) of the point. We shall plot on polar graph paper whenever it is available.

Step 1: Assuming we are **standing at the origin** (the pole) and facing in the positive *x*-direction, we **rotate** through the given angle θ counterclockwise if θ is positive; or clockwise if θ is negative. From this step, we go to Step 2 if r is positive but if r is negative, we go on to Step 3, below.

Step 2: Keeping the direction in which we are facing as a result of our rotation from Step 1, we **count** straight **forwards a distance** r units (r being positive) and then **stop**. We **label** this stopping point (r, θ) if θ is positive or $(r, -\theta)$ if θ is negative.
We then draw r with a **solid line**.

Step 3: If r is negative, then, keeping the direction in which we are facing as a result of rotation from Step 1, we **count** straight **backwards** a distance r units and then **stop**. We **label** this point $(-r, \theta)$ if θ is positive or $(-r, -\theta)$ if θ is negative. Finally, we draw this line segment r with a **broken** (or dotted) **line** to indicate that r is negative.

Example 1 Plot the point having the polar coordinates $(3, 60°)$

Solution

Step 1: Assuming you are at the origin and facing in the positive x-direction (Figure 1.), rotate through $60°$counterclockwise (draw $60°$ counterclockwise since θ is positive).

Step 2: From the origin count 3 units forwards and stop. Place a dot here and label this stopping point $(3, 60°)$. If θ were in radians, we would label this point $(3, \frac{\pi}{3})$

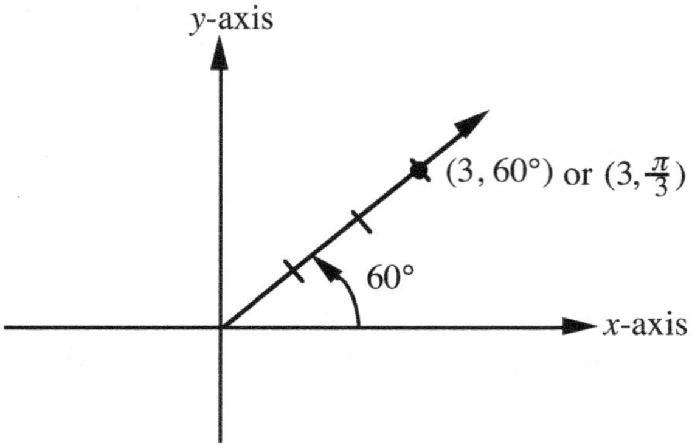

Figure 1 : The graph of the point $P(3, 60°)$

Example 2 Plot the point having the polar coordinates $((3, -60°)$

Solution

Step 1: Since the angle is negative, we draw $60°$ clockwise (Figure 2)

Step 2: From the origin count 3 units forwards and stop. Place a dot here and label this stopping point $(3, -60°)$

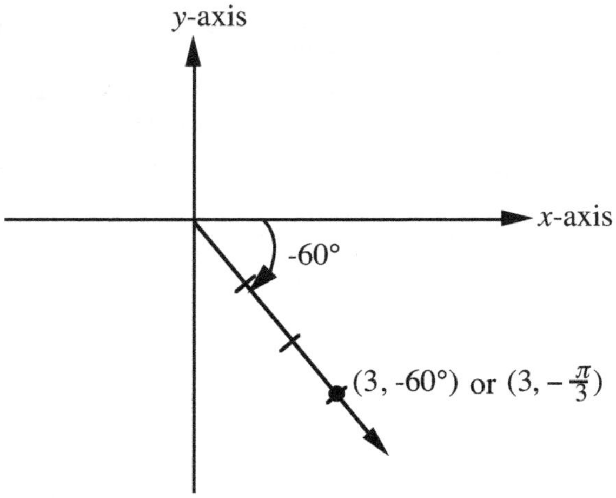

Figure 2 : The graph of the point $P(3, -60°)$

Example 3 Plot the point having the polar coordinates $(-3, 60°)$

Solution

Step 1: Assuming you are at the origin and facing in the positive x-direction (Figure.), rotate through 60° counterclockwise (draw 60° counterclockwise since θ is positive).

Step 2: Still at the origin and keeping the direction in which you are facing, count 3 units straight backwards along OQ and stop. Place a dot here, and label this point $(-3, 60°)$. Using a broken line draw the radius by connecting OQ.

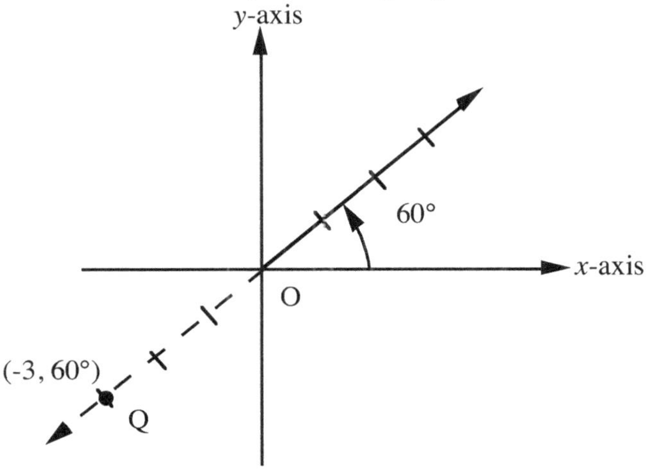

Figure :The graph of the point $Q(-3, 60°)$

Example 4 Plot the point having the polar coordinates $(-4, -60°)$

Step 1: Assuming you are at the origin and facing in the positive x-direction, rotate through 60° clockwise (since θ is negative). See Figure 3

Step 2: Still at the origin and keeping the direction in which you are facing as a result of rotation from Step 1, count 4 units straight backwards along OQ and stop. Place a dot here, and label this point $(-4, -60°)$.

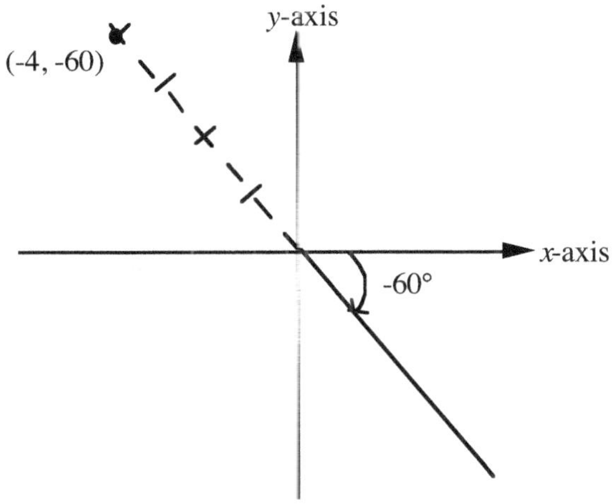

Figure 3 : The graph of the point $(-4, -60°)$

Example 5 Plot the point having the polar coordinates $(-5, -300°)$

Solution

Step 1: Rotate through $300°$ clockwise (since θ is negative).

Step 2: Count 5 units backwards and stop. Place a dot here and label this point $(-5, -300°)$.

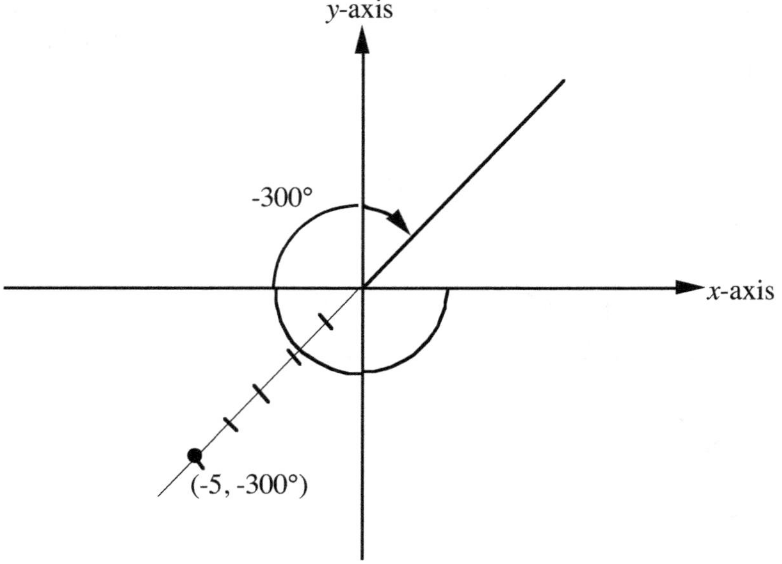

Note: In plotting polar coordinates, we must note that the signs of the ordered pairs (r, θ) depend solely on **how the angles and distances are generated** and do **not** depend on the **quadrant** in which the point happens to be located, unlike the case of the rectangular coordinate system, in which the quadrant determines the signs of the coordinates.

The coordinates (r, θ) and, $(-r, \theta + \pi)$ represent the same point. These two coordinate pairs are the two primary representations of the same given point. There are infinitely many other representations each of which is obtained by adding 2π or (or -2π) successively to the primary representation.

Other representations of the same point: are

$(r, \theta + 2\pi)$ and $(-r, \theta + 3\pi)$.

$(r, \theta - 2\pi)$ and $(-r, \theta - \pi)$.

Determining if a given point is on a curve, given the polar equation

Definition: A given point (except the pole) is on a polar form curve if and only if at least one of the primary representations of the point satisfies the given equation.

Thus if we check for one primary representation and it does not satisfy the given equation, we must also check for the other primary representation.

To check for a point, we substitute the given coordinates in the equation to determine if the left-hand side equals the right-hand side of the equation. If the LHS equals the RHS the point is on the curve, otherwise, it is not on the curve

Lesson 46 Exercises

Plot the following polar coordinates:

1. $(4, 30°)$; **2**. $(-3, 90°)$; **3**. $(-3, \frac{\pi}{3})$; **4**. $(-6, -30°)$; **5**. $(-4, -400°)$

1.

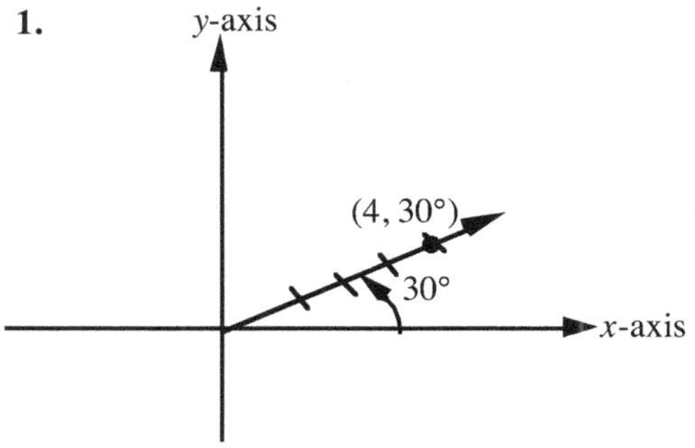

2.

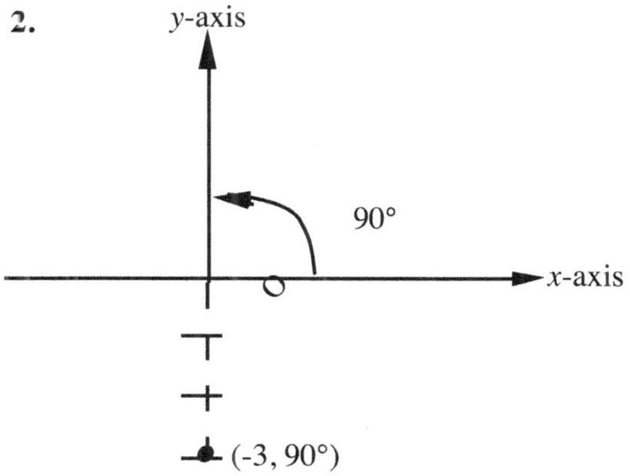

3.

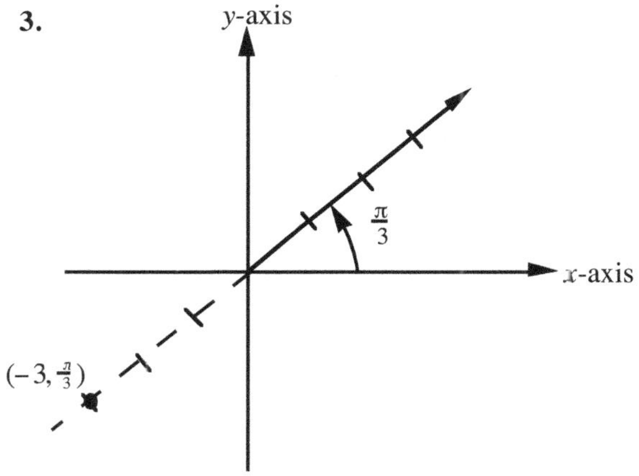

4.

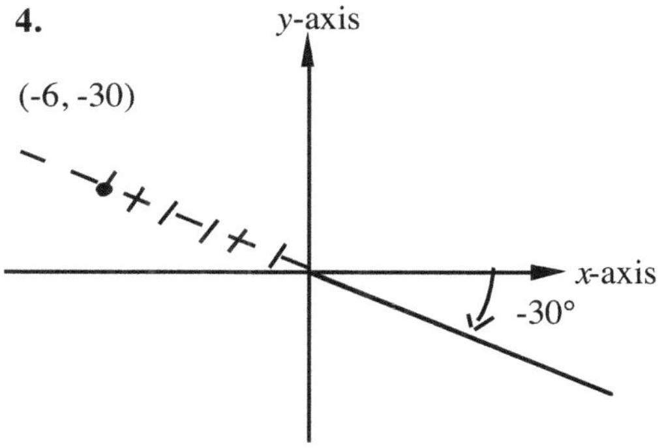

5.

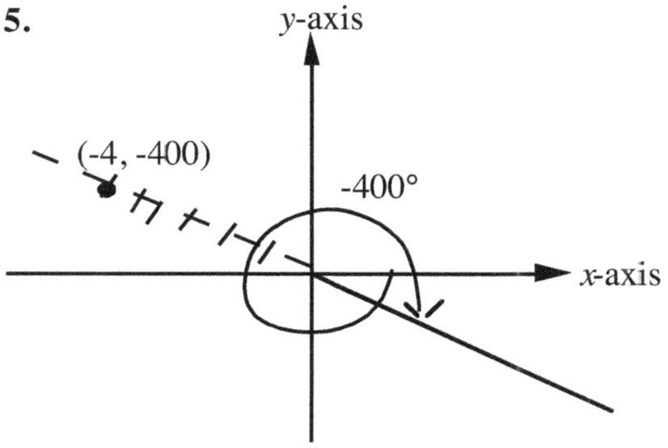

Lesson 47

Graphing Polar Equations

In general, we can sketch the graphs of polar equations by constructing tables for r and θ and plotting points. Consider a table for the polar equation $r = \sin\theta$ (See Table 1). It can be observed from the table that after π radians or 180° the values of r begin to duplicate. We can therefore take advantage of such duplication when sketching polar graphs. We can do this by testing to determine if the graph is symmetric about (a) the x-axis, (b) the y-axis, and (c) the origin.

Table 1

θ	0°	30°	45°	60°	90°	120°	135°	150°	180°	210°	240°	225°	270°	300°	315°	330°	360°
$r = \sin\theta$	0	.5	.71	.87	-1	.87	.71	.5	0	-.5	-.87	-.71	-1	-.87	-.71	-.5	0

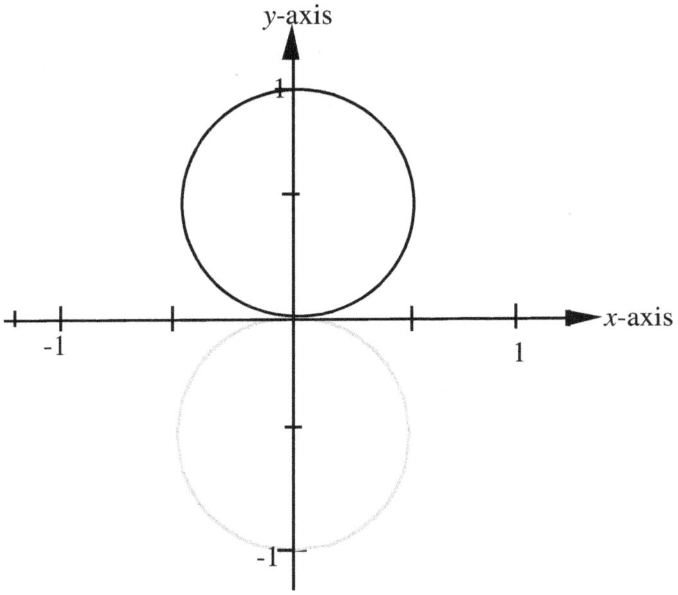

Rules for testing for symmetry about the x-axis

Step 1: Replace (r, θ) by either $(r, -\theta)$, or $(-r, -\theta + \pi)$ and simplify If the equation remains unchanged on simplifying then the curve is symmetric about the x-axis.

Step 2: if $(r, -\theta)$ fails the test, we must check for $(-r, -\theta + \pi)$. If both $(r, -\theta)$and $(-r, -\theta + \pi)$ fail the test, then the curve is **not** symmetric about the x-axis However if either $(r, -\theta)$ or $(-r, -\theta + \pi)$ passes the test first, then we do not have to test for the other.

Although we can test for symmetry before sketching polar curves , for common practical purposes and to save time, we may not test for symmetry for the common polar graphs. We shall familiarize ourselves with or memorize the properties of the common or simple polar curves (the same as we familiarized ourselves with the simple trigonometric functions such as $y = \sin x$.). When we meet an unfamiliar or a more complicated polar equation, then we shall test for symmetry We shall now discuss and sketch some common polar graphs. In the future, we should be able to identify these graphs with the corresponding equations and vice versa.

The following trigonometric identities will be useful when testing for symmetry and graphing polar equations:

1. $\cos(-\theta) = \cos\theta$
2. $\sin(-\theta) = -\sin\theta$
3. $\tan(-\theta) = -\tan\theta$
4. $\cos(\pi - \theta) = -\cos\theta$
5. $\sin(\pi - \theta) = \sin\theta$
6. $\cos(\pi + \theta) = -\cos\theta$
7. $\sin(\pi + \theta) = -\sin\theta$

Also, $\cos(\theta \pm 2\pi) = \cos\theta$; $\sin(\theta \pm 2\pi) = \sin\theta$

The Rose Curves: The equations of the rose curves are

$r = a\sin n\theta$ (where n ia a positive integer)

$r = a\cos n\theta$

If n is odd each curve has n leaves but has $2n$ leaves if n is even.

Example 1 If $n = 1$ (odd number), the sine equation, $r = a\sin n\theta$ becomes $r = a\sin\theta$ and the curve has one leaf (petal). The leaf is circular and has the y-axis as the axis of symmetry (Figure 1).

Example: $r = 2\sin\theta$ (where $a = 2$,)

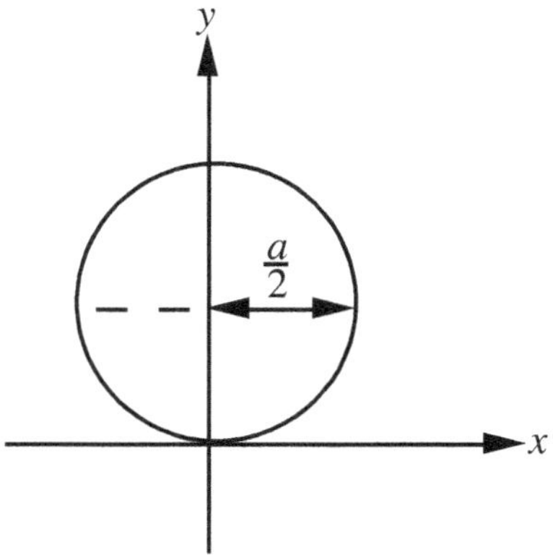

Figure 1 : $r = a\sin\theta$

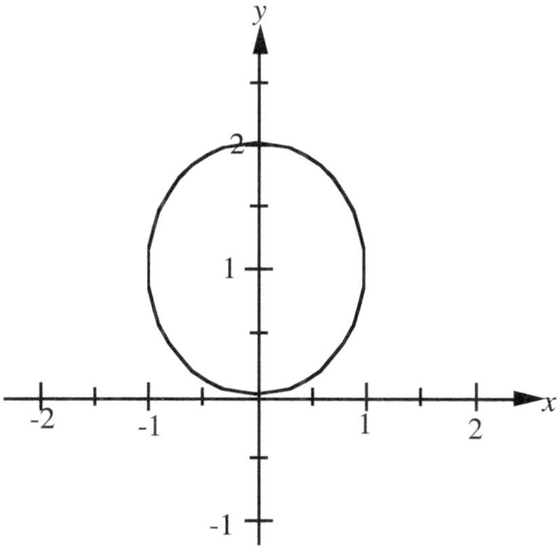

Figure 2 : Graph of $r = 2\sin\theta$

Example 2 If $n = 1$, the cosine equation becomes $r = a\cos\theta$. This curve also like that of the sine has one circular leaf, but the axis of symmetry is the x-axis .
Example: $r = 2\cos\theta$ (where $a = 2$,)

-

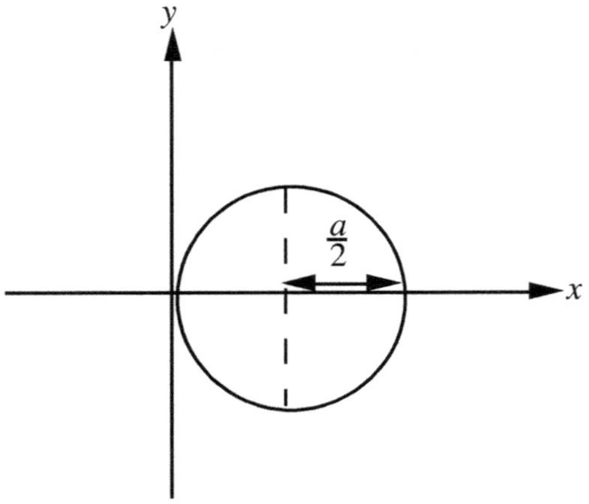

Figure Graph of $r = a\cos\theta$

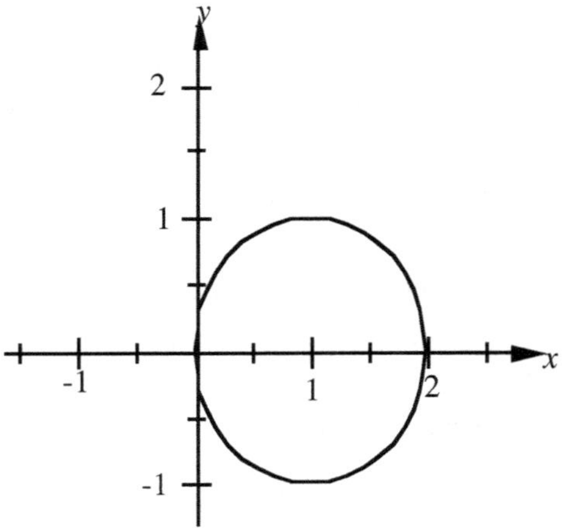

Figure: Graph of $r = 2\cos\theta$

Example 3 Sketch the polar graph for $r = -a\sin\theta$

Solution The graph is the graph of $r = -a\sin\theta$ reflected about the y-axis.
Compare Figure and

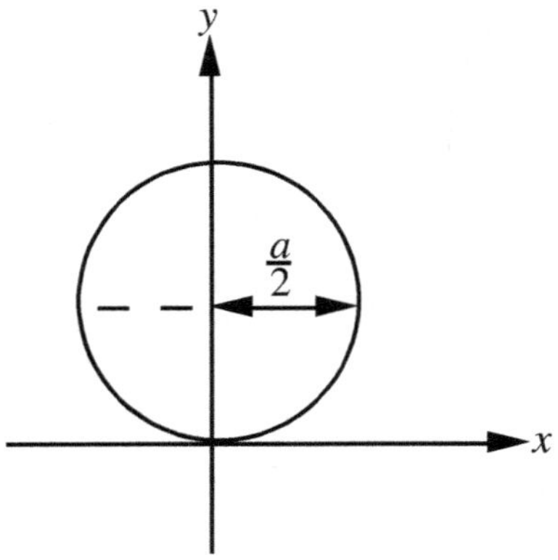

Figure : Graph of $r = a\sin\theta$

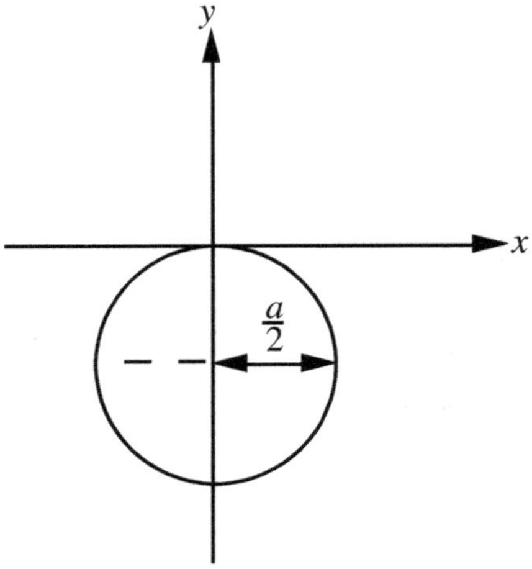

Figure : Graph of $r = -a\sin\theta$

Example 4 Sketch the polar graph for $r = -a\cos\theta$.

Solution The graph of $r = -a\cos\theta$ is the graph of $r = a\cos\theta$ reflected reflected about the y-axis. Compare Figure and

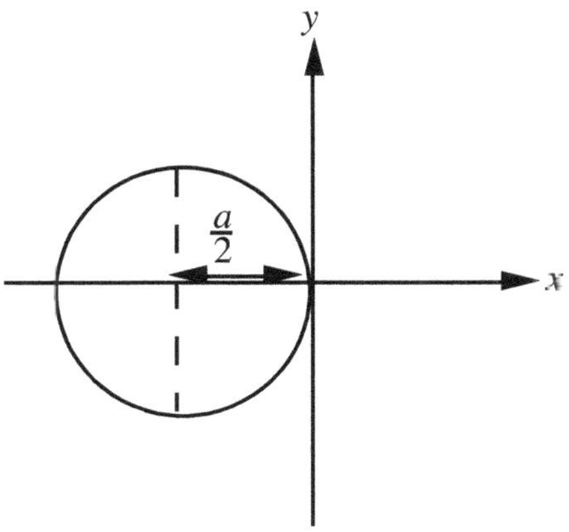

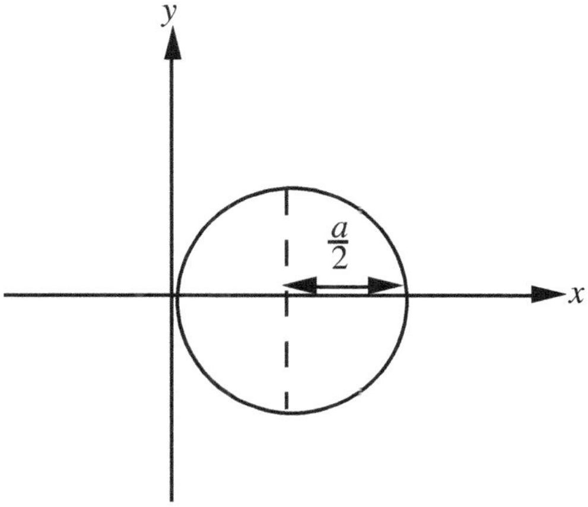

Figure: $r = -a\cos\theta$

Figure: $r = a\cos\theta$

Example 5 Sketch the graph of $r = 3\sin 2\theta$

Solution

Since n is even ($n = 2$), there are 2(2) or 4 leaves or petals. There are $\frac{360°}{4} = 90°$ between the axes of any two leaves. The first leave is in the first quadrant. The leaves are symmetric about the lines $y = x$, and $y = -x$ (Figure 203)

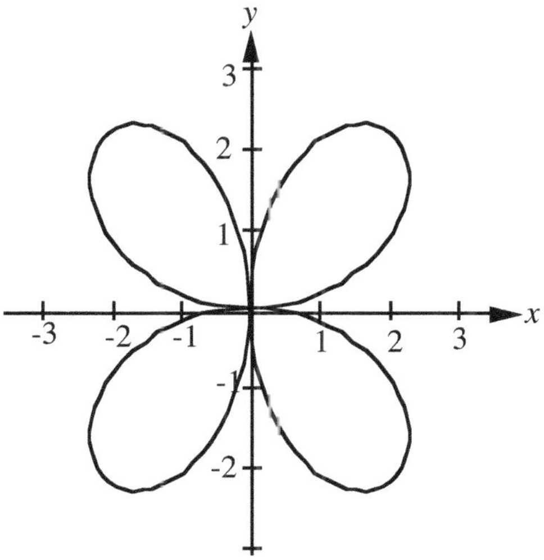

Example 6 Sketch the graph of $r = 3\cos 2\theta$

Solution

Since $n = 2$, there are 2(2) or 4 leaves. The are $\frac{360°}{4} = 90°$ between the axes of any two leaves (as in Example 5 above). However, the axes of symmetry are the x- and y-axes (Figure).

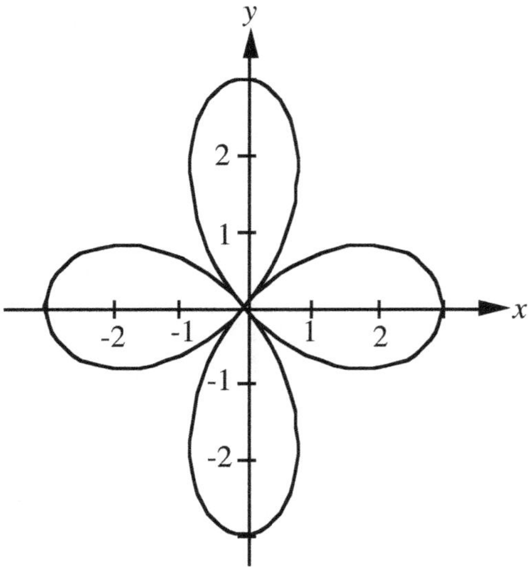

Example 7 Sketch the graph of $r = 4\sin 3\theta$

Solution Since n is odd and $n = 3$, there are 3 leaves. There are also $\frac{360°}{3} = 120°$ between the axes of any two leaves. The first leave is in the first quadrant and it is symmetric about the line $\frac{\pi}{6}$ (Figure).

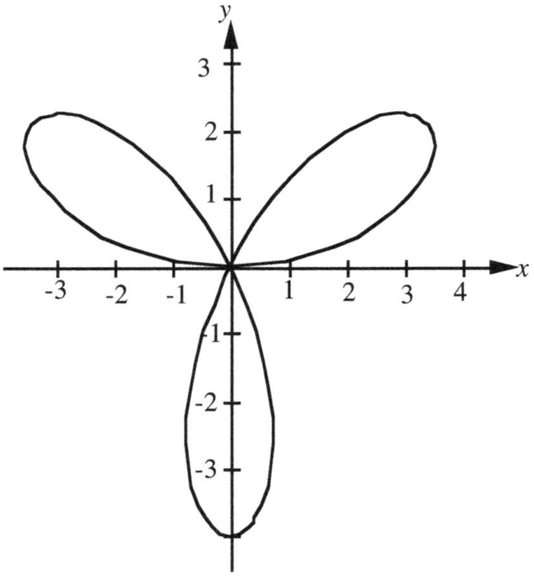

Example 8 Sketch the graph of $r = 4\cos 3\theta$

Solution The graph is similar to that in Example 7, except that there is a difference in the location of the petal (leaf) axes (Figure)

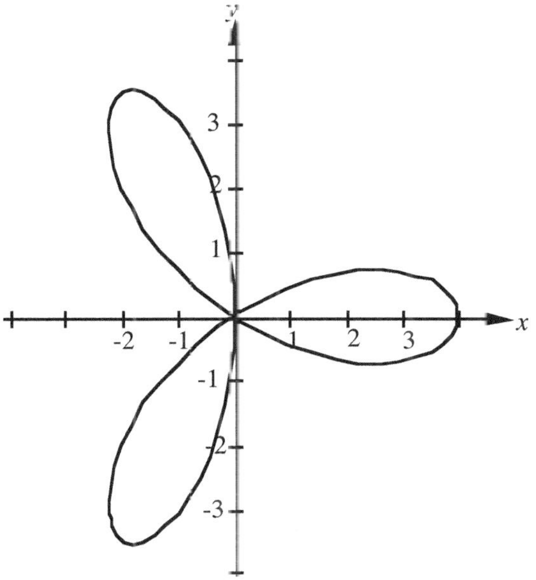

Other polar Graphs

The graphs of the Limacon $r = a \pm b\sin\theta$; $r = a \pm b\cos\theta$; $r^2 = a^2\sin 2\theta$; $r^2 = a^2\cos 2\theta$

Lesson 47 Exercises

Identify (e.g., rose curve) and sketch the graph of each equation.

1. $r = 2\cos\theta$; **2.** $r = 5\sin\theta$; **3.** $r = -4\sin\theta$; **4.** $r = 4\sin 3\theta$; **5.** $r = 5\cos 3\theta$;
6. $r = 4$; **7.** $r = 2\theta$; **8.** $r = 4 - 3\sin\theta$; **9.** $r^2 = 5\cos 2\theta$; **10.** $r = 3 - 3\cos 3\theta$;
11. $r = e^{2\theta}$; **12.** $\theta = \frac{\pi}{6}$.

1.

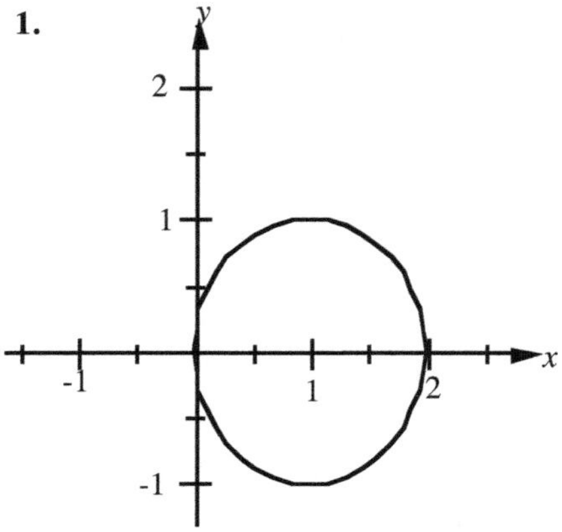

2.

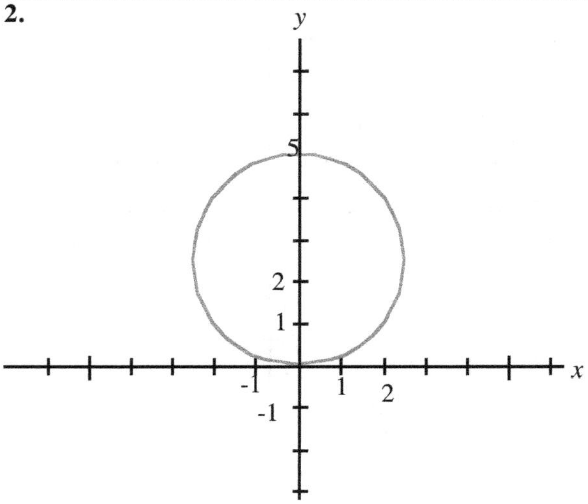

3.

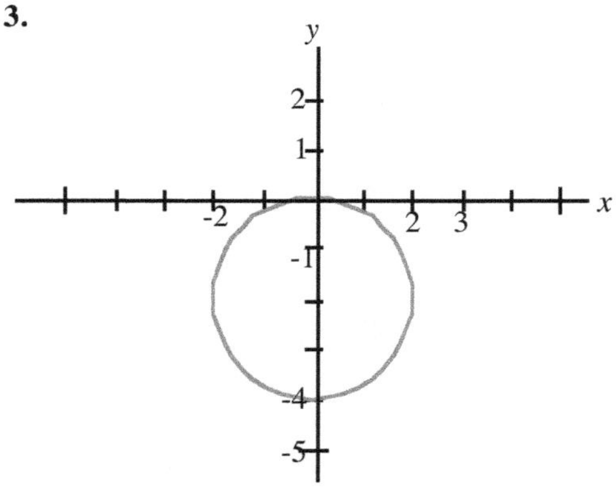

4.

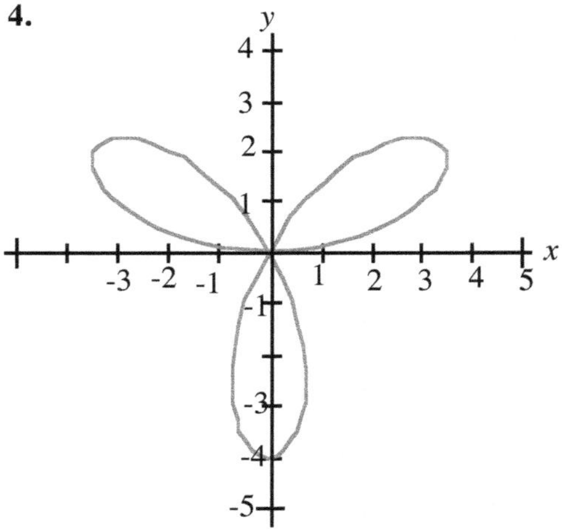

5.

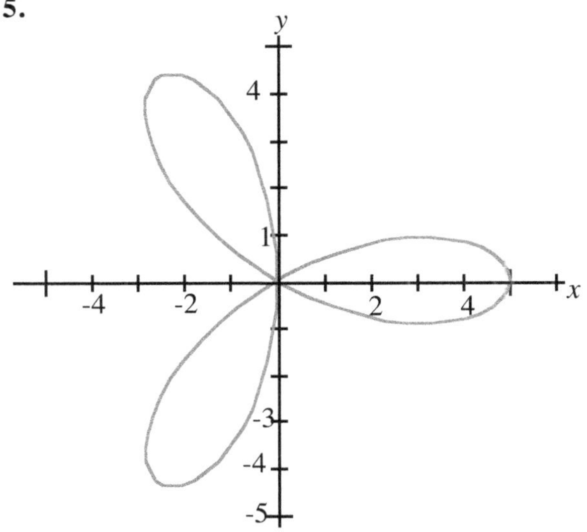

6.

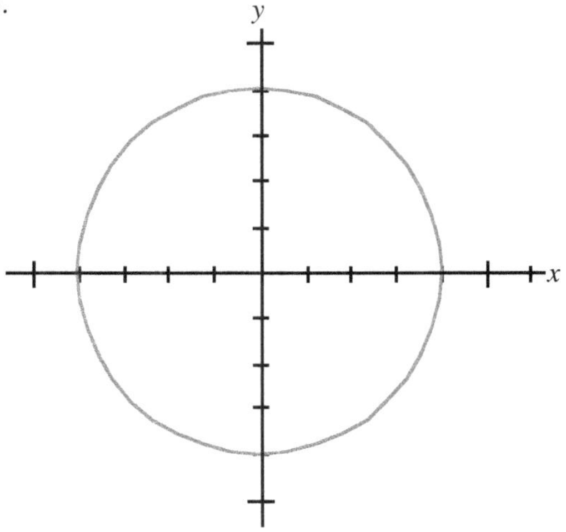

7.

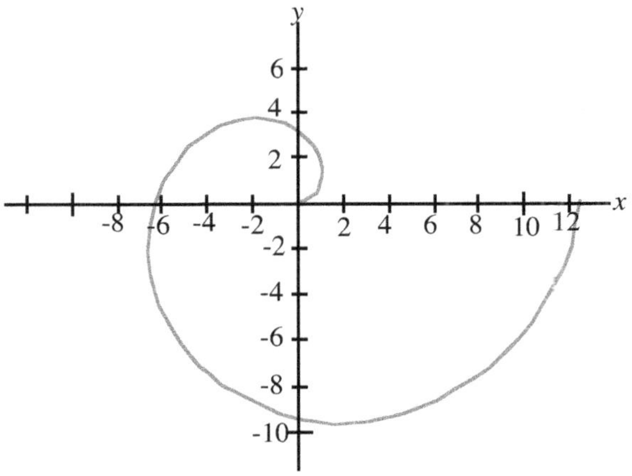

8.

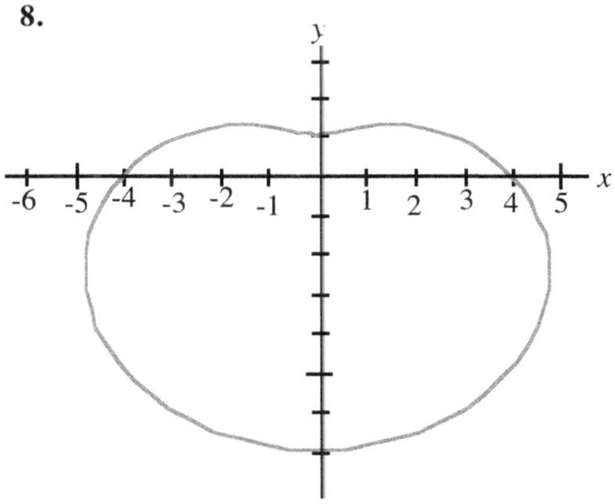

9.

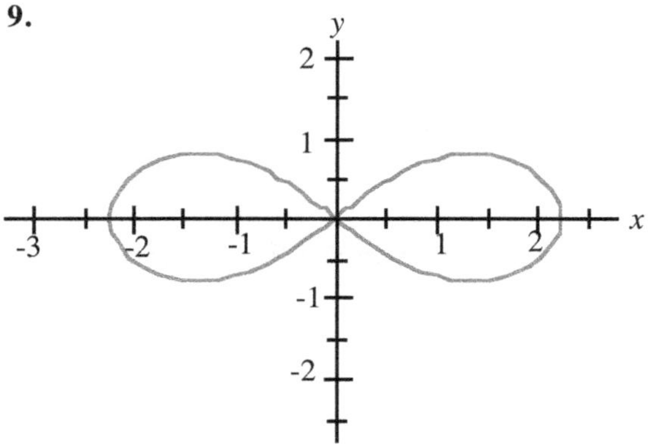
y
2
1
-1
-2
x
-3
-2
-1
1
2

10.

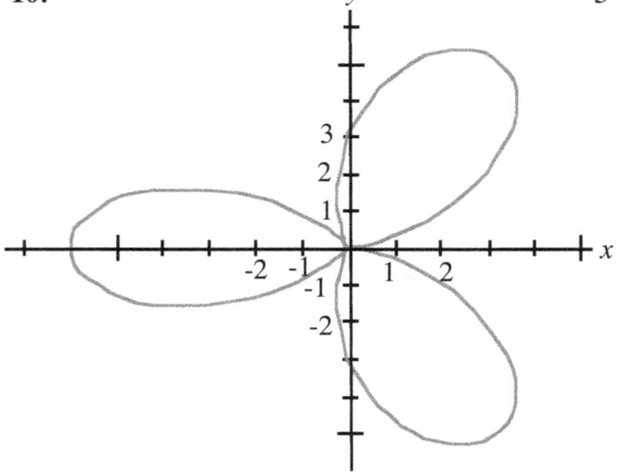
y
3
2
1
-1
-2
x
-2
-1
1
2

11.

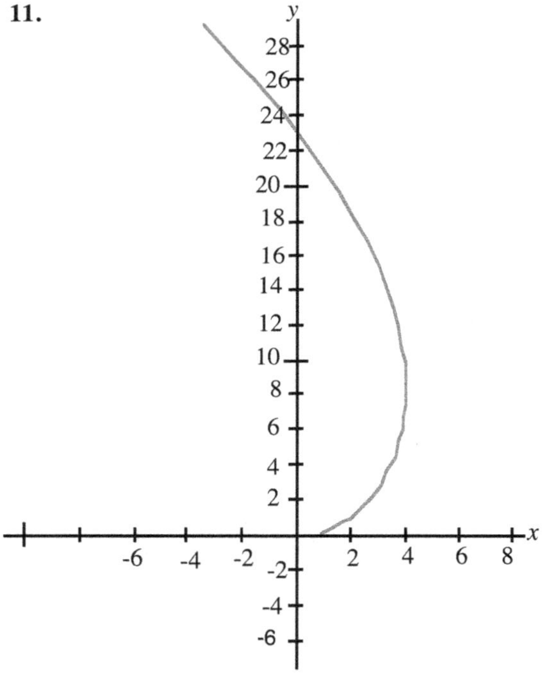
y
28
26
24
22
20
18
16
14
12
10
8
6
4
2
-2
-4
-6
x
-6
-4
-2
2
4
6
8

12

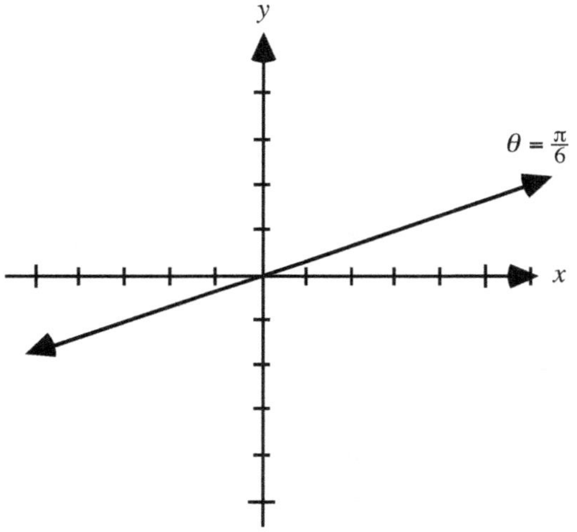
y
$\theta = \frac{\pi}{6}$
x

INDEX

A

Absolute Value 3
Accuracy and Precision 262
Acute triangle 93
Addition
    of vectors 153
amplitude 178
Angle 119
Angle of depression .i.Angles
    angle of depression 113
Angle of elevation 113
Angles
    angle of elevation 113
    complementary angles 103
    in standard position 120, 121
Approximate Numbers 261
arc length 164, 254
Area of a circle 253
Asymptotes 36
    horizontal asymptote 36
    oblique 39
    vertical asymptote 36
Asymptotes, drawing 41
Asymptotic Formula 40

B

Bearing 114

C

Calculators
    using calculators 100
central angle 164
CIRCULAR FUNCTIONS 165
Cofunction Relationships 235
Cofunctions 104
Complementary angles 103, 104
Complementary functions 104
complex numbers 270
Composite Figures 256
Composite Functions 21
composite numbers 1
Composition of functions
    applications 23
Composition of trigonometric functions 220
Concavities of Curves 64
constant function 62
Continuous Functions 33
Contraction of a Curve
    in the x-direction 83
    in the y-direction 84
Coordinate Axes 175
Coterminal Angles 133

counting numbers 2
Critical Points 66
Curves, expansion 82
Curves,contraction 83

D

decimal to a rational number 247
Decreasing Function 62
De Moivre's Theorem 294
Direction
    of resultant 162
Discontinuous Functions 33
Domain 10
Double Angle Identities 234

E

Equations
    of vertical asymptote 36
    trigonometric equations 241
Equilateral triangle 94
equivalent functions 244
Essential dicontinuity 33
Estimation 260
Even function 89
Exact Numbers 261
Excluded Values 10, 12, 34
Expansion of a Curve
    in the x-direction 82
    in the y-direction 84

F

Finite Discontinuity 33
Function 4, 5
Functional notation 8
Functions
    comparison with relations 5
    composite functions 21
    constant functions 62
    continuous functions 33
    discontinuous functions 33
    even functions 89
    function composition applications 22
    graphs of reciprocals 46
    increasing functions 61
    inverse functions 24
    inverse relations 24
    negative functions 60
    odd functions 89
    one-one functions 20
    polynomial functions 12
    positive functions 60
    rational functions 13

G

Graph
of rational function 46
Graphs
of inverse trigonometric functions 217, 223
of InverseTrigonometric functions 223
of y = Arcsin x 223
rational function 52
rational functions 34
shifting or translating 78
sketching inverses 30
vector quantities 151
y = cos x 183
y = cotx 193
y = csc x 190
y = sec x 192
y = sin x 177, 179
y = tan x 187, 188

H

Half Angle Identities 234
Holes,functions 33
Horizontal Asymptote 36
horizontal asymptotes 37
horizontal line test 26
hypotenuse 95

I

Increasing Functions 61
infinite discontinuity 33
integers 2
Inverse Cofunction Identities 252
Inverse Functions 24, 244
Inverse functions,finding 27
Inverse of a function
tests for 25
Inverse Operations
trigonometric functions 220
Inverse Relations 24
inverse trigonometric functions 218, 220, 251
ircular functions 167
irrational number 2
irrational numbers 1
Isosceles triangle 94

L

Law of Cosines 139, 148
Law of Sines 142, 148
Linear Interpolation 115

M

Maximum Point 66
Minimum Point 66
natural numbers 2

N

Negative Functions 60
negative integers 1, 2
Non-negative integers 3
Non-negative real numbers 3
Non-positive real numbers 3
non-real numbers 1
Number Flow Chart 1
Number line
    labeling 176

O

oblique (slant) asymptotes 39
Obtuse triangle 93
Odd functions 89
One-to-One Functions 20
Ordered pair
    definition 4

P

Perimeter (or Circumference) of a Circle 253
periodicity
    of trigonometric functions 177
Point of Inflection 66
Polar form; polar coordinates-286, 302
Polar equations 308
Polynomial functions 12
Positive functions 60
positive integers 2
prime numbers 1
Product or Product as a Sum Formulas 234
Proving Trigonometric Identities 236
Pythagorean Identities 233
Pythagorean Theorem 95, 221
Pythagorean theorem, 148

R

radian measure 100, 163
Range 10
Ratio 2
Rational Expression, improper 34
Rational Expression, proper 34
Rational functions 13
    excluded values 34
    graphs 46
    irreducible 52, 53
    reducible 52

Rational number 2
rational numbers 1
real number line 3
real numbers 1, 165
real numbers. 2
real-valued function 10
reciprocal functions 244
Reciprocal Identities 233
Rectangular components 153
Reference Angle 121
Reference number 168
Reflection of a curve
    equation of 74
    or a line .c3.about the line y = x 73
Reflection of a given curve
    about the origin 72
    about the x-axis 72
    about the y-axis 72
Reflection of a point
    about a vertical line 69
    about the horizontal line 71
    about the origin 69
    about the vertical line 70
    about the x-axis 68
    about the y-axis 67
Reflection of Points 67
Relation 4
Resultant 153
    direction of 161
Right triangle 93
Rounding-Off Numbers 259

S

Scalar Quantity 150
Scalene triangle 94
Sector of a circle 254
Signed Numbers 3
Significant Digits (261
standard notation
    for labeling a triangle 92
standard position 120
    of an angle 121
    of an angle in radians 173
standard-position angle 122
Sum
    of vectors 162
Sum and Difference Identities 233
    applications 238
Sum Identities 234
Symmetry
    about the origin 88
    about the x-axis 87
    about the y-axis 86

T

Terminology 1
Tramslation of Axes 79
Translation of Axes
  applications 80
Translation of Points 77
Triangles 92
  30-60-90 triangle 105
  45-45-90 triangle 106
  acute triangle 93
  classification 93, 94
  equilateral triangle 94
  isosceles triangle 94
  obtuse triangle 93
  right triangle 93, 95
  scalene triangle 94
  solutions of right triangles 107
  standard notation for labeling 92
Trigonometric Equations
  solutions 241
Trigonometric function
  of a real number 167
Trigonometric Functional Value
  given one functional value and no specification of the quadrant of the terminal side.) 137
  given one functional value and the quadrant of the terminal side of the angle 136
  given the measure of an angle 123
  given the oordinates of a point on the terminal side of an angle 125
  of any angle 119
Trigonometric functional values
  of a real number S 173
  of an angle in radians 172
Trigonometric Functions 96
  of Quadrantal Angles 129
trigonometric graphs 174
Trigonometric Identities 232
  proving 236
Trigonometric ratios 96, 111
Turning Points 66
Types of Numbers 1

U

unit circle 165

V

Vector quantities
  graphs 151
Vector Quantity 150
Vectors 150
  addition 157
  resolution into components 153
Vertical Asymptote 36
vertical line test 6, 28

W

whole numbers 2

Mathematical Modeling

Some Reciprocal Relationships

1. Arithmetic If A working alone can do a piece of work in time t_A; B working alone can do the same work in time t_B; C working alone can do the same work in time t_C, and if A, B, and C working together, can do the same work in time t_{ABC}, then

$$\frac{1}{t_{ABC}} = \frac{1}{t_A} + \frac{1}{t_B} + \frac{1}{t_C}$$

That is, the reciprocal of the working-together time equals the sum of the reciprocals of working-alone times (individual times).

2. Geometry: For any triangle, the reciprocal of the inradius (R) equals the sum of the reciprocals of the exradii (r_1, r_2, and r_3).

Thus $$\frac{1}{R} = \frac{1}{r_1} + \frac{1}{r_2} + \frac{1}{r_3}$$

3. Physics (Electricity) For electrical resistances in parallel (in an electric circuit), the reciprocal of the combined resistance, R, equals the sum of the reciprocals of the separate resistances, r_1, r_2, and r_3.

Thus $$\frac{1}{R} = \frac{1}{r_1} + \frac{1}{r_2} + \frac{1}{r_3}$$

4. Physics (Optics)

For two thin lenses in contact, the reciprocal of the combined focal length, F, equals the sum of the reciprocals of the separate focal lengths, f_1 and f_2, .

Thus $$\frac{1}{F} = \frac{1}{f_1} + \frac{1}{f_2}$$

5. Physics (Optics) For spherical mirrors and thin lenses, the reciprocal of the focal length F equals the sum of the reciprocals of the object distance, d_o and the image distance d_i. Thus $$\frac{1}{F} = \frac{1}{d_o} + \frac{1}{d_i}$$

6. Physics (Mechanics). If two bubbles of radii r_1, r_2, coalesce into a double bubble, the radius, R, of the partition is given by

$$\frac{1}{R} = \frac{1}{r_1} - \frac{1}{r_2}$$

www.ingramcontent.com/pod-product-compliance
Lightning Source LLC
Chambersburg PA
CBHW081458200326
41518CB00015B/2297
* 9 7 8 1 9 4 6 4 8 5 1 8 2 *